THE A B C OF WAR MEDALS AND DECORATIONS

BEING THE HISTORY OF THE MANNER IN WHICH THEY WERE WON, AND A COMPLETE RECORD OF THEIR AWARD: THEIR CHARACTERISTICS: HOW THEY ARE NAMED AND HOW THEY ARE COUNTERFEITED

BY

W. AUGUSTUS STEWARD

LIEUT. 1ST CADET BATT. "THE RANGERS"
LATE SQUADRON COMMANDER WEST LONDON MOUNTED RIFLES
OFFICIER D'ACADÉMIE
AUTHOR OF "FROM THE BREASTS OF THE BRAVE," ETC.

With 264 Illustrations in Half-tone and Line

LONDON
STANLEY PAUL & CO
31 ESSEX STREET STRAND W.C.2

First Published in 1915

LA CROIX DE GUERRE

BRITISH NORTH BORNEO, 1900.

NEW NAVAL GENERAL SERVICE MEDAL.

THE MILITARY MEDAL.

THE 1914 (MONS) STAR.

FOREWORDS

If any excuse were needed for penning this, it is to be found in the exceeding interest which was taken in my monograph "Badges of the Brave." Indeed, many readers have requested me to deal, at greater length, with a subject which not only opens up a great historical vista and awakens national sentiment, but, incidentally, serves an educational mission to those who collect and those who sell the metallic records of many a hard-fought field, which, when collated, form an imperishable record of our island story.

The War Medal is a comparatively modern institution, otherwise we might have learned the names of the common folk who fought so tenaciously in the old wars, as, for instance, the Welsh infantry and Irish soldiers who, with the English bowmen, comprised the army of 30,000 which at Crécy routed an army of 120,000; the followers of the Black Prince who captured the impetuous King John at Poitiers, or the English archers whose deadly volleys made such havoc at Agincourt, on that fateful day in October nearly five hundred years ago; the brave seamen who, under Lord Howard, Drake, Hawkins, and Frobisher, fought the "Invincible Armada"; and those who, under Raleigh, vigorously pursued the Spaniards on the high seas. We might have learned something of the men who composed the Royal Scots and the 18th Royal Irish, and helped to vindicate the reputation of the British soldier at Namur, and covered themselves with glory at Blenheim; the gallant Coldstream Guards who did such excellent service under Marlborough at Oudenarde and Malplaquet, as well as the Gloucesters and Worcesters who fought so well at Ramillies, or the

Royal Welsh Fusiliers who served under George II at Dettingen.

When, however, war medals were designed for distribution among successful combatants, a means of decorating surviving soldiers and sailors was established, and at the same time a sentimental and substantial record of a man's labours for his country upon the field of battle. So that, if the veterans of Drake's historic fleet, or Marlborough's dauntless soldiers, were not possessed of badges to distinguish them from the soldiers of industry, we, at any rate, may hold in our hands the medals which were awarded to those who served the immortal Nelson, and be proud to possess the medals which shone upon the breasts of our great grandparents who defied the Conqueror of Europe on that memorable Sunday, and made his sun to set upon the battlefield of Waterloo.

Have you listened to the smart British veteran as he explains the disposition of the troops on that historic occasion—how the French cavalry "foamed itself away" in the face of those steady British squares? How he makes the Welsh blood tingle as he records the glorious deeds and death of Sir Thomas Picton, and the Scotsman's dance through his veins as he explains how, with the cold steel of their terrible bayonets, the Black Watch at Quatre Bras, and its second battalion, the Perthshires, at Waterloo, waited for the charge of the cuirassiers; and how Sergeant Ewart of the Scots Greys captured the Eagle of the 45th, and then, with the rest of the Union Brigade (the English Royals and the Irish Inniskillings), crashed through the ranks of the faltering French, and scattered the veterans of Napoleon's army! Have you seen how the mention of the Guards holding the Château of Hougomont brightens the eye of the Englishman? Yes! Then just think what it is to touch and possess the solid proofs of the deeds that those men did, and to feel that you have in your possession the only recompense those brave and daring men received from a grateful country.

Historical Value.—My collection of medals enables me to cover over a hundred years of history; takes me back to the stirring times when men yet met face to face in the Peninsula and at Waterloo; to the men who founded our Indian Empire. It enables me to keep in touch with sailors who fought in the battle of the Nile, at Trafalgar, and at Navarino, that last of all naval battles in which we British took part—our allies were then the French and Russians—until our battleships met those of the Germans in the great war now waging. It reminds me of the horsemen who made the world wonder ere, with deathless glory, they passed their little day, and of that "thin red line" of Scots, whose cool daring at Balaklava has only been bedimmed by the gallantry of the Light Brigade. It enables me to think more intimately of the men I know who faced the Russians in that terrible winter, and then, like heroes, plodded through the inferno of the Mutiny. It brings back vividly to my mind the days of the Zulu War and the heroism of Rorke's Drift. It reminds me of the daring march to Kandahar and the frontier wars so necessary to hold back the turbulent human surf which beats on the shores of our great Eastern Empire. It enables me to keep closely in touch with those who so quickly dealt with Arabi Pasha and later faced the fanatical hordes of the Mahdi; the young men of this generation who fought so stubbornly at the Modder River, and who stormed the Tugela Heights. It enables me to keep in touch with those "handy-men" and scouts on the fringe of Empire who in Somaliland, Gambia, Benin, Matabeleland, and Bechuanaland uphold the dignity of Britain.

We sometimes read of a man or woman who has shaken hands, sixty, seventy, or eighty years ago, with some great person, or some one whose deeds have made him or her a name in history. The possession of war medals and decorations, or of medals of honour gained by brave deeds in time of peace, brings us in close touch with those who honourably gained them. That is an aspect of medal-collecting which appeals to me, and should to every one who admires pluck, grit, daring, and the willingness to personal sacrifice which these badges of the brave denote.

Finally there is an exceptional feature in the collection of war medals which will also appeal, for, as Sir James Yoxall has pointed out in "The A B C About Collecting," the collector of war medals "has concentrated upon a line which can be made complete." If, however, his inclinations or his means will not permit of the acquisition of a complete set he may specialise in either Military or Naval Medals, or those awarded to special regiments or ships, or to men of his own name, or those earned by boys or nurses.

In order to facilitate the search for bars issued with the various medals, the names inscribed thereon are printed in the text in small capitals: these, of course, must not be taken as representing the type used on the official bars; reference must be made to the illustrations, which, being the same size as the original medals, will materially assist the reader in recognising official lettering.

In conclusion I have to express my sincere thanks for the help afforded and the deep interest taken in my book by Dr. A. A. Payne, whose kindness in providing photographs of examples in his unique collection has enabled me to illustrate many interesting and rare medals; to G. K. J. and F. W. G. for clerical assistance; G. T. F. for sketches; and to Messrs. Heywood & Co., Ltd., for the loan of several of the blocks of medals which had been used in monographs I had written for publication by them.

W. AUGUSTUS STEWARD.

LONDON.

CONTENTS

MILITARY SECTION

NAVAL SECTION

ILLUSTRATIONS

ABBREVIATIONS USED IN THE TEXT

H.E.I.Co. . .	Honourable East India Company.
M.G.S. . .	Military General Service.
N.G.S. . .	Naval General Service.
I.G.S. . .	India General Service.
V.C. . . .	Victoria Cross.
D.S.O. . .	Distinguished Service Order.
D.C.M. . .	Distinguished Conduct Medal.
D.S.C. . .	Distinguished Service Cross.
D.S.M. . .	Distinguished Service Medal.
C.G.M. . .	Conspicuous Gallantry Medal.
G.S.M. . .	General Service Medal.

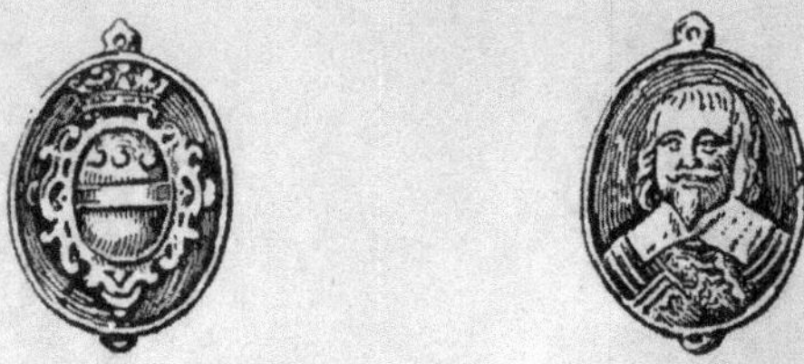

MEDALLION OF EARL OF ESSEX.

SIR THOMAS FAIRFAX MILITARY AWARD, 1645.

BADGE OF CHARLES I AND HENRIETTA.

ROYALIST BADGES.

WAR MEDALS AND THEIR HISTORY

Since the days when woad-clad Britons faced Cæsar's legions we have had a military system, but it was not, so far as we have any substantial evidence, until the days of Elizabeth that personal decorations were awarded for military service, or distinction on the seas or in the field.

In the Middle Ages, a warrior, knighted on the field of battle, was permitted to use a square instead of a swallow-tailed pennon, as a knight's banneret, and to use a war cry, from whence we may trace the origin of the mottoes used with Coats of Arms.

Later, after Sedgemoor (July 6th, 1685), recognition of the lower ranks is recorded in the bestowal of a gratuity of £40 to Sergeant Weems of the 1st Royals (now the Royal Scots) for serving the great guns in an emergency; but these types of award hardly provided that personal note or record which the war medal, as we know it, gives to the recipient.

It is fitting that the Navy, which had existed for centuries prior to the establishment of a standing Army, should take precedence in the bestowal of awards for active service; the rout of the Spanish Armada—in 1588—probably gave the incentive to Good Queen Bess to commemorate the auspicious occasion by the issue of medals in gold and silver, and we may reasonably assume that they were given for personal decoration to the leading officers engaged in the defeat of the Armada, or that the recipients thought that by wearing them they would show respect to the Queen, and thus established the custom of wearing medals, presented by the Sovereign for War Service. A specimen of these medals, with rings and chain for suspension, probably from the neck, is to be seen in the British Museum.

Just over fifty years later, Charles I established the principle for the Army, and thus strengthened the precedent, which was gradually extended, until now every boy and man who has acquitted himself creditably in a campaign, and the nurses also, may rightly claim the medallic recognition and record of their principal services by land or by sea.

Elizabethan Naval Medals.—A silver medal, with an attached half-ring loop for suspension, was apparently given by Queen Elizabeth for naval achievements; it is oval, and bears on the obverse a bust of the Queen, and on the reverse a bay tree on an island, with the legend impressed thereon NON · IPSA · PERICVLA · TANGVNT. This is a splendidly decorative medal. The "Ark-in-Flood medal," though hardly so well designed or so delicately cut, is characteristic. It is generally stated that it was given to the principal officers who fought against the Armada, or to commanders who had distinguished themselves at sea. The medal was struck in gold and silver and measured 2 in. by $1\frac{3}{4}$ in., and was suspended by a fancy loop. On the obverse of the medal is a bust of the Queen facing to the left, with the inscription ELIZABETH ◆ D ◆ G ◆ ANGLIE ◆ F ◆ ET ◆ H ◆ REG.

These, like another medal given by Elizabeth, we may reasonably suppose to have been struck to commemorate the defeat of the "Invincible Armada." James I issued a similar medal in gold and silver, with a ring for suspension. On the reverse is the Ark upon the waters, having above it, like Elizabeth's medal, the symbol indicative of Divine protection, surrounded by the motto STET · SALVUS · IN · VNDIS. There are two kinds of obverse to this medal, one with a portrait bust of the King in armour with a ruff, and the legend FIDEI DEFENSOR encircling the head. This apparently was for military officers, and the one bearing an obverse with the King's head surmounted by a broad-brimmed hat, for his courtiers. Nothing, however, can be stated with any certainty on this point. The motto on the obverse is JACOBUS · D · G · MAG · BRITA · FR · ET · HI · REX; and on the reverse an ark within an oval band containing the motto PER VNDUS SEVAS TRANQVILLA.

Charles I establishes Military Medals.—Having indicated the creation of the campaign or active-service medal for the Navy, we will leave the consideration of naval medals for the section which will be reserved for that purpose and deal with the listing of the military badges of the brave, commencing with those which were struck in the days of Charles I, who established medals for military prowess.

They were, according to the order of the Court, held at Oxford on the 18th day of May, 1643, "to be delivered to wear on the breast of every man who shall be certified under the hands of their commander-in-chief to have done us faithful service in the forlorn hope." These medals were only, it will be seen, given for very distinguished conduct in the field. One bore the Royal image on the obverse, and Prince Charles on the reverse; and the other the bust of Charles on the obverse, with the inscription CAROLUS · D · G · MAG · BRI · FR · ET · HIB · REX, and on the reverse the Royal Arms with the Garter bearing the motto HONI · SOIT · QUI · MAL · Y · PENSE. Both medals were silver and oval in shape, the sizes being respectively 1·7 in. by 1·3 in. and 1·5 in. by 1·2 in.

First Distinguished-conduct Medal.—The first record we have of the bestowal of a medal for conspicuous conduct in the field is in the award made to an Irish commander who distinguished himself at the Battle of Edge Hill—the first battle of the Civil War, fought on October 23rd, 1642—by the recovery of a Royal standard, and certain military accessories. The gallant soldier, who became Sir Robert Welch, was subsequently presented with an oval gold medal specially cut to the King's orders by the Royal "graver of seals and medals," Thomas Rawlins. The King's instructions, given on the 1st day of June 1643, were to the effect that the medal was to have on the obverse his own figure, and that of his son Prince Charles, and on the reverse a copy of the banner Welch saved at Edge Hill, together with the legend PER REGALE MANDATVM CAROLI REGIS HOC ASSIGNATVR ROBERTO WELCH MILITI. Encircling the busts of the King and his son was the inscription CAROLUS REX MAGNAE BRITANNIAE, FRANCAIE ET HIBERNIAE CAROLUS PRINCEPS. The medal, oval in form, was 1·7 in. by 1·5 in.

The First Campaign Medal.—The Dunbar medal—by the famous medallist, Thomas Simon, was struck in two sizes, in gold (1 in. by ·85 in.) and silver (1·35 in. by 1·15 in.), for presentation to officers and men who "did this excellent service," and, to use the quaint words of Oliver Cromwell, in "commemoration of that great mercie at Dunbar" where the Scots Royalists were defeated on September 3rd, 1650. Although these medals, which were worn suspended by a chain from the neck, cannot be regarded as campaign medals in the modern sense, they have the unique distinction of being the first of which there is a reliable record, indicating that the common soldiery and officers equally participated in the receipt of a military decoration for war service. This issue was exceptional, for throughout the existence of the Commonwealth, although medals were struck to commemorate naval victories over the Dutch, they were given to officers only, and during succeeding years this appears to have been the rule also until the beginning of the nineteenth century, when the custom was established to bestow upon all soldiers, from Field Marshal to drummer-boy, the same medal; that was after the battle of Waterloo, June 18th, 1815.

THE DUNBAR MEDAL, 1650.

(Obverse.)

(Reverse.)

COMMEMORATIVE SILVER MEDAL FOR THE BATTLE OF OUDENARDE, 1708.

The Dunbar medal, as the illustration facing page 1 shows, bears on the obverse the bust of Cromwell in armour with the inscription in a semicircle above WORD AT DVNBAR · THE LORD OF HOSTS SEPTEM YE 3 · 1650 On the reverse is quaintly depicted the House of Commons, which referred to the Committee of the Army the consideration of a grant of medals to officers and men who had served in Scotland, and to "set the proportions and the values of them, and their number, and present the estimate to the House." Bronze medals exist, but the reverses of these are plain, despite Cromwell's request that an Army should form the subject, which was only acceded to in so far as the legend is concerned, and a slight suggestion of a fight behind the bust of the Lord Protector.

The Cromwellians voted the sum of £100 to provide a gold medal, for distinguished service, and chain for presentation to Colonel Mackworth, who, as governor of Shrewsbury, had refused to surrender the castle to the Royalists. Of Blake's medal and the other decorations for maritime warfare I shall treat in the section devoted to naval medals, but I might here remark upon the fact that in those Puritan days the seaman appears to have been as much a "handy-man" as in modern times, for, like the present-day marine and seaman, he fought *Per Mare et per Terram*; at any rate Generals Blake and Monk did, for they received naval medals or awards from the Commonwealth for their services against the Dutch.

Various medals were struck during the Civil War by the Royalists, and engravings are extant illustrating medals variously bearing on the obverse the portraits of King Charles I, Prince Rupert, or one or other of his leading supporters and generals, including the Earl of Essex and Sir Thomas Fairfax. The reverse bore either a representation of the Parliament or the arms of the person depicted on the obverse. The victory of Naseby (June 14th, 1645) is commemorated by a silver medal—gilt—with ring for suspension, upon the obverse of which is a portrait of Sir Thomas Fairfax, and the inscription THO: FAIRFAX MILES MILIT PARL DUX GEN, and on the reverse with a circle MERUISTI and POST HAEC MELIORA 1645.

During the reign of Charles II (when the foundation of a standing army was established) and James II, there is no record of military medals being struck, although during the reign of the former, as I shall presently describe, naval medals were issued to commemorate victories over the Dutch, and in the reign of the latter a large silver medal, apparently for presentation to naval men, was issued.

During the reign of William III and Mary, naval medals only were struck, and, as far as is known, only one such during the reign of Queen Anne, for there is no record of any being issued for the famous battles of Blenheim, Ramillies, and Oudenarde (where, as "Prince Elector of Hanover," George III distinguished himself), or the sanguinary battle of Malplaquet, although commemorative medals were struck.

The reign of George I is also barren of military campaign medals, while there were apparently only two issued during the reign of George II—one for the Battle of Culloden (April 16th, 1746), where the Duke of Cumberland well earned the sobriquet of "The Butcher," and the other for the taking of Louisbourg, in Canada (July 27th, 1758), where, after a seven weeks' siege, the French garrison surrendered to General Amherst.

The Culloden Medals.—The Culloden medal was struck to commemorate the crushing of the Jacobite rebellion at the battle of Culloden on April 16th, 1746, when the Scots under Prince Charles Edward and Lord George Murray were defeated by the Duke of Cumberland. The medal, oval in shape, and 1·75 in. by 1·45 in. in size, had on the obverse the head of "The Butcher," with CUMBERLAND above, surrounded by an ornate ribbed border and suspender; a nude figure of Apollo, with a wounded dragon at his feet, occupies the ground of the reverse, with the words ACTUM EST ILICET PERIIT, and in the exergue PRŒL · COLOD · AP · XVI · MDCCXLVI. The medal, by Richard Yeo, was struck in gold, silver, and bronze, and was suspended from a red ribbon with green edges. This is a very rare medal, and few specimens now exist, and it is assumed it was only given to regimental commanders, although as Mr. Grueber, of the British Museum, pointed out there is no evidence of an order that it should be granted.

In the British Museum there is a bronze medal commemorating this same battle; it is possible that it was intended to be worn as a war medal. The striking of the medal is with a scroll suspender at the top with a hole, which leaves no doubt that it was intended for suspension by means of a small ring. There is a small ornamental scroll at the bottom. It is plain on the reverse, the obverse being occupied by the equestrian effigy of the Duke of Cumberland, with the date 1746 in the exergue, and above the group GUL · AUG · DUX · CUM · TERROR · REB.

Capture of Louisbourg.—For the siege of Louisbourg, gold and silver medals were struck for presentation to those officers who had been conspicuous by their gallantry. In gold and silver the medal, 1·7 in. in diameter, bears on the reverse a representation of the bombardment, with LOVISBOVRG TAKEN MDCCLVII, and on the obverse a globe resting upon a figure symbolic of France (some say a fury) dropping fleur-de-lis, and pointing to boats at sea. On the globe, supported by a British Grenadier and a sailor, are the words CANADA and AMERICA, and above all the figure of Fame, with laurel wreath and trumpet flying in front of the Union Jack, and a scroll with the words PARITER IN BELLA; the ribbon for this is half brownish-yellow and light purple. This rare medal is, however, generally placed in the category of historic medals. It is by Thomas Pingo, who was responsible for a series of historical medals, including those commemorating the capture of Goree, December 29th, 1758, the capture of Guadaloupe, 1759, the battle of Minden, July 31st, 1759, and the capture of Quebec (where Wolfe lost his life), September 15th, 1759.

Carib War.—For the Carib War in 1773 the Legislative Assembly of the island of St. Vincent ordered that a medal of silver should be awarded to those who had taken part in suppressing the rebellion which had broken out among the natives. The medal bore upon the obverse the bust of George III in high relief, his hair being tied with ribbon, as was the fashion of the times, but represented as clad in armour. Above the bust is the legend GEORGIVS III MB REX. On the reverse Britannia, characteristically helmeted and resting her left hand upon the Union Shield, offers to a beaten Carib an olive branch in token of peace, while the vanquished is represented as having surrendered his arms, which lie at the feet of Britannia. In the exergue is the date MDCCLXXIII, and above the group PEACE AND PROSPERITY TO ST. VINCENTS. The medal, $2\frac{3}{20}$ in. in diameter, was cast and chased, and bears on the truncation of the King's head the name of the modeller —C. M. Moser. It was worn suspended from the neck by a red ribbon. Among the regular regiments taking part in the campaign were the 14th, 31st, and 70th.

Isle of St. Vincent.—Another medal was awarded for service in the island of St. Vincent. It was given to the militia officers and non-commissioned officers who commanded about five hundred natives—slaves—who helped in the campaign against the Caribs and French troops in 1795. On the obverse of this medal, which was cast and chased in silver and bronze, is a winged figure of Victory, with her right foot planted upon the body of a defeated Carib, who has dropped his musket at her feet. While in her left hand she bears a palm branch, she shows her determination by grasping a sword in the right. Above is the inscription ST. VINCENTS BLACK CORPS; on the reverse is a representation of a black soldier standing at attention with his musket, with fixed bayonet, resting on the ground, encircled by the legend BOLD LOYAL OBEDIENT. The whole of the lettering is in a square type of Roman capitals. This medal is 1⅞ in. diameter.

The next in order we have to consider are those issued by the Honourable East India Company to the officers and men who took part in the Deccan and Mysore Campaigns, between 1778 and 1792, against Hyder Ali and his son Tippoo Sahib, but these, despite the significant character of the wars, were not official or Sovereign awards, and were given to the Company's troops only. Previous to this, the Company had awarded a medal (in 1766) to native officers who had quelled a mutiny among European troops at Morighyr.

The Deccan Medal.—The medal for Guzerat—1778–84—and Carnatic—1780–84—commonly known as the "Deccan" medal, was issued to native troops only, and, as Dr. Payne states, is the earliest Anglo-Indian medal to be awarded to all ranks. On the obverse Britannia is represented resting upon a trophy of arms, and holding a wreath in the right hand, which she extends toward a fort over which the British flag is flying. On the reverse is an inscription in Persian, which explains that the medal is awarded by the Government of Calcutta to commemorate "the excellent services of the brave; year of the Hegira 1199, A.D. 1784." The medal, issued in two sizes, was intended to be worn suspended from the neck by a yellow cord run through a metal loop. It was issued in gold to Subadars,[1] silver gilt to Jemadars[2] and other native officers, and of an inferior character to those who did not hold warrant rank and the ordinary sepoy. A specimen of the smaller medal, in fine condition, has sold for between £7 10*s*. and £8 5*s*.

Mysore Campaign.—During 1791 and 1793 the native

[1] Infantry captain.

[2] Jemadar—of the same standing as a lieutenant in the Cavalry or Infantry, or the Sappers or Miners.

troops were engaged against Tippoo Sultaun in Mysore, during which campaign the city of Bangalore was captured in an exceedingly brilliant manner, thanks to the pluck and daring of British officers. On the 1st day of April 1793 it was decided by an Order in Council to give to the successful troops a medal as a reward for their services. As in the case of the forementioned medal, two grades and sizes were issued. The gold ones were all $1\frac{13}{20}$ in. in diameter, but a smaller silver one was issued $1\frac{1}{2}$ in. in diameter, and while some of the medals have only two cannon-balls at the foot of the sepoy, five are found on others. This is probably due to different men cutting the replicas of the dies, which may have cracked in the striking. On the obverse of the medal a sepoy, standing with his left foot on a dismounted cannon, grasps the British colours in his right hand and the Mysore flag reversed in his left; the background shows the fort of Seringapatam in the distance. On the reverse is the wording, FOR SERVICES IN MYSORE A.D. 1791–1792, encircled by laurel branches, round which is arranged a Persian inscription to the effect that the medal is "a token of the bravery of the troops of the English Government in the war in Mysore in the year of the Hegira 1205–1206." Like the "Deccan" medal, this was intended for suspension from the neck by means of a yellow cord running through a loop. The Subadars were granted gold medals, the Jemadars and Serrangs silver medals—afterwards gilt—and the non-commissioned officers and men silver medals of less value. The order for striking these medals was given to Mr. Mair, a Calcutta silversmith.

Many Counterfeits.—This is a poor class of medal, and doubtless the ease with which it lent itself to the counterfeiter is responsible for the number of cast fakes which one meets with. These, however, are easily detected: in casting, "sandholes" cause little grains on the surface, while dust on the mould causes holes in the cast; the granulations need removing and the holes closing up by the chasing tool, but the "faker" was invariably careless, and the counterfeits may be distinguished from the genuine medals by the presence of grains on or small holes in the surface, and the fact that the "fakes," having been cast from an original, are slightly smaller, owing to the "shrinkage" which always takes place in casting. This shrinkage is responsible for practically closing up the space between the head of the flagstaff and the rim of the medal. The larger medal is $1\frac{3}{4}$ in. and the smaller $1\frac{1}{2}$ in. in diameter. A fine specimen of the larger medal has fetched as much as £18.

Capture of Ceylon.—For capturing Ceylon from the Dutch in 1795–6 it was decided by an Order in Council, dated Fort William, May 15th, 1807, to confer a medal upon the Gun Lascars only who served with the European Artillery. Only two gold medals were struck and presented to native officers, and one hundred and twenty to other ranks. Like those previously mentioned, the medals were intended for suspension from the neck by a yellow cord. One and nine-tenths inch in diameter, this is one of the simplest medals ever issued, and the only one in which lettering suffices for decoration. On the obverse is the inscription, FOR SERVICES ON THE ISLAND OF CEYLON A.D. *1795/6*, and on the reverse, in Persian, "This medal was given by way of acknowledgment of services in Ceylon in the year of the Hegira 1209–1210."

Siege of Seringapatam.—The feeling of animosity on the part of Tippoo Sahib, the son of Hyder Ali, fed by the fact that the Directory in France had designs against British India, led him to make preparations for another conflict, but the Earl of Mornington, who afterwards became Marquis of Wellesley, smartly marched an army into Tippoo Sahib's territories, and on March 27th, 1799, advanced in battle array against the enemy, who, after losing about 10,000 men, retreated in disorder, followed by the British Army, which, pressing on to Seringapatam, reached the city on April 5th, and commenced the famous siege, which terminated on May 8th. Among the slain was Tippoo Sahib, and the British in a magnanimous manner interred his remains in the fine mausoleum which he had caused to be built over the resting-place of his father. Twenty-two officers were killed and 45 wounded; 181 men killed, 624 wounded, and 22 missing; 119 native soldiers killed, 420 wounded, and 100 missing.

The European regiments that took part in the storming were the 33rd Regiment (now the 1st Batt. West Riding Regiment), which, led by the Hon. Arthur Wellesley (afterwards Duke of Wellington), distinguished itself on the road to Seringapatam—on March 27th—by standing its ground in the face of an advancing body of 2,000 men, and then, after firing a volley, boldly charging upon them, and with the aid of cavalry driving them off. The other European regiments included the 19th and 22nd (late 25th) Light Dragoons (both disbanded), the 12th and 73rd Regiments, 74th Highlanders, 75th and 77th Regiments, the Scots Brigade, afterwards the 94th (disbanded in 1818), and the 103rd. For these excellent services the Honourable East India Company issued a medal, of which two kinds exist—one produced in the Soho Mint near Birmingham, the famous place established by Matthew Boulton, and the other from dies cut in Calcutta after strikings of the Soho medal.

Variety of Medals Struck.—Quite a variety were struck. Gold for His Majesty, the Governor-General of India—Lord Melville, the Marquis Cornwallis, certain Nizams, Nabobs, and Rajahs, the Commander-in-Chief and the General Officers on the Staff, and one for the Oriental Museum. (£15 to £20 has been paid at auction for one of these medals.) Silver gilt for the members of the Council of the three Presidencies, the Residents of Hyderabad and Poona, the Field Officers and the General Staff on Service. (£8 has been given for a fine specimen of the gilt medal.) Silver medals were awarded to the captains and subalterns. (As much

HONOURABLE EAST INDIA CO.'S MEDAL FOR SERINGAPATAM, 1799.

HONOURABLE EAST INDIA CO.'S MEDAL FOR EGYPT, 1801.

as £5 and £6 has been paid for one of these.) Bronze for non-commissioned officers of the British Army, and tin for privates. The medals were issued unnamed in 1808, but it was not until August 29th, 1815, that the members of the British Army were permitted to wear them; issued without suspenders, this permission led to the addition of loops and suspenders, and the adoption of a ribbon for suspension, and while some used a dark orange ribbon suggestive of a tiger's skin, in allusion to the victory of the British arms over Tippoo Sahib—"the conquering tiger of God"—the claret-coloured ribbon with dark blue edges, as used with the Peninsular gold medals and crosses and the Waterloo medal, was the recognised ribbon, although some officers wore a watered yellow ribbon.

To the native commissioned officers and non-commissioned officers, sergeants, bandsmen, and trumpeters of European corps, and to others who might be ranked as non-commissioned officers, the bronze medal was also granted, while the tin medals were given to corporals, gunners, and European privates, and to native doctors, guides, and general-utility men with the Army.

The Soho Medal.—This, $1\frac{1}{9}$ in. in diameter, bears on the obverse a representation of the British lion, with a defeated tiger beneath him, a long pennon flying above, and held erect by the lion's tail, bearing near the staff the Union Jack, and in Arabic the defeated Tippoo's title, "Assad Allah El Ghaleb," and in the exergue IV MAY MDCCXCIX. On the reverse is represented the storming of Seringapatam, with the meridian sun indicating that when the sun was in its full glory the successful assault was made and glorious victory was won. Underneath is the legend in Persian, "Seringapatam God conquered 28th day of the month Zikadah, 1213 of the Hegira." The Soho mint was responsible for the striking, in the years 1801 and 1802, of 30 gold, 185 silver gilt, 5,000 bronze, and 45,000 tin medals.

The Calcutta Medal.—Like the medal of Mysore, there are two varieties of the Seringapatam medal owing to mistakes on the part of the die-cutters. The sun at its meridian, a significant and symbolic feature in the Soho medal, is omitted in that struck at Calcutta; moreover the medal is not so large, being only 1⅞ in. in diameter, is thinner, has a loop for suspension by a cord, and, what is very important, of inferior craftsmanship—so that in considering the value or authenticity of a medal for the famous battle of Seringapatam the collector has to remember several important features, since while he may hold in his hand two medals differing somewhat in size, design, and quality of craftsmanship, he may find that both are perfectly genuine. The Calcutta mint issued 83 gold and 2,786 silver medals, so that the collector must carefully consider what he is about when a silver-gilt medal is offered for sale; only 185 *bona-fide* medals of that quality were issued, but of the 850 British-struck silver medals many have been gilded to simulate the rarer medals, while a number of the smaller and inferior Indian-struck medals have been gilded and offered as original gilts. Despite the large number of tin medals issued (45,000), I have seen many fetch £1 each at auction. No bronze or tin medals were struck from the Calcutta die.

Egypt, 1801.—Napoleon, with an eye upon our Indian Empire, invaded Egypt in order "to conquer the East and take Europe in the rear." Fortunate in eluding Nelson, who was cruising in the Mediterranean, he managed to disembark his troops, but he had to suffer the humiliation of Nelson's brilliant victory in the Bay of Aboukir (on August 1st, 1798), by which he lost practically all his fleet. This famous sea battle is known as the Battle of the Nile. It left Napoleon stranded in Egypt with his army, which he marched across the desert. He stormed Jaffa and besieged Acre for two months, but Sir Sidney Smith and his garrison of British and Turks not only held the fort, but in so doing caused the loss of 4,000 of Napoleon's finest troops. Falling back on Egypt he found disquieting news awaiting him, so he returned to France and proclaimed himself First Consul. Meanwhile his troops had to prepare to meet the British expedition which, under Lieutenant-General Abercromby, had been dispatched to evict them. The total British force which landed under the brilliant but fated General was 15,330 men, but of these, it is stated, only about 12,000 could be counted as fighting units. The French had made formidable preparations to resist the disembarkation, and the seamen had to row the troops ashore under a hail of round shot, grape and shell. Many a boat-load of brave men was sent to the bottom, but those who reached the shore rushed through the surf, formed line, fixed bayonets, and, led by Major-General Moore (who found a soldier's grave at Coruña), charged up the slope, drove out the French battalions, and after withstanding a charge of cavalry, compelled the French to retreat along the road to Alexandria, whither they were ultimately pursued by the British army, many daring exploits taking place *en route*. One is particularly noteworthy. The Perthshire Regiment, wearing at the time brass helmets, were mistaken by the French cavalry for dragoons, and the Scots were consequently charged in an impetuous manner by them. The affair, which took place at Mandora, was a precursor of the brilliant action of the 93rd Highlanders at Balaklava, for, standing firm, the Perthshires coolly awaited their mounted adversaries until they were within fifty yards of them, and then, firing a volley, caused the main body to wheel off by their left and retreat in confusion. Sir Ralph Abercromby narrowly escaped death in this affair, but, as Archibald Forbes states, he "was rescued by the devoted bravery of the Perthshire Regiment."

The Highlanders at Alexandria, 1801.—At the battle of Alexandria, March 21st, 1801, Generals Moore and Oakes were wounded, but continued to lead their men, who did prodigies of valour ere the French retreated to their entrenchments before Alexandria. Abercromby had, during the day, kept close to the Highlanders, whom he cheered with the words, "My brave Highlanders, remember your country! Remember your forefathers!" ere ordering them to pursue the enemy; but having sent his staff officers on different missions, he was left a solitary and conspicuous figure. A couple of French cavalrymen, seizing the opportunity, endeavoured to capture the gallant Commander-in-Chief, but, refusing to yield, he made a brilliant fight until a corporal of the 42nd (Black Watch) ran up and shot one of the men, when the other made off, but was bayoneted by another of the 42nd. Meanwhile it was noticed that he dismounted with difficulty from his horse, but as he exhibited no sign of suffering it was not until blood was observed on his thigh that any idea was entertained that he had been wounded. It was then found that a musket-ball had lodged in his hip joint. This caused his death on board the *Foudroyant* on March 28th, 1801. General Hutchinson, who took command of the force, ultimately invested Alexandria, and after a three days' armistice the terms of capitulation were signed, and arrangements made for the departure of the armies. The French, who were commanded by General Menou, left Egypt with about 27,500 men, and the British with 22,350.

H.E.I. Co.'s Medal for Egypt.—The Military and General Service medal, with bar for Egypt, was not issued to the survivors of the series of strenuous fights which concluded with the capitulation of Alexandria, until February 11th, 1850, almost forty-nine years after. Meanwhile a large number of the army of about 5,000 men, which sailed from India under the command of Sir David Baird, had received a special medal from the Honourable East India Company, which, by a General Order, was granted on July 31st, 1802. The promptitude of the Company in rewarding soldiers stands out in pleasant relief against the thoughtlessness

PENINSULAR GOLD MEDAL.

Awarded to Lieutenant-Colonel Alex. Gordon, 83rd Foot, who was killed at Talavera.

or procrastination of the home Government. This army of 2,000 Europeans, 2,000 sepoys, and 400 artillerymen, marched across the Nubian Desert to Keneh, and after descending the Nile encamped at Rosetta, but saw no fighting. The British regiments which formed part of the army to cross the desert were the 10th and 61st Regiments of foot, and detachments of the 80th, 86th, and 88th.

Silver medals, $1\frac{1}{9}$ in. in diameter, were struck for about half the force, and sixteen gold ones. The medal, illustrated facing page 12, bears on the obverse an energetic figure of a sepoy, carrying the Union Jack, with a background suggesting a camp. In the exergue is a Persian inscription to the effect that the medal is presented to commemorate the defeat of the French armies in Egypt by the bravery and ability of the victorious army of England. On the reverse, suggestive of the transportation of troops, is a ship in full sail, and on the land which forms the background the Pyramids and an obelisk indicative of Egypt. In the exergue is the date MDCCCI.

Order of the Crescent.—Selim III presented a series of gold medals to the officers who took part in the operations. These varied in weight and size, some being set with diamonds, while a few of the larger ones were enamelled crimson in the centre. The medals, being struck from different dies, have the crescent either on the right or left of the star. The principal naval and military officers received medals measuring $2\frac{1}{10}$ in. in diameter; those ranking next received a lighter and smaller medal, $1\frac{9}{10}$ in. in diameter; while a third medal, given to captains, was still lighter and only $1\frac{7}{8}$ in. in diameter. The non-commissioned officers received a silver medal $1\frac{4}{10}$ in. in diameter, so that one may gauge by the size the rank of the officer who received it. The medals were suspended from an orange-coloured ribbon by means of a hook and chain. I have seen specimens without attachment for suspension.

The army which landed under Abercromby included the 1st Battalion of the Coldstreams and the Scots Guards; the 1st Royal Scots; 2nd Queen's Royal; 8th King's; 13th, 18th Royal Irish; 19th, 28th; 23rd Welsh Fusiliers; 28th, 30th, 40th (flank companies), 42nd Black Watch; 44th, 50th, 1st and 2nd Battalions of the 54th, 58th; 79th Cameron Highlanders; 89th, 90th Perthshire Light Infantry; 92nd Gordon Highlanders; one squadron of the 11th Light Dragoons; the 12th and 26th Light Dragoons (unmounted); Artillery and Engineers; also Dillon's and De Rolle's Regiments; a detachment of Baron Hompesch's mounted riflemen, and the Corsican Rangers. The army was further strengthened by a squadron of the 8th Light Dragoons; the 10th and 27th Foot; 1st and 2nd Battalions of the 89th, and the Queen's German Regiment.

The 22nd Light Dragoons; two battalions of the 20th Regiment; the 24th, 25th, and 26th Regiments; as well as the Ancient Irish Fencibles, the Chasseurs Britanniques, and De Watteville's Regiment, did not reach Egypt until July 1801. The medals of men attached to the regiments which first landed and drove the French off to Alexandria, following up their retreat, and particularly those who took part in the most furious fighting, are the most sought after; those of the Black Watch and Gloucesters being especially and rightly valued, likewise those of the 8th and 13th Foot.

The Red Heckle.—There is even now a common notion that the Black Watch (which lost 54 officers and men killed, and 262 wounded) had its white heckle changed to red because of its prowess against the French "Invincible Legion" in the historic ruins outside Alexandria, where, it is alleged, the heckles in the fallen bonnets became dyed with the blood of the slain, and in honour of their part in the battle of Alexandria they were granted this signal favour. Such, however, is not the correct story, for it is now generally conceded that the red heckle was granted for their courageous conduct in retaking some guns, at Geldermaslen on January 4th, 1795, which the 11th Light Dragoons had abandoned. The Dragoons wore a red feather in their helmets, and this was passed on, so to speak, to the Black Watch, and the "red heckle" has distinguished the gallant regiment ever since.

Highland Society's Medal.—For the "distinguished and brilliant manner in which the 42nd acted at Alexandria," the Highland Society of London had a medal struck to commemorate the capture of the "Invincibles'" standard, and one was presented to each officer and private, and to the relatives of the dead men. The medal is 2 in. in diameter, and bears on the obverse the bust of Sir Ralph Abercromby and the inscription ABERCROMBIUS DUX IN EGYPTO CECIDIT VICTOR 28 MAR 1801, and on the reverse a Highlander capturing a French standard, with the inscription in Gaelic NA FIR A CHOISIN BUAIDH' SAN EPHAIT 21 MAR 1801, which being interpreted means "These are the heroes who achieved victory in Egypt." On the edge of the medal is engraved in Gaelic O'N CHOMUN GHAELEACH D'ON FHREICEADAN DUBH NA XLII RT ("From the London Highland Society to the Black Watch 42nd Regiment"). In this battle the 28th (the 1st Gloucester Regiment) gained the distinction of wearing their badge at the back and front of their helmets, they having—while in line—turned about to meet an attack in the rear by cavalry while repelling an attack on their front by infantry.

THE MAHRATTA WAR

Dowlat Rao Scindiah, aspiring to the conquest of Hindustan, entered into negotiations with other Mahratta (Máráthá) leaders to attack the British and the Peishwah, who having allied himself to the British had been restored to his ancient position. The treaty of Bassein, in which the British agreed to restore him, has been described as "the greatest diplomatic triumph the world has ever witnessed." The war did not last long.

Allighur.—The first serious engagement was the storming of Allighur, "a bloody struggle, an hour in duration," on September 4th, 1803. The place was defended by General Perron, who had risen from the rank of a non-commissioned officer of the French Marine to be Commander in Chief of the Scindiah's army. General Lake, the British Commander-in-Chief, was his opponent, and he took the fortress by blowing in the gate, an operation described in the Wellington dispatch as "one of the most extraordinary feats I have heard of in this country." The 76th Regiment, led by Colonel Monson, was mainly instrumental in this daring feat; 2,000 of the defenders fell, and 17 officers and 261 men of the British army were killed and wounded. The 27th and 29th Dragoons were the only other European regiments present.

Delhi.—Within a few days the battle of Delhi was fought and won. A portion of the force was dispatched, and on the 11th day of September, 1803, encamped near the Jehna Nullah within 6 miles of Delhi. Here Lake was, with 4,500 fighting men, confronted by an army of 13,000 infantry and 6,000 horsemen, with 79 pieces of artillery, under the command of a French officer, Louis Bourquien. Lake, despite the fatigued condition of his men, decided to give battle, and the result was a crowning victory for the British General, who with a native army—he had only two of the King's regiments with him, the 76th Foot and 27th Dragoons

GOLD MEDAL FOR MAIDA, 1806.
Awarded to Sir Henry Edward Bunbury, Bart., K.C.B.

MEDAL FOR AVA, 1824-6.

(since disbanded), to stiffen it—utterly routed the Mahrattas, who lost 3,000 men. The British losses were severe, for 585 men, including 17 European officers, fell ere the victorious army bivouacked on the field of battle, after being seventeen hours under arms. On September 13th the gallant Lake and his men entered Delhi. The result of this victory was the restoration of the Great Mogul, the blind Shah Alum, who had been a prisoner of the Scindiah's for years.

Assaye.[1]—Meanwhile Major-General the Hon. Arthur Wellesley had besieged and taken Ahmednuggur, possibly at the time the strongest stone fort in India. This he made his base of operations, and hastened to get into touch with Colonel Stevenson in order to make a combined attack on the enemy on September 24th, but while Wellesley was on the march he discovered, when at Naulmah on the 23rd, that he was within 6 miles of the Mahratta army; he decided, therefore, to take what he afterwards termed "a desperate expedient." With a force of 6,400 infantry and 1,600 cavalry (of these only 1,500 were British) and only 17 guns, the man who was to become the hero of Waterloo offered battle to the armies of the confederate chiefs, which included 30,000 horse and 128 cannon in position behind the steep banks of the River Kaitna. The result of the battle was as brilliant as the commander's resolve was daring, for the Mahratta army was routed by one only a fifth its size, and left 2,000 dead on the field; but the cost to the British was a third of the army killed or wounded, including 170 officers. Wellesley had one horse killed under him by a cannon shot, and another by a bayonet wound in the breast—sufficient evidence of the part the General took in the attack. The battle of Assaye, Wellington's first victory, is excellent evidence that he was as bold in attack as he was brilliant in defence. The 19th Light Dragoons (since disbanded), the 74th and 78th Foot were the British regiments engaged in

[1] Spelt ASSYE on bar of Army of India Medal.

the battle, and while the latter bore the brunt of the conflict, the Dragoons largely contributed to it by their brilliant charge.

Siege of Agra.—The Mahrattas, who had evacuated Delhi after the surrender of their European officers, fell back on Agra, where the garrison, desirous of keeping the great treasure the city held, had shut the gates to their Mahratta countrymen. General Lake, having ordered his army in pursuit, had therefore to deal with these before laying siege to Agra. After a stiff fight, a charge of sepoys was all that was necessary before proceeding to take the city, which when it fell, on October 18th, was found to contain treasure worth £280,000. For this battle no recognition was given, as we shall see when listing the bars for the medals granted for the wars and battles in India between September 4th, 1803, and January 1826.

Asseerghur.—At the battle of Asseerghur (October 21st, 1803) there was only one British regiment present, the 94th Foot (the Scotch Brigade). On October 16th Colonel Stevenson took possession of Burhampoor, and then proceeded to Asseerghur, where he assaulted the Pettah, or Citadel, on the 17th, and on the 21st, after cannonading the fort, received the surrender of the place. The battle practically cleared the Scindiah out of the Deccan, but did not humiliate him.

Laswarree.—The decisive defeat of the Scindiah's army at Laswarree on November 1st, 1803, might be called the crowning glory of Lake's life, for the battle was a particularly desperate one, fought against great odds, and with a tenacity which had not characterised any previous battles, hard-fought though they had been, the enemy contesting the ground inch by inch. The Mahratta force comprised 9,000 infantry, 4,000 cavalry, with 72 guns, and of this fine army 7,000 were left dead on the field of battle, while 2,000 were taken prisoners; the British force, numbering only 4,000, had 172 killed and 652 wounded. The British regiments present were the Royal Irish Hussars (8th);

the 27th and 29th Light Dragoons; and the 76th Regiment, which lost heavily in the battle, but covered itself with glory after a night march of nearly 25 miles; indeed, the whole of the British army had been under arms for sixteen hours. General Lake, who it was said had "carried bravery to the very verge of rashness," was created Baron Lake of Delhi and Laswarree.

Argaum.—The battle of Argaum, fought on November 29th, 1803, broke the Mahratta power, Wellesley routing the enemy, who abandoned 38 pieces of cannon and all their ammunition. In this battle the 74th and 78th Highlanders were particularly praised by Major-General Wellesley for repulsing and practically destroying a body of Persian or Arab cavalry. The other infantry regiment present was the 94th, and the only British cavalry regiment the 19th Light Dragoons.

Gawilghur.—Pressing on, Wellesley arranged with Colonel Stevenson to besiege the fort of Gawilghur, the event taking place on December 15th, 1803. This was only possible after considerable hardships and endurance on the part of the army, which had to make roads along which it could drag its guns by hand. Their labours, however, were rewarded by a comparatively bloodless victory for themselves, in which the enemy lost a large number of men during the storming, among them being Killadar and Bery Sing. Two days later the treaty of peace between the Rajah of Berar and the Honourable East India Company was ratified, and the Scindiah also ratified a treaty of peace on January 5th, 1804, and thus a campaign of five months' duration was brought to an end by the astuteness and daring of Lake and Wellesley. The 74th and 78th Highlanders, and the 94th Foot (then the Scotch Brigade, now 2nd Battalion Connaught Rangers) were the only British regiments present at Gawilghur.

Defence of Delhi.—Holkar yet remained to be dealt with, and a series of unfortunate circumstances made him somewhat bold, so he set about besieging Delhi, which was held by Indian troops and four companies of Europeans under British officers, who, numbering only 2,500, had to defend a city 10 miles in circumference from the attacks of an army of 70,000 men with 130 guns. Meanwhile General Lake, learning of the trouble, marched rapidly up from Agra, and reached the city on October 18th, 1804. Holkar thereupon raised the siege, crossed the Jumna, and proceeded to lay waste the land.

Battle of Deig.—Lake, however, with his usual daring, made after him with his European cavalry and artillery, accompanied by the Delhi garrison. General Frazer was detached to march on Deig, where in the battle outside he was mortally wounded. In this battle the 76th (the only British regiment present) again distinguished itself, and drove the left flank of the enemy into the morass, where many were drowned. Besides the 76th the H.E.I. Co.'s 101st (Bengal Fusiliers) and the Bengal Native Infantry took part. During this action, in which the Mahrattas lost about 2,000 men, Holkar was closely pursued by General Lake, and was ultimately compelled to fly with a few adherents before the surprising onslaught of Lake, who, finding that the Rajah of Bhurtpoor had assisted with his cavalry in the battle of Deig, and not only sheltered Holkar's retreating troops but had fired on their pursuers, determined to act promptly and firmly.

Capture of Deig.—On December 13th he invested the fortress, but not until the night of the 23rd was the assault made—ten days having been occupied in making a breach—and then within one hour the place was taken after a short but desperate resistance; the citadel, however, still remained in the hands of the enemy, who evacuated the town and citadel on Christmas Eve 1804, and fled to Bhurtpoor, whither Lake resolved to pursue their army. The British regiments present at the capture were the 8th and 29th Light Dragoons, and portions of the 22nd (the Cheshire Regiment) and the 76th Foot. The H.E.I. Co.'s

1st European Regiment, 6th Light Cavalry, and 8th Native Infantry were also present.

Lord Lake made the one mistake in his Indian warfare when he sat down before Bhurtpoor, because *matériel* necessary to a successful issue was lacking. After several unsuccessful assaults, Lake deemed it advisable to withdraw, and on the 24th took up a position 6 miles off, there to await fresh supplies of stores and ammunition. On April 9th he renewed operations, and by the 21st had settled accounts with the Rajah. The Scindiah, having joined Holkar, was pursued by Lake, and on November 23rd, 1805, a treaty was ratified with him; later Holkar was also glad to sign a treaty, whereby the peace of Hindustan was secured for twelve years, and here we must leave for a period the tale of Indian wars, which went unrewarded until April 1851—forty-eight years after the battle of Allighur—and take up the thread with the war in Nepaul.

Maida, 1806.—The first official war medal for military service was awarded to the seventeen superior officers who were engaged in the decisive and brilliant victory of Sir John Stuart—then Major-General—at Maida in Calabria, on July 4th, 1806, when the French were defeated, leaving about 700 dead on the field. The order for striking the medals was issued in 1808. On the obverse is a laureated head of King George with the legend GEORGIVS TERTIVS REX, and on the reverse the figure of Britannia hurling a spear with the right hand, while guarding herself with a circular shield whereon the Union Jack is delineated, and just underneath this the inscription MAI DA IVL. IV MDCCCVI. Above is a small flying figure of Victory, crowning her with a wreath of laurel, and behind, the ancient symbol of Sicily—the triguetra, or triple leg—like the Manx arms; in the exergue are two crossed darts tied with a knot. This medal, 1½ in. in diameter, framed and glazed, was worn from the buttonhole of the uniform, suspended by a claret-coloured blue-edged ribbon from a gold swivel and loop. The names and rank of recipients are inscribed on the edge; it was cut by Pidgeon. The medal illustrated facing page 20 was awarded to Lieut.-Gen. Henry Edward Bunbury, Bart., K.C.B., who was Under Secretary of State for War, 1809-16. It was kindly photographed for use in this book by his grandson, Sir Henry Bunbury, Bart.

THE PENINSULAR WAR

Napoleon, having by a series of brilliant battles gained supremacy over Central Europe, issued a decree against all commercial intercourse with Great Britain, but having lost any semblance of maritime power by the crushing defeat of Trafalgar (October 21st, 1805), this was a somewhat idle procedure. Portugal declining to carry out his Imperial decrees, Napoleon alleged that the house of Braganza no longer held power, and by cajoling the Spaniards into a false security led them to believe that they would participate in the spoils of Portugal; consequently they permitted an army under Marshal Junot to over-run Portugal, but Napoleon's ulterior object was to place his brother Joseph, King of Naples, on the throne of Spain, and in order to effect this purpose the weak-minded Charles IV and his son Ferdinand were induced to go to Bayonne, where Napoleon extracted from them the renunciation of the throne of Spain; meanwhile the Royal family of Portugal had fled to Brazil. On July 20th, 1808, Joseph Bonaparte entered Madrid, but the high-spirited Spaniards refused to accept him, and established a "junta" at Seville, declared Ferdinand VII king, and drove Joseph out of Madrid within a fortnight. The Spaniards, however, felt unequal to dealing with the usurper, and appealed to England, and not in vain —for an army of 10,000 men was speedily dispatched, under Sir Arthur Wellesley, who fought his first successful battle against the French in Spain on August 17th, 1808.

Roleria.—This was the battle of Roliça, erroneously called Roleria through a mistake in dispatches, and maintained on the colours ever since, when General Laborde was driven from the position he held. The following regiments took part in the battle—5th, 6th, 9th, 29th, 32nd, 36th, 38th, 40th, 45th, 60th, 71st (known with the 43rd as the "Glasgow Light Infantry," owing to the number of Lowland Scots in the regiments), 82nd, 91st, and the Rifle Brigade, which, as Wellesley stated in his dispatch, had just been formed, but the brunt of the fighting fell on the 5th, 9th, 29th, Riflemen of the 60th and the 95th, and also the flank companies of Major-General Rowland (afterwards Viscount) Hill's Brigade. Meanwhile Marshal Junot, the French Commander-in-Chief, decided to brook no delay, and advanced to attack Wellesley, who was at this time at the Vimiera heights near the Maceria River.

Vimiera.—In this battle (August 21st, 1808) Wellesley gave some indication of that power in organising defence which was to prove so valuable at Waterloo. The French attacked with considerable spirit, but the British retaliated with equal vigour and drove them off. In this battle the 50th, known as the "Blind Half Hundred" owing to ophthalmia having attacked the men, and as the "Dirty Half Hundred" owing to the black coming off their cuffs when rubbing their faces, made brilliant and determined charges, which, with those made by the 43rd and the 71st, compelled the French to retreat. In this battle a piper of the 71st, being wounded in the leg, sat upon his knapsack, and declared as he placed his pipes to his lips, "Diel hae me, lads, if ye shall want for music." There were no Victoria Crosses in those days, or George Clark would have had one. Wellesley, who had been superseded in the command by Lieutenant-General Sir Hew Dalrymple and Sir Harry Buzzard, now returned to England, disgusted, it is said, with the Convention of Cintra, by which the French army was permitted to evacuate Portugal and leave Lisbon in September for

Brittany, from whence it soon proceeded to re-enter the Peninsula by way of the Pyrenees. Lieutenant-General Sir John Moore, who arrived in Spain after the battle of Vimiera, was placed in command of the troops, consequent upon the recall of Sir Hew Dalrymple and the resignation of Sir Harry Buzzard. His glorious death at Coruña, after a brilliant retreat, is an imperishable memory.

Sahagun.—The story of the advance against Marshal Soult is marked by many brilliant episodes, among them the daring charge of the 15th Hussars on December 20th, 1808, at Sahagun, where, unexpectedly finding a broad ditch to pass, they leapt it in "a true fox-hunting spirit," and put the enemy to flight, with the loss of ten colonels and 160 men prisoners. As a result of this action, Sir John Moore established his headquarters at Sahagun. Though the 10th Hussars also took part in the affair, the 15th is the only cavalry regiment which won the right to carry the name on its battle-roll; the 7th, 18th Hussars, and the 3rd Hussars of the King's German Legion, were also present, and were later deemed worthy of the bar, likewise two batteries of Horse Artillery. It is noteworthy that only fifteen bars for Sahagun were issued, those who took part in the battle of Benevente as well, receiving a single bar inscribed with the two names.

Benevente.—On December 24th Sir John Moore thought it advisable to retire from Sahagun, and on the 28th reached Benevente in very inclement, snowy weather, where another brilliant cavalry affair took place on the 29th, when the 10th Hussars, under General Lord Paget, made a most dashing charge against the cavalry of the Imperial Guard, led by General Lefebvre-Desnouettes, who, together with 70 cavalrymen, were made prisoners. The English cavalry lost 50 killed and wounded, and the French 150 killed. Dr. Payne points out that the single bars for this and the previous battle, as well as the bar SAHAGUN-BENEVENTE, were issued so indiscriminately

PORTUGUESE PENINSULAR GOLD CROSS FOR COMMANDERS.
Awarded to Lieutenant-Colonel Richard Brunton.

that it is difficult to distinguish to which bar the recipient was entitled. There is a medal with the single bar BENEVENTE in the Chelsea Hospital.

Retreat to Coruña.—Then began the awful yet masterly retreat to Coruña. Through rain and snow, and over almost impassable roads, the British army of about 19,000 retreated with the French army of about 60,000, with 91 guns, hanging on without daring to do more than press the retreat, which became so demoralising to Moore's army, that not only had baggage to be abandoned, but guns and treasure to the amount of £25,000 were thrown over a precipice near Nogales to prevent any possibility of their falling into the hands of the enemy. And so, with his army of comparatively raw troops, Sir John Moore resolutely and defiantly made his way to Coruña, pursued by the veterans, under Ney and Soult, only to find on arriving at the port that the transports which were to bear them to England had not arrived! There was nothing to be done but to accept battle, and this he did in such a splendid manner that it makes his death in the hour of victory an imperishable memory, while the recollection of the valour of his much-tried and distressed soldiers will last as long as men respect courage, especially that which triumphs in the face of adversity. Sir John Moore was buried on the ramparts of Coruña by his staff officers and a few men of the Black Watch and other regiments he loved, but—

"No useless coffin enclosed his breast,
Nor in sheet nor in shroud we wound him;
But he lay like a warrior taking his rest,
With his martial cloak around him."

Over his soldier's grave the French Marshals placed a plain tomb, while Marshal Soult, who commanded the French, had placed upon a rock near which Sir John Moore fell a Latin inscription recording the fact. He was not allowed, however, to sleep in his soldier's grave, for the Spaniards removed his body to a more conspicuous position, and covered it with a magnificent monument eloquent of his deeds, but hardly in keeping with his simple spirit.

In the battle of Coruña, fought on January 16th, 1809, the 50th Regiment particularly distinguished itself by its use of the bayonet, as did also the 42nd, who made a daring and victorious dash into "the key of the fight," Elvina village—indeed, the battle is said to have been won by Lord William Bentinck's brigade; Sir David Baird's division, consisting of the 4th, 42nd, and 50th, which, as mentioned in general orders issued by Lieutenant-General Hope, "sustained the weight of the attack." The 23rd Royal Welsh Fusiliers were the last to embark.

The regiments engaged at Coruña also included the Grenadier Guards; 1st, 2nd, 4th, 5th, 6th, 9th, 14th, 20th, 26th, 28th, 32nd, 36th, 38th, 43rd, 50th, 51st, 52nd, 59th, 71st, 76th, 79th, 81st, 82nd, 91st, 92nd, 95th, and the Rifle Brigade; 7th, 10th, 15th, 18th Light Dragoons; 1st and 2nd Light Battalions King's German Legion, and the 3rd Dragoons of the same corps. Three batteries of the Royal Horse Artillery were also present.

Martinique.—This island was taken by the British in February, 1762, and on several other occasions, but a bar was only given for the campaign in 1809, which had been concluded, despite very bad weather, by its surrender to Lieutenant-General Beckwith on February 24th, 1809. The French were under Admiral Villaret-Joyeuse. It was restored to France in 1814. For this campaign a bar was granted when the decision was made to issue the Military General Service medal, commonly but erroneously known as the Peninsular medal, since not only were bars for Guadaloupe and Java included, but also for Fort Detroit, Chateauguay, and Chrystler's Farm in America.

For the capture of Martinique the superior officers were awarded a gold medal, similar to that I shall shortly describe as having been conferred upon the officers who served in the Peninsula.

The following regiments took part in this expedition—

60th, 63rd, and 90th; 1st West India Regiment and the flank companies of the 7th Royal Welsh Fusiliers; 8th, 13th Light Infantry, 15th, 23rd, and 25th Regiments.

Talavera.—In the spring of 1809, Sir Arthur Wellesley left England to take command of the Anglo-Portuguese army, and, landing at Lisbon on April 22nd, placed himself at the head of 25,000 British and Portuguese, crossed the Douro, drove Marshal Soult out of Oporto, advanced into Spain, and giving battle to the French at Talavera, defeated them after a long-drawn conflict lasting over the 27th and 28th July. The French were commanded by the intrusive King Joseph Bonaparte, assisted by Marshal Victor. The French losses in the first day's fighting were about 1,000 officers and men, the British losing about 800. For the two days the French had a total of 7,394 placed *hors de combat*, including Generals Lapisse and Morlot killed, and Generals Sebastiani and Boulet wounded; the killed numbered 944, and the wounded 6,274, while 150 were taken prisoners. The British losses amounted to 6,268; Generals Mackenzie and Langewith, 857 officers and men being killed; 3 generals, 192 officers, and 3,718 men wounded, 9 officers and 643 men missing. The Spaniards gave their losses as 1,200 killed and wounded. This battle, the victory in which was claimed for both sides, gained for the victorious General the titles of Baron Douro of Wellesley and Viscount Wellington of Talavera, with a pension of £2,000 per annum.

The following regiments were engaged in the battle of Talavera, those marked with an asterisk being specially mentioned in dispatches. 1st Batt. Coldstream Guards; 1st and 3rd Batts. Scots Guards; 2nd, 3rd, 4th, 5th, 6th, 2nd Batt. 7th Regiment,* 2nd Batt. 24th, 1st Batt. 29th,* 2nd Batt. 31st,* 1st Batt. 40th, 1st Batt. 45th,* 1st and 2nd Batts. 48th, 2nd Batt. 53rd,* 5th Batt. 60th,* 1st Batt. 61st, 2nd Batt. 66th, 71st, 79th, 82nd, 2nd Batt. 83rd, 2nd Batt. 87th, 1st Batt. 88th, 91st, 92nd, 1st Batt. 97th,* and the Rifle Brigade. 3rd Dragoon Guards, 4th Dragoons, 14th (now Hussars), 16th (now Lancers), and 23rd Light Dragoons; Royal Artillery, Engineers, and Staff Corps; 1st and 2nd Light Battalions, 1st Light Dragoons, and 1st, 2nd, 5th, 7th Line Battalions of the King's German Legion.

Peninsular Gold Medals.—Following this brilliant victory it was notified in the *London Gazette*, September 9th, 1810, that a gold medal would be granted to commemorate the battles of Roleria, Vimiera, Coruña, and Talavera. It was, however, only given to officers who had taken part in these engagements, or who had participated in any brilliant cavalry affair. It was also decided that the medal to which an officer would have been entitled, had he survived, should be given to his next of kin. The medals vary in size, but are all of the same design. The larger size was conferred upon General Officers, and the smaller ones upon colonels and senior officers. The medal was only given to those who were actually in command during an engagement, or succeeded thereto owing to the commander being killed or placed *hors de combat.*

The gold medal bears on the obverse the figure of Britannia, wearing a rather squat Roman-looking helmet, seated upon a globe, and bearing in her extended right hand a laurel wreath and in her left a palm branch, which rests upon an oval shield bearing the Union Jack; to Britannia's right is the British lion. On the reverse is a simple wreath of laurel, with the name of the battle and the date engraved in Roman capitals. The medal is attached to the ribbon by a simple suspender, and a gold buckle was frequently worn in the middle of the ribbon. General and Commanding Officers were ordered to wear the medal suspended from the neck by means of a red ribbon with blue edges—familiarly known as the Peninsular ribbon—and below the rank of lieutenant-general by the same ribbon from a buttonhole of their coat. The senior officers complained that it was very inconvenient to

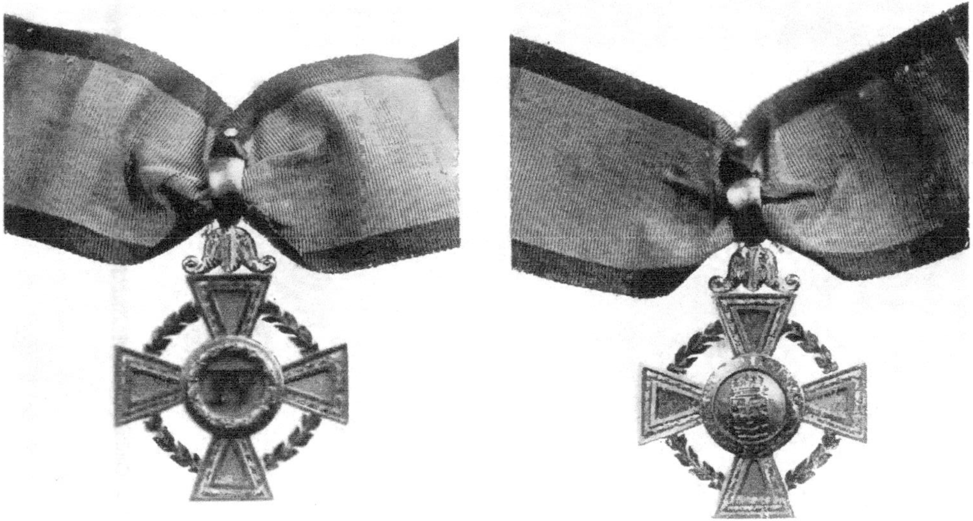

PORTUGUESE OFFICER'S CROSS FOR THE PENINSULAR CAMPAIGN.
Awarded to Lieutenant-Colonel Richard Brunton.

wear a medal dangling from the neck while on horseback, and they were supported in their plea by Wellington. This medal is generally found glazed, to prevent its surface being damaged, the name and rank of the officer being engraved upon the edge of the medal itself.

Guadaloupe.—General Beckwith, who had captured Martinique, attacked the West Indian Island of Guadaloupe, and captured it on February 6th, 1810. The French, who lost nearly 600 men, were commanded by General Ernouf (see **Martinique** for regiments engaged).

Busaco.—Wellesley's Spanish allies had failed him, as Sir John Moore with his sorry experience had predicted, and so after the battle of Talavera de la Regna he retired into Portugal, whither Napoleon sent Marshal Masséna to "drive the English Leopards into the sea," but the "Sepoy General," as Napoleon had dubbed Wellesley, was alive to the possibilities of the situation, and in face of the overwhelming numbers of the French, decided to prepare the impregnable lines of Torres Vedras, stretching from the Tagus to the sea; but before he could entrench himself he was compelled to check the pursuit of Masséna on the Busaco Sierra on September 27th, 1810, where his army of British and Portuguese resisted all attempts to dislodge them. In this battle both the British and Portuguese troops fought with daring determination, so that it would appear invidious to single out special regiments for notice, but the 38th, 45th, and the 8th Portuguese Regiments were mentioned in dispatches. The 88th Connaught Rangers repulsed a division of French infantry, while the 74th Highlanders did a like service. General Simon, who had bravely led his brigade up the mountain side, was captured by two men of the 52nd Regiment, and the French losses amounted to 2,000 killed and as many wounded. The British losses were 197 killed and 1,072 wounded. Marshals Ney, Masséna, and Regnier commanded the French, the English and Portuguese being under Wellington.

At the battle of Busaco the following regiments took part: the Coldstreams and Scots Guards; 4th, 14th, and 16th Light Dragoons, and three batteries of artillery; 1st Royal Scots 3rd Batt. 3rd, 5th, 7th, 9th, 11th, 24th, 27th, 29th, 31st, 34th, 38th, 39th, 40th, 42nd, 43rd, 45th, 48th, 50th, 52nd, 53rd, 57th, 60th, 61st, 66th, 74th, 79th, 83rd, 88th, 97th, and the Rifle Brigade; there were also present the following units of the King's German Legion; 1st and 2nd Light Battalions, and the 1st, 2nd, 5th, and 7th Line Battalions.

Rodrigues and Isle of France.—To keep our order chronologically we must leave for a moment the army in Spain to call attention to the fact that it had become necessary, in order to protect British shipping from the French ships harbouring in the Isle of France, or the Mauritius, that the place should be taken from them. Previous to this, as part of the same plan, it had been decided to also take the Island of Rodrigues, as a preliminary to the capture of the Isle of Bourbon. For the action which was fought at Rodrigues it was decided to award a medal to those who had taken part in the expedition in the then French islands, and on September 10th, 1811, a General Order was issued stating that a medal would be awarded to those who had taken part in the action at Rodrigues, the defeat of the French on the Isle of Bourbon, and the capture of the Isle of France. The medal was awarded to native troops only, although the 86th and 96th Regiments bear BOURBON on their colours. 2,201 were struck, 45 gold and 2,156 silver. The obverse bears a figure of a sepoy standing beside a field gun, holding the Union Jack unfurled in his right hand, and in his left his musket with bayonet fixed; with his left foot he tramples on a French colour and Eagle, whilst the British fleet at anchor forms the background. On the reverse is a Persian inscription surrounding a laurel wreath, stating that "this medal was conferred in commemoration of the bravery and accustomed fidelity exhibited by the sepoys of the English company in the capture of the Mauritius Islands in the year of the Hegira 1226." Outside the wreath is the record RODRIGUES VI JULY MDCCCIX. BOURBON VIII JULY & ISLE OF FRANCE III DEC. MDCCCX. The medal, $1\frac{9}{10}$ in. in diameter, was intended to be worn suspended from a silken cord.

It is noteworthy that a corporal of the 86th planted the regimental colours on the redoubt, and in the fight for the Isle of Bourbon Lieutenant Munro of the 86th was killed, also 17 men killed and 59 wounded. At the taking of the Isle of France only 29 of both services were killed, and less than 150 wounded and missing; besides the fleet of 70 sail, an army of 10,000 men under Major-General Abercromby took part in the expedition.

Barrosa.—On September 29th Wellington retired with the allied armies of 25,000 British and 30,000 Portuguese behind the famous lines of Torres Vedras, where his force was augmented by the arrival of English troops. Masséna made a series of ineffectual attacks, which had the effect of keeping Wellington's army from getting stale, and then retired to Santarem. Meanwhile a force of 4,500 British, under Lieutenant-General Graham, and 7,000 Portuguese under General Dom Manuel le Lapeña, reached Tarifa to operate against the rear of the French army under Marshal Victor at Chaclana. Graham had unfortunately contented himself with the command of his own troops, leaving Lapeña to assume the position of Commander-in-Chief. On April 5th, 1811, Lapeña halted his army on the Cerro de Puerco, now known as the heights of Barrosa. The Spaniards, lacking in discipline and devoid of *esprit de corps*, straggled into a situation which gave the keen-eyed French Marshal an opportunity he was not slow to take advantage of, for the blundering Lapeña had left a position which could have been held on the heights of Barrosa, with the result that his Spanish army was routed and driven towards the sea. Lapeña's stupidity had placed Graham in a very awkward situation. He had obeyed the command to move to Te Bermeja, but, apprised of the danger that threatened the Spanish General, he countermarched, only to find that Lapeña had bolted with his rabble of soldiery, and that the French held the key of the position. Graham determined to take the heights, and it is said there was never a shorter, more violent, or bloodier conflict; it lasted one hour and a quarter, and resulted in a loss to the French of about 3,000 killed and wounded, including General Ruffin and Brigadier Rousseau, who were mortally wounded, and six guns. The British out of their small army lost in killed and wounded about 1,243 men, including 50 officers and 60 sergeants. The 87th Regiment, Royal Irish Fusiliers, distinguished themselves by charging the French line, and capturing the Eagle of the 8th Infantry, the first to be taken during the war; two companies of the Coldstream Guards accompanied the 87th in its charge.

The following British regiments took part in the battle: 2nd Batts. of the Grenadier, Coldstream, and Scots Guards; flank companies of the 1st Batt. 9th Regiment; 1st Batt. 28th; flank companies 2nd Batt. 47th; 2nd Batt. 67th; flank companies 2nd Batt. 87th; 2nd Batt. 87th; 2nd and 3rd Batts. Rifle Brigade; Royal Artillery and Royal Engineers, also the 2nd Hussars of the King's German Legion.

The British had been under arms for twenty-four hours, and fought the battle with Lapeña and his men acting the part of spectators, except for the Spanish Walloon Guard and the regiment of Ciudad Real, who joined in the fray "impelled by the instinct natural to brave men." Napier, in comparing the two leaders at Barrosa, states: "The contemptible weakness of Lapeña furnished a surprising contrast to the heroic vigour of Graham, whose attack was an inspiration rather than a resolution—so wise, so sudden was the decision, so swift, so conclusive was the execution."

SPANISH CROSS FOR ALBUHERA.

SPANISH CROSS FOR CIUDAD RODRIGO.

SPANISH GOLD CROSS FOR VITTORIA.

Fuentes d'Onor.—On Sunday, May 5th, 1811, the battle of Fuentes d'Onor (Fountain of Honour) was fought. The British and Portuguese armies under Marshal Beresford had invested Almeida, upon which Marshal Masséna marched his army, and the British light division and cavalry retired as the French advanced upon the village, which gave its name to the battle, and together with the heights behind was occupied by the 1st, 2nd, and 3rd division. The French attack upon the village on May 3rd necessitated the aid of reinforcements, and the 71st (Highland Light Infantry), the 79th (Cameron Highlanders), and the 24th (South Wales Borderers) were dispatched, the first-named regiment charging the enemy and driving them out of that portion of the village which they had occupied. The contest raged until nightfall, and over 250 officers and men fell on each side in what was practically a hand-to-hand conflict. The 4th was spent quietly, but on the 5th Masséna decided to try the fortune of the day, which at the commencement seemed to smile in his favour, for Junot carried the village of Posa Velha, while the French cavalry not only drove in the cavalry of the allies, but the whole movement caused Wellington to make a daring change of position, during which an episode unique in the annals of warfare took place. In the retirement, two guns, under Captain Norman Ramsey, had been left behind, and the French and British also were astounded to see Ramsey charge through the French masses at the head of his battery, "his horses breathing fire and stretching like greyhounds along the plain, his guns bounding like things of no weight, and the mounted gunners in close and compact order protecting the rear." Shortly after the 2nd Battalion of the Black Watch again distinguished itself by repulsing a charge of French dragoons.

While the fighting was proceeding on the vast plain, a furious attack was made on Fuentes d'Onor, which was desperately held by the 24th, 71st, and 79th Regiments, but the unequal contest was determined by the bayonet charge of the 88th Connaught Rangers. Nightfall compelled the cessation of hostilities, and both armies lay down to rest, looking to the morrow to decide the battle, but the French gave no sign of battle, and on the 8th retreated on the road to Ciudad Rodrigo. Masséna claimed Fuentes d'Onor as a victory, despite the fact that his object, to relieve Almeida, was unattained. In this sanguinary battle the British lost 200 killed, 1,028 wounded, and 294 missing. It is said that Masséna's total loss was about 3,000; of the French dead 200 were found in the lower village of Fuentes d'Onor. This was Masséna's last battle, Napoleon recalling the *enfant chéri de la victoire,* who returned to France broken-hearted, and when opportunity offered gave his allegiance to Louis XVIII.

The regiments engaged in the battle were the 2nd Foot Guards (Coldstreams), 3rd Foot Guards (Scots Guards), 1st Dragoons; 14th and 16th Light Dragoons; 1st, 4th, 5th, 9th, 24th, 30th, 38th, 42nd, 43rd, 44th, 45th, 50th, 51st, 52nd, 60th, 71st, 74th, 79th, 83rd, 85th, 88th, 92nd, 94th, and the Rifle Brigade; Horse Artillery; 1st Hussars; 1st and 2nd Light Battalions, and the 1st, 2nd, 5th, and 7th Line Battalions of the King's German Legion.

Albuera.—"Modern history," wrote General Picton, "presents no example of an action so obstinately disputed as that of Albuera." It was there that the "Die-hards" gained their coveted name, and when the roll-call of the "Buffs" was called, after the battle, only three privates and a drummer answered to their names. It was the fight for the regimental colours which accounted in a degree for the grave loss of life, so Lieutenant-General Cust states. Ensign Thomas refused to yield the colours and was killed. Lieutenant Latham, when severely wounded, tore the precious flag he had received from Ensign Walsh from the broken staff, and concealed it under his body, where it was found after the battle by a sergeant of the 7th Fusiliers.

Latham Gold Medal.—Latham, who had a wonderful

recovery, was presented by the officers of the "Buffs" with a gold medal commemorating the event, and a royal authority was given him to wear it. The "Buffs," as I have indicated, were almost annihilated—, 4 officers, 4 sergeants, and 208 rank and file were killed; 13 officers, 11 sergeants, 1 drummer, and 222 rank and file wounded; 1 officer wounded and taken prisoner; 2 officers, 15 sergeants, 1 drummer, and 161 rank and file missing—a loss of 644 officers and men in one battle. The "Buffs" had already lost 27 killed and 208 wounded, and 7 missing, at Talavera. The 57th, Middlesex Regiment, who were called upon by their dying Colonel Inglis to "die hard," went into action 570 strong, but when the battle was over, 23 officers and 400 rank and file "were lying as they had fought, in ranks, with every wound in the front."

The battle was fought on May 16th, 1811, the Spanish leaders agreeing that Marshal Beresford should have supreme command, and to receive battle in the village of Albuera. The allied army consisted of 32,000 infantrymen and 2,000 cavalry, with 38 guns; half the force was Spanish, and the other half equally composed of British and Portuguese. The Spaniards, as usual, were slack, and did not get into position properly. The battle opened by the French attacking the heights on the right of the Spaniards, and though they fought stubbornly at first they gradually gave way before the onslaught of the French. Matters had become very serious when General Stewart sent the leading brigade of his division up the hill under General Colborne; the advance was made in a drizzling rain. The French made several brilliant charges, and in the *mêlée* the commanding officer, Marshal Beresford, was attacked by a Polish lancer, whom he seized bodily and threw from his saddle. Those were heroic days indeed! The bad weather, which had so far favoured the French, actually helped to save the day for the English, for the drizzling rain prevented Marshal Soult from appreciating the sad condition of the British. On the heights the 31st had resolutely held its ground; other regiments had been cut to pieces, and Beresford thought of retreat, when the cool daring of Colonel (afterwards Lord) Hardinge, who took upon himself to order General Cole to advance, and, taking General Abercromby's brigade to the ground, saved the day—to which, by the way, the Portuguese regiments of the fourth division contributed—General Cole leading the 7th and 23rd Fusiliers in person. Cole drove off the Lancers and recovered the lost guns, but ere the enemy were driven from the ridge, they had fired a terrific volley from all the guns, which killed Sir William Myers, wounded the intrepid Cole and three officers, while the Fusiliers "reeled and staggered under the iron tempest like sinking ships, but nothing could stop them." Soult, rushing into the thickest of the fight, tried to encourage his brave soldiers to stand; it was useless. "Suddenly and sternly recovering, they closed on their terrible enemies, and then was seen with what a strength and majesty the British soldier fights." To again quote Napier, "Nothing could stop that astonishing infantry . . . their measured tread shook the ground, their dreadful volleys swept away the head of every formation . . . the French reserve, mixing with the struggling multitude, endeavoured to sustain the fight, but the effort only increased the irremediable confusion; the mighty mass gave way, and like a loosened cliff went headlong down the steep. The rain flowed after it in streams, discoloured with blood, and 1,500 unwounded men, the remnant of 6,000 unconquerable British soldiers, stood triumphant on the fatal hill." The dead and wounded lay in two distinct lines on the ground, and in such compact masses that 7,000 bodies occupied so small an area that the artillery advancing into action had to pass over the living and the dead, friend and enemy.

The Immortals.—The "Die-hards," the "Buffs," the 7th Fusiliers, and the 23rd Welsh Fusiliers immortalised themselves in that bloodiest of all battles,

ALCANTARA MEDAL, 1809. (Obverse.)

MILITARY GENERAL SERVICE MEDAL.

ALCANTARA MEDAL, 1809. (Reverse.)

where in four hours no less than 15,000 men were placed *hors de combat.* Well might M. Thiers say that "a sort of fatality rendered the heroic bravery of our troops powerless against the cold courage of the English." The 23rd lost 347 men, 80 killed, and 257 wounded and missing, and it is noteworthy that one company was brought out of action by a corporal. The 29th mustered only 2 captains and about 100 junior officers and men; the 57th was little better off, likewise the 48th when the French were driven across the Albuera River. Of the 7,000 British soldiers engaged, 4,300 were killed or wounded; the dead included Major-General Houghton, who was killed while leading his division, 33 officers, 33 sergeants, and 917 men. The Spanish and Portuguese allies lost about 2,000 men killed or wounded, and as the allied army consisted of about 34,000 men, the brunt of the fighting was borne by the British troops. The French, who numbered about 33,000, lost about 1,000 more men than the British, including 5 of their General Officers killed or wounded; 2,000 dead and 800 of their worst wounded were left on the field when the French drew off, leaving the British uncertain as to what the morrow might bring forth, but reinforcements coming for the British, Soult moved off in the direction of Badajoz, followed by Beresford.

The following regiments took part in the battle: 3rd Dragoon Guards; 4th Dragoons; 13th Light Dragoons; 1st, 3rd Foot (the "Buffs"); 1st and 2nd 7th (Royal Fusiliers); 1st 23rd (Welsh Fusiliers); 1st 27th Inniskillings; 2nd 28th; 29th; 2nd 31st; 2nd 34th; 2nd 39th (Dorsetshire); 1st 40th; 1st and 2nd 48th (Northamptonshire); 1st 57th (West Middlesex); 1st and 2nd 60th and four companies of the 5th Battalion; 2nd 66th (Berkshire); 97th Queen's Own; 1st and 2nd Light Battalions of the King's German Legion, and one company of Brunswick Oels.

Java.—The capture of the East Indian island of Java, which the Dutch called the most precious jewel in their diadem, was effected after operations lasting from July 27th, 1811, when Lieutenant Edmund Lyons—after Admiral Lyons—made a daring landing with a few seamen, until August 26th, ten days after the battle fought at Serondel near Samarang. During this expedition the storming of Fort Cornelis took place, when the enemy lost 1,000 killed; and three Brigadiers, 30 Field Officers, 70 Captains, 140 Subalterns, and 5,000 rank and file prisoners; 400 cannons were captured. A million sterling was received by the British force as prize money. The British loss (naval and military) was 150 killed, 788 wounded, and 16 missing. By the Treaty of Vienna Java was restored to the Dutch in 1814.

H.E.I. Co.'s Java Medal.—A medal, 1·9 in. in diameter, was awarded by the Honourable East India Company to the sepoys who took part in the expedition. On the reverse is depicted the attack upon Fort Cornelis, with a British flag flying above the Dutch, and above all CORNELIS. The following is inscribed in Persian upon the reverse: "This medal was conferred in commemoration of the bravery and courage displayed by the sepoys of the English company, in the capture of the Kingdom of Java in the year of the Hegira 1228." "Java conquered XXVI August MDCCCXI." 133 gold and 6,519 silver medals of this type were struck.

King George III conferred gold medals upon the officers engaged, and when the Naval and the Military General Service Medals were awarded, the surviving participants in the capture received a medal with bar inscribed JAVA. Four distinct medals were issued for this expedition.

The military present at the capture of Java were the 14th (now West Yorkshire Regiment), 59th (now 2nd East Lancashire Regiment), 69th (now 2nd Welsh Regiment), 78th Highland Regiment (now 2nd Seaforths), 89th (now Royal Irish Fusiliers), Bengal Volunteers and sepoys of "John Company."

Ciudad Rodrigo.—The siege of Ciudad Rodrigo followed the "soldiers' battle" of Albuera. The army, after considerable manœuvring, marching, and counter-marching, besieged Ciudad Rodrigo, which, together with the siege of Badajos, had to be undertaken before Wellington's plan of campaign could be carried out. On January 1st, 1812, the order for the siege of the fortress was suddenly issued, but some days elapsed before the order could be put into execution, owing to the inclement weather, snowstorms preventing any possibility of operations. On the 8th the investment of the fortress was commenced, and the earth redoubt of Teson Grande stormed and taken by Colonel Colborne with three companies of the 52nd Regiment. On the 13th the Convent of Santa Cruz, one of the two convents which had been fortified by the French, was stormed and carried by the light companies of the brigade of Guards, and on the night between the 14th and 15th the 40th Regiment escaladed and carried the fortified San Francisco Convent. Meanwhile preparations had been progressing for the general assault, which took place on the night of January 19th. Wellington had written in his orders "Ciudad must be stormed this evening," and the order was carried out in a brilliant manner, but alas! two British Generals were killed, General Crauford being mortally wounded whilst leading the light division, and General Mackinnon killed on reaching the ramparts of the greater breach. In this assault General Picton adjured the Connaught Rangers, who passed with the forlorn hope to storm the breach, to "spare powder and trust to cold iron," and it was by the use of the bayonet that the 43rd and 95th drove the French from the "fausse braye." The city was taken after a siege lasting twelve days, but the troops "committed frightful excesses," which some have condoned as a prescriptive right which successful besiegers may enjoy. The British losses were 9 officers and 217 rank and file killed, 84 officers—including Major-General Vandeleur and Major (afterwards Sir George) Napier, who lost an arm—and 1,000 men wounded. For the successful issue to his generalship, Wellington received the patent of an earl, and was made a Spanish duke, the British Government increasing his annuity to £4,000 per annum.

The regiments engaged in the siege were the 2nd Foot Guards (the Coldstreams); 3rd Foot Guards (Scots Guards); 5th, 7th, 24th, 30th, 40th, 42nd, 43rd, 45th, 48th, 52nd, 60th, 74th, 77th, 83rd, 88th, 94th, 97th, and the Rifle Brigade.

Siege of Badajoz.—Before bringing to a successful issue the siege of Ciudad Rodrigo, Wellington had been busily engaged in preparing for the third siege of Badajoz. On March 16th, 1812, it was invested, but the weather for some time was particularly unfavourable to the besiegers, rain falling in torrents so that the men in the trenches were knee-deep in mud and water. By March 23rd, however, the weather had improved, and the ground was in better condition for placing the guns, which by the 25th were in position at Picurina, the place being taken by storm on that day. On the 26th breaching batteries opened fire on the city, and by April 6th it was deemed possible to storm the castle by escalade, Picton's division being ordered to do so. The bastion of La Trinidad was to be stormed by Colville's division, and the Santa Maria by the light division under Colonel Barnard. while the lunette of San Rocque was to be carried by the 48th under Major Wilson. At 10 o'clock at night the attack was made: General Kempts, leading his brigade to the foot of the castle, was there wounded, but, his men dashing on, the castle was won in an hour and a half. "At the breaches the tumult was such as if the earth had been rent asunder . . . the carnage was frightful. It is doubtful whether, since the invention of gunpowder, any mass of men had ever been more fearfully exposed to all its murderous power. The dying were piled upon the dead in mounds, which the living could not pass; and the French soldiers, undisturbed in their avocation, raised the deriding cry

PENINSULAR GOLD CROSS.

PENINSULAR GOLD MEDAL WITH BARS.

'*Vive l'Empéreur!*'" By midnight 2,000 men had fallen outside the city, and Wellington ordered the troops to withdraw; but this was as impossible as to go forward. Meanwhile some of the 5th division had entered the town (the 4th Regiment being in first), which by 6 o'clock the next morning was surrendered by the Governor. The besiegers lost 131 officers and 1,707 rank and file killed, and 564 officers and 6,083 men wounded. Napier states that "when the havoc of the night was told to Wellington, the pride of conquest sunk into a passionate burst of grief for the loss of his gallant soldiers," for in the assault alone the British casualties were 59 officers and 744 men killed, and 258 officers and 2,600 rank and file wounded. No wonder one of the Connaught Rangers exclaimed with an oath, "Och! Boys, Soudradrodrago was but a flay-bite to this." The 43rd Monmouth Light Infantry lost more men than any other regiment; 20 officers, including its chief, Colonel M'Leod (who was only twenty-seven), 335 sergeants and privates killed and wounded. Alas! the Saturnalia of Ciudad Rodrigo was repeated, and the disciplined men who had fought with such valour gave way to demoniacal passions which we of to-day can scarce believe. The indignity of the threatened gallows was the only way in which Wellington could curb his men, but not before several of the worst plunderers had been executed.

The following regiments were engaged at the siege and storming of Badajoz: 2nd Foot Guards; 1st, 4th, 5th, 7th, 23rd, 24th, 30th, 38th, 40th, 43rd, 44th, 45th, 48th, 52nd, 60th, 74th, 77th, 83rd, 85th, 88th, 94th, and 95th Regiments, the Rifle Brigade, and the 13th and 14th Light Dragoons.

Salamanca.—The battle of Salamanca, fought on Sunday, July 22nd, 1812, is noteworthy as the first general action in the Peninsular War in which Wellington attacked; his decision was due to Marmont making a sudden movement which threatened to cut the British lines of communication with Portugal, but although the position appeared to be a critical one for the British General, he turned to the Spanish General Alva and said, "Mon cher Alva, Marmont est perdu!" In this battle the 3rd division, under Major-General Pakenham, were the first in action, and brilliantly carried everything before them. In this division were the Connaught Rangers, who, incensed by the death of a favourite officer, Major Murphy, could hardly be kept in hand. General Pakenham, noticing their impatience, ordered them to "be let loose," when they charged with an impetuosity which nothing could withstand, and together with the 45th and 74th broke through the masses of infantry, bayoneting all they could overtake. Composed of such material, no wonder the 3rd was called "the fighting division." It took just forty minutes to defeat Marmont's left wing, which in his endeavour to close the Ciudad road had got separated from the right, and into the gap made thereby Wellington poured his men. The right, however, reinforced by those who had escaped the conflict on the left, made a very determined resistance, and the shades of evening were falling ere the enemy made a last brave effort to retrieve the day, and, indeed, appeared to be on the way to doing so when Wellington ordered Clinton's division, numbering 6,000 bayonets, to advance, which, after a furious struggle, compelled the French to give way in confusion. Night had set in, and the remnants of Bonnet's division, which Clinton had put to rout, had the good fortune to find a means of escape through the abandonment of the ford of Alba de Tormes by the Spaniards. For six hours the battle of Salamanca raged with unabated fury, and with varying fortune on the right, so that the divisions which had been actively engaged, on an exceedingly hot day, were glad to bivouac on the ground where they had fought a battle, which again demonstrated the brilliant commandership and the keen, quick eye with which Wellington detected an opportunity when it offered. Salamanca was, in the opinion of experts, the most skilful of any of his victories, though the mistakes which Marmont made would have been obvious to men of lesser capacity. A decisive battle, it would have proved much more fatal to the French had darkness not precluded the possibility of any considerable effort at pursuit on the part of the victors. As it was, however, Wellington pressed the retreating French to the ford of Huerta, and ordered the cavalry to follow the fugitives; continually pressing on the rear, Wellington forced his way on to the romantic city of Valladolid, which he entered in triumph on July 31st. On August 12th he entered Madrid, to the plaudits of the people, who hailed him as the deliverer of their country.

In this battle the allies' losses exceeded 5,000. The British lost General Le Marchant, who was killed while leading a brilliant and successful charge of the heavy brigade, 24 officers, and 686 rank and file killed. Field-Marshal Beresford, Lieutenant-Generals Cotton, Cole, Leith, and Major-General Allen were wounded, besides 182 officers of inferior rank; 4,270 of the rank and file were wounded. The Portuguese lost about 304 killed and 1,552 wounded; the Spaniards lost—4! The French, who had at the battle about 42,000 men and 74 guns, lost about 7,000 men, besides Generals Desgraviers, Ferrey, and Thormières killed. Marmont, the Commander-in-Chief, was badly wounded early in the action by a shell; Bonnet was severely wounded, and Clausel, who commanded after the disablement of Marmont and Bonnet, slightly; 130 officers and 7,000 men were taken prisoners, and 2 Eagles, 6 standards, and 11 pieces of artillery were captured by the victors.

Essentially a General's victory, with exceedingly far-reaching consequences, Wellington was rewarded with a marquisate, an augmentation of his coat of arms to commemorate his services, and a grant of £100,000 to maintain the dignity of his rank; while the Spanish Regency presented him with the Order of the Golden Fleece. The vanquished Marmont, Duke of Ragusa, on the other hand, was told by Napoleon that he had "sacrificed to vanity the glory of the country, and the good of my service."

The following regiments were present at Salamanca, although some were in reserve: 2nd and 3rd Foot Guards; 1st, 2nd, 4th, 5th, 7th, 9th, 11th, 23rd, 24th, 27th, 30th, 32nd, 36th, 38th, 40th, 42nd, 43rd, 44th, 45th, 48th, 51st, 52nd, 53rd, 58th, 60th, 61st, 68th, 74th, 79th, 83rd, 88th, 94th, and 95th Regiments; artillery and Rifle Brigade; 5th Dragoon Guards; 3rd, 4th, 11th, 12th, 14th, 15th, and 16th Light Dragoons. The King's German Legion was represented by the 1st Hussars; 1st and 2nd Light Battalions; 1st, 2nd, and 5th Line Battalions.

Fort Detroit.—From that unfortunate day in 1773, when the youth of Boston boarded the ships and cast the cargoes of tea into the sea, until the signing of the Treaty of Ghent in 1814, the relations between the English-speaking peoples were continuously strained; the centenary of that auspicious day, and the happy continuance of peace, we celebrate this year. With the events which led up to the outbreak of hostilities I am not here concerned, my province being merely to record such facts as will help in the appraisement of the intrinsic, historical, or sentimental value of medals awarded to those who have carried the burden of war. The capture of Fort Detroit was an episode which followed the declaration of war by the American Senate on June 18th, 1812. Early in July 2,500 Americans, under General Hull, crossed the Detroit and invaded Upper Canada, but, opposed by about 1,330 Canadian militia and regulars under Major-General Brock, they were forced back to Fort Detroit, on the American side of the St. Laurence, which Brock proceeded to invest, but having constructed his batteries and opened fire, before making the assault he sent his aide-de-camp to summon the United States General to obviate bloodshed by surrendering. This he did on August 16th.

The regulars represented in this affair were 30 Royal

Knight's Badge, Military Order of San Fernando.
M.G.S. Medal.
Officer's Gold Cross for Peninsular.
King John VI's Jewelled Badge, for Special Service.
Knight's Badge, Order of the Tower and Sword.
Badge and Star (Knight Commander's) of the Order of St. Bento d'Aviz.
Officer's Badge, Order of the Tower and Sword.

Artillerymen, about 250 men of the 41st (the Welsh Regiment), which, together with its linked battalion, the 69th, bears DETROIT on its colours; 50 of the Royal Newfoundland Regiment; 400 Canadian militiamen, and 600 Indians.

Vittoria.—Following the abortive siege of Burgos, September 19th, 1813, Wellington retired to Frenada and Coria on the frontiers of Portugal, and went into winter quarters to await reinforcements. They having arrived, and the army being well equipped and in a good disciplinary spirit, he commenced the second part of the Peninsular campaign, which was destined to be a series of successes for the British commander, who commenced the final campaign in May 1813, by bidding adieu to Portugal, and marching into Spain with 70,000 men. He swept everything before him to the Pyrenees. "Neither," says Napier, "the winter gulleys, nor the ravines, nor the precipitate passes among the rocks, retarded even the march of the artillery—where horses could not draw, men hauled; when the wheels would not roll, the guns were let down or lifted by ropes—six days they toiled unceasingly, and on the seventh (June 20th) they burst like raging streams from every defile, and went foaming into the basin of Vittoria." There Joseph Bonaparte and Marshal Jourdan had taken up the position from which they were to be driven by the forceful Wellington, who spared nothing to achieve his object. "Never," wrote Napier, "was an army more hardly used by its commander, and never was a victory more complete." It was a terrific struggle; "the hills laboured and shook, and streamed with fire and water," ere the pseudo-king, Joseph Bonaparte, and his army fled in confusion, leaving all his personal belongings, and the rich pictures and plunder he had taken from every part of Spain, strewn on the road, for whosoever had a mind to acquire; 151 guns, 415 caissons of ammunition, 4,000 rounds for guns, and 2,000,000 musket cartridges were also abandoned, and a full military treasure-chest. Indeed, never had such an accumulation of military stores and private wealth been abandoned by a routed army.

The French General Gazan, who took part in the battle, recorded that the French "lost all their equipage, all their guns, all their treasure, all their papers, so that no man could prove how much pay was due to him." Joseph Bonaparte's carriage was abandoned in the street in his haste to evade Captain Wyndham, who made a bold dash to secure him; the Sword of State, emblematic of the kingship which the brother of Bonaparte had claimed, but which he had lost for ever, and the marshal's baton belonging to Jourdan, together with the Eagle of the 100th Regiment, were among the spoils. It is also recorded that "a perfect herd of women, including General Gazan's wife, and a number of the wives, mistresses, actresses, and nuns, belonging to officers and men of the French army, were abandoned." The Field Marshal's baton, which was taken by the 87th Royal Irish Fusiliers, commanded by Lieutenant-Colonel (afterwards Viscount) Gough, was sent to the Prince Regent, who responded by sending Wellington the Field Marshal's baton of Britain.

The French loss in killed and wounded was estimated by M. Thiers at 5,000; the allies' losses were, British 500 killed, 2,300 wounded, and 266 missing; the Portuguese loss was about 150 killed and 900 wounded; the Spanish losing 89 killed and 460 wounded.

The British regiments present were the 1st and 2nd Life Guards and Horse Guards, represented by two squadrons; 1st, 3rd, and 5th Dragoons; 3rd, 4th, 10th, 11th, 12th, 13th, 14th, 15th, 16th, and 18th Light Dragoons; 2nd and 3rd Foot Guards; 1st Royal Scots; 2nd Queen's Royal; 4th King's Own Royal; 5th Northumberland Fusiliers; 6th Royal 1st Warwickshire; 7th Royal Fusiliers; 9th East Norfolk; 20th; 23rd Royal Welsh Fusiliers; 24th; 27th Inniskilling; 28th North Gloucestershire; 31st, 34th, 38th Staffordshire; 39th Dorsetshire; 40th; 43rd Monmouth Light Infantry; 45th Sherwood Foresters; 47th Lancashire; 48th Northamptonshire; 50th Queen's Own; 51st King's Own Light Infantry; 52nd Oxford Light Infantry; 53rd Shropshire; 57th West Middlesex; 58th, 59th, 60th King's Royal Rifle Corps; 66th Berkshire; 68th Durham Light Infantry; 71st Highland Light Infantry; 74th Highland Regiment; 82nd Prince of Wales Volunteers; 83rd; 87th Royal Irish Fusiliers; 88th Connaught Rangers; 92nd Gordon Highlanders; 94th; 95th Derbyshire.

Pyrenees.—The rout at Vittoria gave Wellington possession of the passes of the Pyrenees, and when in 1849 it was decided to decorate the surviving soldiers of the Peninsular War, all those who had taken part in the series of extraordinary actions in the Pyrenees, Roncesvalles, Maya, Santarem, and Buenzas, from July 28th to August 2nd, 1813, were awarded the medal with bar inscribed PYRENEES, and those who were entitled to the medal for other actions received the additional bar. Napoleon ordered his brother to retire into private life and instructed the officials not to pay him the respect due to a monarch. Soult was ordered to take up the command, and did so with a zest and ability that has called forth the admiration of all who recognise in the dread game of war that sportsmanlike character which thoughtfully dares all and takes its rebuffs with a smile.

Siege of Sebastian.—Meanwhile, Wellington instructed the Spanish Generals to blockade the fortress of Pampeluna, and besieged San Sebastian, which was blockaded by the British Navy on July 3rd, and invested by the troops under Sir Thomas Graham on the 9th. During this, the first siege of San Sebastian, a Frenchman states that on the 19th and 20th for fifteen hours the British fired 350 shots per gun. On the 24th General Graham ordered the assault, and the way was led by Colin Campbell (afterwards Lord Clyde) of the 9th, and Machet the engineer. Campbell, doing many a daring deed, saw all his friends dead round him, while he, seriously wounded, was spared to achieve greater things. In this assault 44 officers of the line and 500 men were killed, wounded, or taken prisoners. The battalion of the Royal Scots present alone lost 87 killed and 246 wounded in the unsuccessful assault. In the meantime Marshal Soult, who had reached Bayonne on July 13th to take command of the united French army, attacked Major-General Byng's brigade at Roncesvalles on the 25th, while Count D'Erlon proceeded to attack Sir Rowland Hill in the Pass of Maya, where very desperate fighting ensued. The fighting at Roncesvalles and Maya continued until nine in the evening, when it was deemed advisable to abandon the Pass of Maya to the enemy.

Gallantry of Gordons at Maya.—In the day's fighting the allies lost 1,600 men and 4 guns. In the Pass of Maya the British troops had been engaged for ten hours, and it is noteworthy that the 92nd Gordon Highlanders, being without ammunition, although ordered by General Stewart not to charge, could not be restrained, and not only charged, but led a charge against the enemy. In this engagement the 92nd lost 1 officer and 34 men killed, and 18 officers and 268 men wounded, one wing being practically annihilated. Napier states in his history of the war that "so dreadful was the slaughter, especially of the 92nd, that it is said the advancing enemy was actually stopped by the heaped dead and dying. The stern valour of the 92nd, principally composed of Scotsmen, would have graced Thermopylæ." In this sanguinary combat the 82nd also particularly distinguished itself.

On the 28th the British were attacked in position on a mountain ridge between the valleys of the Lanz and the Guy, with Byng's brigade in reserve on the second ridge of Huerta, and the 6th division adjoining; the latter had scarcely got into position, when General Clausel attacked it from the Sorauren side; the fight

PRUSSIAN MEDALS FOR NAPOLEONIC WARS.

BRONZE MEDAL TO BRITISH GERMAN LEGION FOR PENINSULAR WAR.

SILVER MEDAL TO HANSEATIC LEGION, 1813-15.

(See section, Some Continental and Foreign War Medals, p. 357.)

became general, and 27th and 48th Regiments, charging, "rolled back the enemy in disorder, and threw them headlong down the mountain side." With admirable heroism the French soldiers returned three times to the charge, but their efforts could not avail against the dogged determination of the allied troops. In this affair every regiment in the 4th division, the 40th, 7th, 20th, and 23rd, charged four times, while Major-General Ross had two horses shot under him. On the 29th the rival armies were inactive, Soult apparently considering his plans for the relief of Pampeluna and San Sebastian. He decided to abandon any advance on the former, and proceeded to relieve San Sebastian, but by masterly manœuvres Wellington checkmated his opponent, and compelled him to abandon the endeavour, and on August 2nd the French troops evacuated Spain at all points.

For nine days the armies had confronted one another, and the allies had lost in the different actions no less than 7,096 officers and men killed and wounded, while the aggregate loss of the French is put down at 15,000. At the conclusion of this series of conflicts Sir William Stewart was wounded, and Wellington narrowly escaped being taken prisoner by a French detachment which surprised him whilst studying a map; the French Commander-in-Chief, Soult, also narrowly escaped with his liberty.

St. Sebastian.—While Soult and Wellington had been busily engaged, General Rey, the Governor of San Sebastian, had employed himself in strengthening the city and castle, so that when the second siege was renewed on August 5th, 1813, he was in a position to make a very desperate defence. Many inactive days were spent, to the chagrin of Wellington, who complained of the exceeding inactivity of the navy, which compelled him to lose more than half of August. However, by the 26th he had breaching guns and mortars in position, and with these he opened fire upon the city, and as a result 250 yards of the walls were reduced to ruins; but it was not until the night between August 30th and 31st that the place was assaulted. Then, again, the British soldier gave proof of his cool bearing, not once, but again and again. At 8 a.m. the town was bombarded by all the batteries, and the fire was kept up until 11 a.m., when the order for the assault was given. A storming party of 750 volunteers was asked for from the 1st, 4th, and light divisions, "men who could show other troops how to mount a breach." They successfully passed one danger which destroyed a party of thirteen that had rushed to cut the *saucisson* of a mine, to die in mounting the great breach. At the sea wall hundreds of the British were killed by the explosion of a mine, and then "began a frightful slaughter," for there was hardly another conflict so desperate and so sanguinary as that which took place at the storming of the great breach; "the appearance of the breach was perfectly delusive; nothing living could reach the summit; no courage, however desperate, could overcome the difficulties, for they were alike unexpected and unsurmountable." The officers encouraged their men by word and action, "crowd after crowd were seen to mount, totter, and to fall, and at length the whole mass sank to the bottom of the breach, but remained stubborn and immovable on this lower part." General Graham, beholding this, resorted to an unparalleled expedient by ordering all the breaching batteries to fire over the stormers' heads on to the inner wall. This daring inspiration saved the situation, for the terrible storm of missiles, passing over the heads of the British troops, did their deadly work by demolishing the defences, and killing or driving away the defenders. Another furious effort was made upon the breach, and when this seemed as though it must fail, explosions occurred all along the enemy's defences, and the British soldiery, breaking into the first traverse, poured through the town, pushed past the barricades, which the defenders could not hold even for a few moments in face of the impetuous stormers, who were, however, compelled to leave the castle unattacked. In this terrible conflict the chief engineer, Sir Richard Fletcher, was killed; Burgoyne, the second engineer, wounded; also Generals Sir James Leith, Oswald, and Robinson. The total loss of the besiegers on the day of the capture was 761 killed, 1,697 wounded, and 45 missing. The castle was pluckily defended, but on the 8th was reduced to ruins, and the garrison, surrendering, were allowed to march out, the brave Rey at their head, with all the honours of war. The siege had lasted seventy-three days, during which time nine assaults had been made. The disgraceful scenes of Ciudad Rodrigo and Badajoz were surpassed at San Sebastian by the addition of revolting cruelty to the excesses of drunkenness, and the crimes of rape and murder.

The regiments engaged in the siege and capture were 200 of the Guards; 1st, 2nd, 4th, 7th, 9th, 11th, 20th, 23rd, 24th, 27th, 36th, 38th, 40th, 43rd, 47th, 48th, 51st, 52nd, 53rd, 59th, 68th, 82nd, 85th, 87th, 88th, and 95th, but only the Royal Scots, 4th, 9th, 38th, 47th, and 59th Regiments, are permitted to carry ST. SEBASTIAN on their colours.

Chateauguay.—This battle, fought on October 26th, 1813, occurred during the military campaign in America. Having achieved a series of successes, the Americans concentrated about 18,000 regulars and 10,000 militia at Sackett's Harbour. Sir George Prevost, thinking it expedient, quitted Kingston and proceeded to the capital, whilst Sir R. Skeaffe assembled as many troops as he possibly could, 8,000 militia readily responding to the call. Major-General Hampton, commanding the eastern division of the United States army, crossed the Canadian border with 8,000 men on October 21st, having arranged to co-operate with General Wilkinson, who was in command of about 10,000 men, but through some misunderstanding the two forces were unable to combine, and on October 26th General Hampton found himself faced by a body of 800 men, under Lieutenant-Colonel De Salaberry, who disposed his little army so admirably that he not only caused the enemy a severe loss, but compelled Hampton to fall back to Plattsburgh, from which he "had not the resolution again to return to the frontier."

Those engaged in this affair were the Canadian Fencible Light Infantry, Canadian Militia, two companies of Voltigeurs, some Indians, and Royal Artillery.

The Peninsular Gold Cross.—The long-drawn character of the Peninsular War, and the frequent general engagements which called for recognition, occasioned the recipients of the gold medals inconvenience in wearing them; in consequence the names of battles were sometimes engraved on the medals as suggested in Wellington's dispatch of October 1st, 1811, dated from Richoso. Subsequently an order was issued from the Horse Guards, on October 7th, 1813, to the effect that amended regulations would be adopted in the grant and circulation of marks of distinction, *i.e.* medals. The regulations set forth that (1st) only one medal should be borne by each officer, (2nd) that for the second and third events a gold clasp attached to a ribbon, from which the medal is suspended, inscribed with the name of the battle or siege to which it relates. These bars, which were unconnected, were 2 in. long by $\frac{3}{5}$ in. broad, with laureated edges, the names of the battles being soldered on. They were pierced and lapped bright. (3rd) That upon a claim being admitted to a fourth mark of distinction, a cross shall be borne by each officer with the names of the four battles or sieges respectively inscribed thereupon, and to be worn in substitution of the distinctions previously granted. (4th) That upon each occasion of a similar nature, that may occur subsequent to the grant of the cross, the clasp shall again be issued to those who have a claim to the distinction. The following officers were eligible for the distinction: General Officers, Commanding Officers of Brigades, Commanding Officers of Artillery and Engineers, Adjutant-General and Quartermaster-General, also Assistant Adjutant-General and

(Obverse)
WATERLOO MEDAL.

ARMY OF INDIA MEDAL.

(Reverse.)
WATERLOO MEDAL.

Quartermaster-General ranking with Field Officers; Military Secretary having the rank of Field Officer, Commanding Officers of Battalions or Corps equivalent thereto, and Officers who may have succeeded to the actual command during an engagement, owing to the death or removal of the original Commanding Officer.

The cross (see facing page 44), Maltese in form, is $1\frac{1}{2}$ in. square, with a bright lapped double border framing a laurel mount. In the centre in bold relief is the British Lion statant, and on the four arms of the cross are arranged the names of the battles the recipient was engaged in. The name of the first battle is arranged in the space above the Lion. The obverse and reverse are similar; the cross was suspended from a crimson ribbon with blue edges—the regulation ribbon at this time—$1\frac{7}{10}$ in. wide, by means of a gold swivel and laureated ring, connected to the cross by means of a gold loop and ornamentation. The names and regiments of recipients were engraved on the edges of the arms of the cross.

Number Issued.—The Duke of Wellington received the only cross with 9 bars, representing 13 engagements; two were issued with 7 bars; three with 6 bars; seven with 5 bars; eight with 4 bars; seventeen with 3 bars; eighteen with 2 bars; forty-six with 1 bar, and sixty-one the cross alone. Eighty-five large gold medals were issued, and 599 small gold ones. One hundred and forty-three bore 1 bar, and seventy-two 2 bars, leaving four hundred and sixty-nine medals without bars or clasps.

Nivelle.—For nearly six cold and inclement weeks after the fall of San Sebastian, Wellington remained inactive in the Pyrenees, the troops suffering considerably in the bleak situation. Then began the operation which has been described as one of the boldest of the war. Wellington determined to seize the great La Rhune mountain, 2,700 ft. high, which stood between the Nivelle and Bidassoa valleys and the dependent heights; by so doing he could menace the centre of the French line, and cross the border. Soult lost his head, and called upon the French peoples of the provinces to take up arms and war "to the knife." Wellington coolly responded by informing the inhabitants that England did not make war upon the people, but upon their ruler, who did not allow others to remain at peace, and impressed upon his soldiers that any acts of violence or marauding would be punished by death. Early on the morning of October 7th, 1814, the British made their first movement by crossing the mouth of the Bidassoa River—a feat classed among the ablest and boldest of Wellington's operations; the enemy were taken by surprise, and the British carried everything before them. At 3 o'clock in the afternoon the famous exploit of Lieutenant William Havelock, "El Chico Blanco," took place. He had been sent by General Baron Alten to see the progress that Marshal Giron's Spaniards had made, and finding them irresolute, despite their previous successes, his fiery spirit could not brook the check. Taking off his hat, he called upon the Spaniards to follow him, set spurs to his horse, and cleared the abattis at a bound. "Then," wrote Napier, "the soldiers, shouting for 'El Chico Blanco' (the fair boy)—so they called him, for he was young and had light hair—with one shock broke through the French, and at the very moment when their centre was flying under the fire of Kempt's skirmishers from the Puerto de Vera." (Lieutenant-Colonel Havelock, elder brother of Sir Henry Havelock, was born on January 23rd, 1793, and killed while leading the 14th Light Dragoons —"450 sabres against an army amounting to more than 15,000 men with heavy cannon"—at Ramnuggur, November 22nd, 1848.) In three days the allies lost 814 men, and the French 1,400, but the victorious allies were now on French soil, and working well together, the Spaniards rising to the occasion as success succeeded success. For some days the armies were inactive; Wellington was getting ready to resume the offensive, while Soult was preparing to attack. The enemy, however, attacked, and carried a redoubt in the camp of Sarre held by the Spaniards. They then attacked, on the morning of the 13th, the advance posts of the army of Andalusia, under the command of Marshal Giron, but were easily repulsed. Ultimately the French retreated in confusion towards the bridge of Nivelle, and on November 10th the battle of La Nivelle was fought. It was a beautiful morning, and it must have been a grand sight to see the army of 90,000 men descend to the battle. "Three guns pealed from the mountain heights of Achubia . . . and the battle of the Nivelle commenced." Driven from the centre, many of the French troops crossed the Nivelle at St. Pè, while Major-General Colville, with the 3rd division, and General Le Cor with the 7th, drove off those who held the heights above, and established the allied army on the rear of the enemy's right. Night ended the battle, and under cover of the darkness Soult withdrew his army and abandoned the position he had been fortifying for three months. On the 12th he took up his position in front of the camp at Bayonne. In this engagement the allies took 1,400 prisoners, 51 pieces of cannon, and 6 tumbrils of ammunition. The French also lost 2,000 men killed and wounded.

In his dispatch Wellington particularly referred to the gallant conduct of the 51st and 68th Regiments in the attack on the heights above St. Pè. The allies lost 26 officers, including Lieutenant-Colonel Lloyd of the 94th, 28 sergeants, 289 rank and file; 155 officers, including Major-General Kempt, and 2,146 men wounded; 3 officers and 70 men missing.

The following regiments were present in the engagements leading up to the battle of Nivelle. 1st and 2nd Batts. Foot Guards; 1st, 2nd, 3rd, 4th, 5th, 6th, 7th, 9th, 11th, 20th, 23rd, 24th, 27th, 28th, 31st, 32nd, 34th, 36th, 38th, 39th, 40th, 42nd, 43rd, 45th, 47th, 48th, 50th, 51st, 52nd, 53rd, 57th, 58th, 59th, 60th, 61st, 62nd, 66th, 68th, 71st (did not take active part), 74th, 76th, 79th, 82nd, 83rd, 84th, 85th, 87th, 88th,

91st, 92nd, 94th, and 95th; the Rifle Brigade; 12th, 13th, 14th, and 18th Light Dragoons, and the 1st and 2nd Light Battalions, also the 1st, 2nd, and 3rd Line Battalions of the King's German Legion.

Chrystler's Farm.—The day following the battle of Nivelle in France saw a victory for British arms in Canada at Chrystler's Point, commonly called Chrystler's Farm. This battle was a sequel to that at Chateauguay. I have referred to the fact that the American General Wilkinson, with his army of 10,000 men, had failed to combine with Major-General Hampton, but by November 3rd he landed a portion of his troops near Point Iroquois, to which place the British General dispatched a force of 800 regulars and militia, under Colonel Morrison, to stop the advance. Meeting 3,000 men who had been landed under Major-General Boyd at Chrystler's Farm, Colonel Morrison's little army routed them after a fight lasting two and a half hours, with a loss of 250 killed and wounded, besides 100 prisoners and one gun. The effect of the action at Chrystler's Farm was to compel the enemy to evacuate Lower Canada.

Men of the 49th and 2nd Battalion 89th Regiments, the Royal Artillery, Canadian Militia, Fencibles and Voltigeurs, likewise a few Indians, were engaged in this battle. As comparatively few regulars took part in the American battles, Fort Detroit, Chateauguay, and Chrystler's Farm, and because they were awarded so many years after, medals to British regiments bearing these bars are very rare, as indeed are those of the Colonials who took part.

Nive.—Soult, having lost the Nivelle, withdrew his army to Anglet, and formed a line of defence with his unfinished fortified camp in front of Bayonne. Wellington, owing to the bad weather—the 12th was very foggy, and rain fell in torrents from the 11th until the 20th November—was unable to move, and was consequently constrained to maintain the allied army in very close quarters. The bad weather abating, the

NASSAU MEDAL FOR WATERLOO.

HANOVERIAN MEDAL FOR WATERLOO.

British commander determined to enlarge the productive area for his army by forcing the line of the Nive, and taking up a position on the left bank of the Adour, where reasonable subsistence could be found for the allied army. The time had not passed tamely, for in pressing the posts slowly forward, Generals Wilson and Vandeleur were wounded, while the light division alone lost 100 men. On December 8th the troops made a forward movement, and on the 9th the Nive was crossed near Cambo by the right of the army under Lieutenant-General Sir Rowland Hill, while the 6th division, under Lieutenant-General Sir H. Clinton, successfully passed at Ustaritz, and by this manœuvre the enemy were driven from the right bank of the river, and retired towards Bayonne by the great road of St. Jean Pied de Port. Those opposite Cambo were nearly intercepted by the 6th division, one regiment being driven from the road and compelled to march across country.

On December 10th, 11th, 12th, and 13th, Soult made a series of determined and often desperate attacks upon the British divisions, but without success, and was ultimately forced to fall back upon Bayonne, leaving the courses of the Nivelle, Nive, and Adour in the possession of the allies, who were thus enabled to obtain plentiful supplies for their needs, while the enemy were harassed by having their communications threatened and supplies restricted. During the conflicts between the 9th and 13th the French lost about 6,000 killed and wounded, and two guns. The British losses were 650 officers (including 6 Generals) and men killed; 233 officers, 215 sergeants, and 3,459 rank and file wounded; 17 officers, 14 sergeants, and 473 rank and file missing.

The battle on December 13th, which determined the series of conflicts, was practically fought and won by the corps under Sir Rowland Hill—indeed the glory of the day was frankly given him by Wellington, who, taking him by the hand, said, "My dear Hill, the day is your own." There could be little doubt about that, for on the heights of St. Pierre—with only 16,000 men and 14 guns—he defied and drove off 35,000 French bayonets and 22 guns which assailed him in front, and the corps (8,000) of General Paris, with the Light Cavalry under Pierre Soult, which threatened his rear. The battle was regarded on both sides as "one of the most sanguinary that the French Army of Spain had fought, and that there was not one where so many deaths took place on the battlefield." No wonder the enemy never again took the offensive!

The following regiments were present in the series of engagements: 1st, 2nd, and 3rd Foot Guards; 7th, 12th, 13th, 14th, 16th, and 18th Light Dragoons; 1st, 2nd, 3rd, 4th, 5th, 7th, 9th, 11th, 20th, 23rd, 27th, 28th, 31st, 32nd, 34th, 36th, 38th, 39th, 40th, 42nd, 43rd, 45th, 47th, 48th, 50th, 52nd, 53rd, 57th, 59th, 60th, 61st, 62nd, 66th, 71st, 74th, 76th, 79th, 83rd, 84th, 85th, 87th, 88th, 91st, 92nd, 94th, 95th, and the Rifle Brigade. Many of the regiments particularly distinguished themselves on the day of victory; notably the 92nd, which, scattering light troops that should have checked them, charged and repulsed a column; re-forming behind St. Pierre, they again advanced with colours flying and pibroch sounding, as if going to a review, and offered battle to a French column five times their superior: the challenge was, however, declined, for like Napier the French commanding officer recognised that men who could act so understood war.

Orthes.—On February 27th, 1814, the battle of Orthes (Orthez) was fought. A period of inaction had again been forced on Wellington by the severity of the weather, but by the second week in February the conditions were such as to make the roads passable, and the British commander proceeded to follow up his successes. Several minor, but in some instances obstinate, engagements, were fought, among them that of Garris, where Wellington, determined to force the passage of the Bidouse, ordered the 29th, with the 28th in support, to carry a bold hill, occupied by 4,000 of the enemy, before dark. This they did in a brilliant manner. Soult, finding the enemy pressing him rather forcibly, crossed the Gave d'Olèron in the night, and marching rapidly to Sauveterre, took up a position on the left of the Gave, while Wellington, having strongly established his right upon the Gave d'Olèron, returned to St. Jean de Luz to superintend the putting together of the remarkable and "wondrous bridge," which had been devised to enable the army to cross the Adour, 3 miles below Bayonne. On the 25th Sir John Hope crossed with his whole corps for the investment of the fortress; meanwhile Wellington was acting vigorously at Gaves, and General Hill was ready to force the passage of the Gave d'Olèron, which he did on the 24th, while Picton crossed between Montfort and Lass; other divisions having made equal progress, Marshal Soult withdrew his army across the Gave du Pau, and determined to make a final stand at Orthez. There, on the semicircular heights, the army of about 40,000 stood at bay, while the allies with about 37,000 men advanced to the attack. They were irresistible, but Soult retreated in a masterly manner, defending his divisions with splendid valour at each ridge they passed. Once forced on to the plain, they fell into disorder, and, charged and pursued by the British cavalry, they hastened to the river, over which they passed in scattered parties, and Soult retreated towards Toulouse, where the last and most unnecessary battle of the campaign was to be fought. It is estimated that 8,000 men were lost to the French in killed and wounded, while the allies lost 18 officers, 25 sergeants, and 234 men killed; 134 officers (including the Duke of Wellington), 89 sergeants, and 1,700 rank and file wounded; 1 officer, 5 sergeants, and 64 men missing.

The following regiments were present: 3rd, 7th, 10th, 13th, 14th, 15th, and 18th Light Dragoons;

2nd, 5th, 6th, 7th, 10th, 20th, 23rd, 24th, 27th, 28th, 31st, 32nd, 34th, 36th, 37th, 39th, 40th, 42nd, 45th, 48th, 50th, 51st, 52nd, 57th, 58th, 60th, 61st, 66th, 68th, 71st, 74th, 82nd, 83rd, 87th, 88th, 91st, 92nd, 94th, 95th, and the Rifle Brigade. The 7th Hussars (Light Dragoons) distinguished themselves in the neighbourhood of Sault de Navailles, and made a charge under Lord Edward Somerset which Wellington described as "highly meritorious." The 52nd Oxford Light Infantry were also mentioned in dispatches.

Toulouse.—On April 10th, 1814, the sanguinary and unnecessary battle of Toulouse was fought, for the allies had entered Paris on March 31st, and Napoleon had abdicated. It is alleged that Soult fought the battle out of personal vanity or pique, since he had news—on April 7th—of what had happened in Paris. It was perhaps his last desperate effort for his master. However, this last battle of the Peninsular War was fought with terrible determination on both sides; but when the range of heights had been carried by the British, and the redoubts of the city won, Soult abandoned the town of Toulouse with 3 Generals and 1,600 men "prisoners at the generosity of the conqueror." The allied British, Portuguese, and Spanish armies lost in killed and wounded about 4,650, the latter including Generals Brisbane, Pack, Mendizabel, and Espelette, while the French had 5 generals and 3,234 men placed *hors de combat.* These numbers include, British losses, 31 officers and 563 men killed; 248 officers and 3,898 men wounded; 3 officers and 15 men missing.

In storming the redoubts the 42nd Black Watch, which had the honour of leading the attack, displayed their usual courage, but the withering fire from the defenders was such that in a short time it would have annihilated the regiment; indeed, out of the 500 who went into action, scarcely 90 reached the redoubt, from which the enemy fled, but these "leapt over the trenches like a pack of hungry hounds in pursuit." Two officers and sixty men of inferior rank were all

PRUSSIAN JUBILEE MEDAL GIVEN BY WILHELM I IN 1863 TO VETERANS OF 1813-14-15 CAMPAIGNS.

JUBILEE MEDAL GIVEN BY THE CITY OF HANOVER TO VETERANS OF WATERLOO.

"ST. HELENA MEDAL" GIVEN BY NAPOLEON III TO SURVIVORS OF WATERLOO AND THE NAPOLEONIC WARS.

that remained unwounded of the right wing of the regiment that entered the field in the morning. During the battle the standard of the regiment had passed through the hands of three wounded officers, until it was borne by a sergeant and defended by a few men. The 42nd, likewise the 36th, 79th, and 61st, who also lost considerable numbers, were mentioned in Wellington's dispatches. The 13th Light Dragoons, the 14th and 15th Dragoons, and the 18th Hussars, were also mentioned for gallant conduct during the advance on Toulouse.

The regiments represented at the battle were: 2 squadrons of the 1st and 2nd Life Guards and Horse Guards; 1st Dragoons; 3rd and 5th Dragoon Guards; 3rd, 4th, 7th, 10th, 13th, 14th, 15th, and 18th Light Dragoons; 1st, 2nd, 3rd, 5th, 7th, 11th, 20th, 23rd, 27th, 28th, 31st, 32nd, 34th, 36th, 39th, 40th, 42nd, 43rd, 45th, 48th, 50th, 52nd, 53rd, 57th, 60th, 61st, 66th, 71st, 74th, 79th, 83rd, 87th, 88th, 91st, 92nd, 94th, and 95th.

Rewards for Generals.—On May 30th, 1814, a treaty of peace was signed between Great Britain and France, Louis XVIII having been restored, and Napoleon permitted to retire to the Isle of Elba, where the allies allowed him to reign as sovereign. The Duke of Wellington was raised to the rank and dignity of a Duke and Marquis of the United Kingdom; Parliament suggested an annuity of £10,000, to be "paid annually out of the consolidated fund for the use of the Duke of Wellington, to be at any time commuted for the sum of £300,000 to be laid out in the purchase of an estate," but the sum was objected to as being too small, and an additional £100,000 was voted. Wellington's lieutenants—Generals Hope, Graham, Cotton, Rowland Hill, and Beresford—were also rewarded by pecuniary grants, and raised to the peerage.

Nothing for Officers and Soldiers.—The majority of the officers and all the men who had participated in this seven years' campaign were, notwithstanding the generosity of the nation to the leaders, left undecorated, and, irony of ironies, the Commander-in-Chief, who had shed tears at the loss of brave soldiers, and who had been so well rewarded, raised his voice against the bestowal of decorations upon the men who had fought and bled with him!

CONTINENTAL PENINSULAR MEDALS

Alcantara Medal.—For the battle of Alcantara a medal of gold was given to some of the officers, and Dr. Payne has in his collection one awarded to Brigadier-General William Mayne, K.T.S., late Colonel Loyal Lusitana Legion, which by his kindness I am able to illustrate. It is $1\frac{7}{20}$ in. in diameter, enclosed in a gold-rimmed glass case, suspended from a gold-hinged waved bar $1\frac{7}{10}$ in. long by a salmon-pink ribbon $1\frac{1}{2}$ in. wide, with a gold buckle attached thereto. On the obverse is engraved an outstretched sleeved arm and hand, below is the word CUIDADO, and above an eye, the whole within a wreath of palm and olive. On the reverse is the inscription, in bold Roman capitals, AL MERITO, and underneath, in script, "De la Juntade Gobierno y guerra de la Villa de Alcantara MDCCCIX." On the rim the name of the recipient is engraved in Spanish: AL CORONEL MAYNE DE LA LEAL LEGION LUZITANA.

Gold Cross for Albuhera.[1]—On March 1st, 1815, it was decided by the Spanish Government to issue a cross to Generals, officers, and men of the army of Estremadura, who had distinguished themselves at the battle of Albuhera on March 16th, 1811. This is an oblong, four-pointed cross of gold enamelled red, the points terminating in gold knobs, having red enamelled gold-edged flames issuing between the arms of the cross. In the centre of the cross, on a white enamelled ground, is the letter F in script, with VII in gold, surrounded by a frosted gold oval band bearing

[1] Should be Albuera, but is spelt ALBUHERA on cross, also bars.

the inscription in bright gold letters AL BUHE RA. The cross is suspended from a squat green enamelled laurel wreath, to which is fixed a gold loop and ring for attachment to the ribbon, which is $1\frac{1}{2}$ in. wide, and red with blue edges.

Gold Cross for Vittoria.—On April 22nd, 1815, a gold enamelled cross was instituted in Spain for presentation to officers to commemorate the battle of Vittoria, June 21st, 1813. The battle was won by the division of the 4th Corps of the Army, under the command of the Captain-General, the Duke of Ciudad Rodrigo (Wellington), and Field-Marshal D. Francisco Thomas de Longa. The cross is $1\frac{1}{5}$ in. in diameter, and enamelled white with a red four-pointed star (see facing page 36), and between the cross, which is enamelled the same both sides, is a green enamelled laurel wreath; the cross is suspended from a gold Spanish crown surmounted by an orb ensigned with a gold fleur-de-lis; the ring for attachment to the ribbon being run through the ball. On the obverse is a white enamelled circle with gold borders encircling a red irradiated ground, whereon are three gold crossed swords, with a gold ribbon (inverted) bearing the motto IRURAG BAT. On the reverse, on a domed gold ground, is the legend in raised stamped letters, arranged in three lines, RECOMPENSA DE LA BATALLA DE VITTORIA. This cross depends from a $1\frac{1}{2}$ in. ribbon composed of equal stripes of black, crimson, and pale blue.

Gold Cross for Commanders.—This gold cross, as the illustration facing page 28 shows, is star-like in form, but owing to the number of engagements recorded the number of arms varies. This particular cross, weighing nearly 2 oz., was awarded to Lieutenant-Colonel Richard Brunton, of the 13th Light Dragoons, late of the Portuguese Service, and forms part of the great collection of Dr. A. A. Payne. The cross was instituted by King John VI of Portugal, on July 26th, 1816, and was given to British officers who served on the Portuguese Staff during the Peninsular War, and who were in command of divisions, brigades, or regiments. The arms of this cross, enamelled white, have double points terminating in gold beads, resting on a blue enamelled band bearing a gold wreath of laurel; on each of the arms is a green enamelled torpedo-shaped overlay, each recording the name of an action in which the recipient was engaged: a gold scroll forms the attachment for the loop which carries the broad red ribbon, edged with blue, by which the cross was suspended from the neck. On the obverse a small gold bust of King John occupies the centre, which is enamelled light blue surrounded by a conventional continuous border of leaves. On the reverse, within a similar band, but on a white enamelled ground, are the initials of the recipient. Sir Denis Pack was awarded the cross with eleven actions recorded on it.

Officers' Gold Crosses.—King John VI also instituted, on July 26th, 1816, an officer's cross for bestowal upon those officers who had participated in all or any of the six campaigns which were fought during the Peninsular War. The cross of gold was awarded to all who fought in at least three campaigns, crosses of silver being given to those who had only been engaged in one or two. The cross illustrated was given to Lieutenant-Colonel Richard Brunton, for service in four campaigns. It has four ornamental arms, $1\frac{11}{20}$ in. long, resting upon a green enamelled laurel wreath. In the centre of the obverse are the arms of Portugal in gold on a convex ground, surrounded by a blue enamelled circle, with the inscription GUERRA PENINSULAR. On the reverse is a continuous gold wreath of laurel, enclosing the pierced Roman numerals indicative of the total campaigns in which the recipient was engaged. The loop to carry the broad red blue-edged ribbon is attached to the cross by means of an inverted husk of gold. The gold cross, awarded to Lieutenant-Colonel Sir Edward Brackenbury, differs somewhat from that awarded to Colonel Burton. The arms on the obverse of the cross are $1\frac{7}{20}$ in. long, and the coat-of-arms is of a somewhat different character, and is not surmounted by a crown. On the reverse the single gold pierced figure 4 is within a blue enamelled circle. A gold ball and swivel bar are used for attachment to a blue watered ribbon $1\frac{3}{5}$ in. wide.

Officer's Silver Cross.—This was of the same size as Lieutenant-Colonel Brackenbury's, and suspended from a gold swivel bar, but the ribbon for suspension was blue silk ribbed, with broad pale yellow borders; the width $1\frac{3}{5}$ in.

Crosses for Ciudad Rodrigo.—These were instituted on December 6th, 1819, for bestowal upon those who had distinguished themselves at the sorties on July 10th, 1810. They were given in gold to officers and in silver to privates who comprised the garrison. The cross has four arms, with blue enamelled edges tipped with gold beads. On the obverse is a gold tower, with rays issuing therefrom, upon a red enamelled ground enclosed by a white enamelled oval band. On the reverse is an inscription in gold letters upon a white enamelled ground, VALOR ACREDITADO EN CIUDAD RODRIGO. (See facing page 36.)

Medal for Brunswick Contingent.—On October 30th, 1824, Duke Charles II decided to award a medal to those who formed the Brunswick Contingent, which came to England in 1809 with the Duke Frederick William, and took part in the Spanish and Portuguese war against Napoleon. The Contingent fought in nearly all the battles between 1810–14. The officers received a silver medal, and sub-officers and privates one of bronze. The medal, $1\frac{3}{16}$ in. in diameter, has a plain double border, and is suspended from a crimson ribbon by means of a loop and ring. On the obverse is a wreath of laurel and oak and PENINSULAR; on the reverse is a military trophy with a studded shield, arranged sideways, on which the Duke's initials, C. C., are arranged in monogramic form.

Medal for British German Legion.—King Ernest Augustus founded, on May 11th, 1841, a medal for distribution to the survivors of the British German Legion, who had served in the Peninsular War between 1803–14. The medal, having a double rim, was of bronze made from captured cannon. One and seven-twentieths inch in diameter, it bears on the obverse a cross *patée* with double raised edges, in the centre of which is the King's monogram, E. A. R., surmounted by a crown; on the reverse within a laurel wreath is the inscription TAPFER UND TREU (Brave and Loyal), and within the rim KÖNIGLICH-DEUTSCHE LEGION (Royal German Legion). The medal is suspended from a $1\frac{2}{5}$ in. white ribbon, with yellow stripe near the edge, by means of a ring and loop. I have in my possession one without the inscriptions, a laurel wreath and 1813 entirely filling the reverse. (See facing page 58.)

Eve of Waterloo.—On February 27th, 1815, Napoleon, untamed by adversity, escaped from Elba, and on March 1st landed at Cannes in Provence with 1,000 determined followers, but so rapidly did the people rally to his standard that by the 20th he had reached Paris with an army of such proportions that Louis was compelled to retire to Ghent, and the Bourbon dynasty was again forgotten. Early on the morning of June 12th Napoleon quitted the capital, on the 14th joined the army collected on the frontier, and commenced his last campaign, which was to terminate on the 18th at Waterloo. Wellington had been sent to take charge of an army which he described in a letter as "very weak and ill-equipped, and a very inexperienced staff," but we now know that, though it was "a green army," the right material was there, and only required the moulding of opportunity to make good form. Wellington's army comprised British, Hanoverian, and Belgian troops, with contingents of Nassau and Brunswick Oels. Of the 78,500 men comprising it, only about 43,000 were British, German, or Hanoverians. By the 18th it had been reduced to 74,040. The Prussian army was 115,000 strong. The French army, estimated at not less than 154,370, with 296 pieces of cannon—and variously at 127,000, 122,000, and 115,000, with 350 guns—was mainly composed of men "whose trade was war, and whose battles were as many as their years"; so that against the heterogeneous mass of men of various tongues commanded by Wellington, his great opponent was to hurl an army of veterans who could well be called upon to repeat the deeds of Marengo, of Friedland, and of Austerlitz. As it happened, the raw levies of Britain were to emulate the deeds of their brethren at Badajoz and Albuera.

On June 15th Napoleon's army crossed the Sambre, and on the morning of the 16th at Ligny he forced back the Prussians, but Blucher—falling back upon Wavre instead of Ligny—upset the French General's calculations. Meanwhile Wellington, having made arrangements with Blucher, was with his officers in the celebrated ballroom of the Duchess of Richmond at Brussels. Two hours after midnight the gaiety of the brilliant ballroom was subdued, and following the "sound of revelry by night" came "the cannon's opening roar" at Quatre Bras.

Quatre Bras.—By 8 o'clock the British had left Brussels, the 42nd and 92nd Highland Regiments having assembled to the sound of the pibroch, "Come to me and I will give you flesh"—an invitation to the wolf and raven, for which, as Forbes relates, the coming day did in fact spread an ample banquet. At 2 o'clock, after a march of 20 miles in sultry weather, they arrived at the hamlet of Quatre Bras, to find the Prince of Orange pluckily but feebly endeavouring to check the enemy, who had already gained Le Bois de Bossu. There, at the intersection of the Charleroi-Brussels and Namur-Nivelle roads, 7,000 Dutch-Belgians with 16 guns and no cavalry stood against Marshal Ney with 16,000 infantry, 1,700 cavalry, and 36 guns. It was, indeed, an unequal contest until the arrival of Pack's and Kempt's brigades. Wellington at once decided the wood must be taken, and then ensued an exceedingly sanguinary conflict ere, with the ultimate arrival of the Guards, the French were forced to retire. The Guards (mostly young soldiers) had arrived from Enghein after a march of 27 miles, having started at 3 o'clock in the morning to join the main body. They had halted at Nivelles to rest and feed, but the arrival of a Staff officer, urging them to hurry on, induced them to abandon their meal, and resume their march. They arrived on the field of battle at an opportune moment, and then, despite their fifteen hours' march without food or drink, the first brigade of Guards, having loaded their muskets and fixed bayonets, pushed their way into Le Bois de Bossu, and within half an hour regained possession of it, but they could do no more than hold it. Meanwhile a series of splendid struggles had been taking place. When the 95th had been ordered to clear Le Bois de Bossu, the Royals and the 28th (Gloucesters) were severely engaged on the left, while the 44th, the 42nd and 92nd Highland Regiments were hotly assailed on the right.

The 28th, after gallantly standing its ground under a furious cannonade, was suddenly, and on three different sides, assailed by French cavalry. Two faces of the square were charged by the lancers, while the cuirassiers galloped down upon another. It was, as Maxwell states, a trying moment. "There was a death-like silence; and one voice alone, clear and calm, was heard. It was their colonel's (Sir Philip Benson), who called upon them to be 'steady!' On came the enemy! The earth shook beneath the horsemen's feet, while on every side of the devoted band the corn bending beneath the rush of cavalry disclosed their numerous assailants. The lance-blades approached the bayonets of the kneeling front rank—the cuirassiers were within forty paces—yet not a trigger was drawn; but, when the word 'Fire!' thundered from the colonel's lips, each face poured out its deadly volley—and in a moment the leading files of the French lay before the square, as if hurled

BRUNSWICK WATERLOO MEDAL.

SAXE-GOTHA-ALTENBURG WATERLOO MEDAL.

by a thunderbolt to the earth. The assailants, broken and dispersed, galloped for shelter to the tall rye, while a stream of musketry from the British square carried death into the retreating squadrons."

42nd and French Lancers.—The 42nd were not so fortunate, although their difficulty gave them the opportunity of again demonstrating the stern stuff of which they were moulded. The 42nd and 44th, the two foremost regiments posted in line on a reverse slope on the top of the Charleroi road, were suddenly and unexpectedly attacked in the rear by the French lancers under Wathier. The 42nd had almost formed square in the tall rye grass, all but the two flank companies having run in to form the rear face, when the leading squadron of lancers drove in the incompleted square, carrying along with it, by the impetus of their charge, several men of these two companies, and by spearing a number created considerable confusion, during which those who had been detached in the *mêlée* fought back to back until the fire of their comrades drove off the cavalry, and enabled them to rejoin the ranks, where the body of the Black Watch, coolly standing firm, gradually closed up its faces, and bayoneted the daring lancers who had penetrated their square. These brave fellows, however, died hard; they killed the commanding officer, Sir Robert Macara, and wounded the three officers upon whom the command had devolved in the space of a few moments. The 42nd lost at Quatre Bras 298 officers and men killed and wounded.

The 44th were still more unfortunate, for, like the 42nd, they had found a difficulty in forming up in the corn, which was up to their shoulders, and having less time faced their ranks about and waited for the French lancers to get in close before firing a volley. The foremost went down, but those lancers were veterans who did not easily pale before the leaden storm, and individual troopers dashed on; they were, however, repulsed by the rear of the 44th, and galloped away eastward under a fire from the left company of the line. As a result of the cavalry attacks the 42nd and 44th were so reduced in numbers that at 5 p.m., when the first main attack had concluded, the remnants of the two regiments formed in a single square. Pack's Brigade, formed of the 42nd, 44th, 92nd, and 95th, lost in this strenuous engagement 800 out of 2,000 men.

The next cavalry charge fell upon Halkett's division, from which, in response to an appeal from Pack, whose men were out of ammunition, the 69th Regiment had been dispatched to assist. The 69th were later ordered by Halkett's aide-de-camp to form square to receive the advance of Kellerman's cavalry, which had trotted in from Charleroi, but the Prince of Orange unwisely countermanding the order, Kellerman and Guiton, leading the 8th Cuirassiers, passed round the flank of the 42nd and 44th square, and charging the unfortunate 69th in line, completely rolled them up; and within a few moments, of the 580 composing the 69th, 150 were dead or dying. During this unfortunate episode, Mr. Clarke, a volunteer, fought with remarkable desperation, killing three cuirassiers and preserving the colours he was carrying, although he had received twenty-six sabre cuts. Kellerman's success was short-lived, for the British artillery fire had begun to play havoc with his cavalry, while the fierce charge of the squadrons of cuirassiers upon the square of the 28th was repelled by deadly discharges of musketry, and the French cavalry were compelled to flee, and in their wild course, 2 miles from the field, they swept 2,000 dismounted horsemen into the stampede. Another incident should be recorded: the 92nd, having repulsed an attack of cavalry and infantry in an advanced movement, were retiring to the wood, when a French column halted and fired upon the Highlanders; already assailed by a superior force, this notwithstanding they held their ground until relieved by a regiment of the Guards, when they retired to their original position, but not before the regiment had lost 28 officers and nearly 300 men.

Losses at Quatre Bras.—Night compelled the cessation of hostilities, as at Ligny, and the French, having retired to their position held in the morning, left the British in possession of Gemioncourt and the southern end of the Bois de Bossu. The allied army lost at Quatre Bras 3,750 men, and of these 327 killed and 2,156 wounded were British; the French losses were at least 4,500. All those who took part in this battle, where the bull-dog tenacity of the British soldier was shown at its best, were awarded the Waterloo medal. In his dispatch Wellington stated, "I must particularly mention the 28th, 42nd, 79th, and 92nd regiments and the battalion of the Hanoverians." In addition to these the following British regiments were present at Quatre Bras: 2nd and 3rd Batts. 1st Foot Guards; Royal Scots; 32nd, 33rd, 44th, 69th, and 95th Regiments.

WATERLOO

After Quatre Bras Marshal Ney fell back upon Frasnes, and the wearied British, with their allies, piled arms, lighted their camp fires, and laid down to rest upon the field which they had so hardly held. The Prussians, after an intensely obstinate struggle, had been forced to retire to Ligny, when Blucher determined to unite with his 4th Corps and concentrate on Wavre. Wellington, recognising the importance of keeping in communication with the Prussian General, decided to fall back upon a position fronting the village of Waterloo, which, by the way, had already been selected by the British commanders as a suitable battleground. In retiring from Quatre Bras, the British cavalry were attacked by the French, and the 7th Hussars and squadrons of the 11th and 23rd Light Dragoons charged without success. The Life Guards, however, under Lord Uxbridge, compelled the enemy to retreat, and the British were allowed to quietly take up their position, which they did in decidedly inclement weather, rain falling heavily most of the day, and the night closing wild and stormy. Violent winds, heavy rain, thunder and lightning added to the discomfort of the cheerless bivouac of the allies. The Duke of Wellington and his staff occupied the village of Waterloo, while Napoleon with his brother Jerome and his generals passed the night of the 17th in a farm-house half a league from the Château de Hougomont, which, in advance of the right centre of the British, was held by detachments of the Brigade of Guards and about 300 Nassau Riflemen. This was considered by Wellington to be the key to the position, and how well the Guards maintained it is a matter of glorious history. It is estimated that the allied forces numbered 74,400, and the French 90,000, when just before noon Joseph Bonaparte directed the 2nd Corps to advance against Hougomont. "The British batteries opened on the French masses as they debouched—their own guns covered their advance—and under the crashing fire of 200 pieces of artillery—a fitting overture for such a field—Waterloo opened, as it closed, magnificently."

The Guards at Hougomont.—The Nassau light troops were soon forced from the wood surrounding the Château by overwhelming numbers, and the enemy pressed on to the Château, but the vigorous and disconcerting fire of the Foot Guards arrested them while daring charges compelled the French to give up possession of part of the wood. They were not, however, to be denied, and with great determination repeatedly assaulted the Château. During one of these attacks so closely did they press that the Guards not only lost their position in the garden, but falling back rapidly on the main building could not securely close the gate of the yard, and a number of daring French soldiers entered; the fire of the defenders, however, followed by a courageous sortie from the Château forced the French from the yard, but only after a most gallant struggle between the combatants. Then Lieutenant-Colonel (afterwards Sir James) Macdonnell, with the assistance of Captain Wyndham, Ensigns Gooch and Hervey and Sergeant Graham of the Coldstream Guards, closed the gate, but not until, as Captain Siborne relates, the French soldiers had fallen "a sacrifice to their undaunted and conspicuous gallantry." Failing to take the Château by assault, the French artillery was turned upon the old building; soon the tower was in a blaze, and, the fire reaching the chapel, a number of wounded—friend and foe—perished in the flames, which strangely enough are stated to have ceased at the feet of a wooden image of Christ. But when the shades of evening fell, the Coldstreams and Scots Guards still held Hougomont, despite the repeated attacks of 30,000 men, and when the day had been decided it was found that in the sanguinary conflict which had raged so long around this advanced position 6,000 men had been killed. 1,500 were killed within half an hour in the four-acre orchard; 600 French fell in the attack on the Château and farm; 1,100 British were killed in the orchard and meadow; 400 near the farm-garden; 25 in the garden; and 2,000 of both armies behind the great orchard. Opposite the gate of the Château 300 British soldiers lie buried.

The Death of Picton.—The second attack was made by D'Erlon, with the whole of his corps, against the left and centre; fortune favoured them temporarily when they obtained possession of Papelotte and La Haye Sainte. In repelling this attack the gallant Picton fell. The 5th division, waiting until the heads of the enemy's columns were within 40 yards, delivered a terrible volley, which annihilated the leading sections and put the main body into confusion. Picton, seizing the opportunity, called upon the men to "Charge!" but the word had hardly thundered from his lips ere a bullet pierced his forehead, and he fell from his saddle mortally wounded. The loss of a commander frequently means defeat, but "as he fell he heard the Highland lament answered by the deep execrations of Erin, and while the Scotch slogan was returned by the Irish hurrah his fading sight saw his favourite division rush on with irresistible fury. The French column was annihilated, and 2,000 dead enemies told how desperately he had been avenged. This was probably the bloodiest struggle of the day; when the attack commenced—and it lasted not an hour—the 5th division exceeded 5,000 men; when it ended they reckoned scarcely 1,000!"

Charge of the Union Brigade.—Following this affair, the famous charge of the Union Brigade took place. Lord Anglesea, observing that the French lancers and cuirassiers were preparing to make a flank attack upon the British infantry, wheeled the Royals, Royal North British Dragoons (Scots Greys), and Enniskilleners into line, charged and overwhelmed the French cavalry, and falling upon the disorganised infantry completed the brilliant work of the 5th division. It was in this charge that Sergeant Ewart of the Scots Greys captured the Eagle of the 45th Regiment. The Eagle of the 105th Regiment was captured by Captain Clarke of the Royal Dragoons—securing the right for these regiments to wear the Eagle badge—and 2,000 men were taken prisoners. The impetuosity of the British Dragoons carried them into the rear of the French position, and they were driven back by the French horsemen, their brave leader, Sir William Ponsonby, being killed in the retirement. Meanwhile the incessant attacks which Napoleon commanded had played havoc with several of the British regiments, necessitating the bringing into action of Wellington's reserves—indeed, the position had become so serious that the left wing of the army, though only partially engaged, had suffered so severely that it could not afford to send assistance to the right or centre.

Great Regimental Losses.—The loss in individual regiments was awful. Four hundred men of the 27th were mowed down in square without drawing a trigger (their medals in fine condition have fetched as many pounds as the number of the regiment represents).

The 92nd, reduced to 200 men, made a daring attack upon a French column 2,000 strong, and with the aid of their countrymen, the Scots Greys, routed it. The 33rd, reduced to a skeleton, asked for support, and the commanding officer was told to "stand or fall where he was," and of the 28th and 73rd it is related that Wellington asked, pointing to a mass of killed and wounded men of these regiments, "What square is that so far in advance!" But the impoverishment of the regiments by incessant attacks and murderous artillery could not break the indomitable tenacity of the British; in vain had Milhaud's Cuirassiers, forty squadrons strong, thrice attacked the British squares. In vain did the remnants of these valorous squadrons, assisted by Kellerman and Guyot—making a total of seventy-seven squadrons, make a desperate effort to pierce those stubborn and thinning squares of British infantry. In vain did Ney repeat his attack upon the centre, even though he annihilated what remained of the German Legion, which had with such admirable courage held the farm of La Haye Sainte, and broke the formation of the British troops, causing Wellington to ask the question relating to the 28th and 73rd. In vain did the ten battalions of the Imperial Guard, led by Marshal Ney, push their way up the slopes between Hougomont and La Haye Sainte, for the Foot Guards —with the 52nd, 71st, and 92nd Regiments—offered such a murderous reception that "the Guard turned and fled." No wonder Napoleon ejaculated, "*A present c'est fini—sauvons nous!*" for, Wellington ordering the whole line to advance, the weary, hungry, and even wounded soldiers rushed forward with a joyous cheer and forced the retreat, which, as the British leader, with the 42nd and 95th, threw himself on Ney's flank, and the allied cavalry charged the enemy's columns, became an utter rout.

The Price of Victory.—But the Old Guard, true to their traditions, made a last desperate stand in square against the British cavalry; it was desperate and grand, but ineffectual, and again they turned and fled, and "the finest army, for its numbers, that France had ever embattled in a field was utterly defeated, and the dynasty of that proud spirit for whom Europe was too little was ended." Over the terrible carnage which followed I must draw a veil, for the unrelenting animosity of the Prussians to the French led them to retaliate in a most vindictive manner, and thousands who had bravely fought and survived the day lived only to fall ignominiously under the revengeful sabre or thrust of Prussian lance. When Wellington recrossed the battlefield, where the destiny of Europe had been changed by the defeat of the greatest of the world's generals, he could have had little stomach for supper, for he had to pass over that 2 square miles of Belgian territory where 50,000 dead or wounded men and horses lay—the terrible price of victory.

The following British regiments were represented at Waterloo: 2 squadrons 1st and 2nd Life Guards; 2 squadrons Royal Horse Guards (Blue); 1st Dragoon Guards; 1st Royals,* 2nd Royal North British Dragoons * (Scots Greys); 6th Inniskillen Dragoons *; 12th, 13th, 16th Queen's and 23rd Light Dragoons; 7th, 10th Royal; 15th King's; 18th Hussars; 2nd and 3rd Batts. 1st Foot Guards (Grenadiers); 2nd Batt. 2nd Foot Guards (Coldstreams); 3rd Batt. 3rd Foot Guards (Scots Guards); 3rd Batt. 1st Royal Scots; 1st Batt. 4th; 3rd Batt. 14th; 1st Batt. 23rd Royal Welsh Fusiliers; 1st Batt. 27th Inniskilling; * 1st Batt. 28th (Gloucesters); 2nd Batt. 30th; 1st 32nd; 33rd; 1st Batt. 40th; 1st Batt. 42nd Black Watch *; 2nd Batt. 44th; 51st; 1st Batt. 52nd; 2nd Batt. 69th; 1st Batt. 71st Highland Light Infantry; 2nd Batt. 73rd Perthshires (2nd Batt. Black Watch); 1st Batt. 79th Cameron Highlanders; 1st Batt. 92nd Gordon Highlanders; 1st, 2nd, and Prov. Batts. 95th; 8 Troops Royal Horse Artillery; 6 Brigades Royal Artillery; Corps of Royal Artillery Drivers; Royal Foot Artillery; Royal

PISTRUCCI'S WATERLOO MEDAL.

(See page 373.)

Engineers; Royal Sappers and Miners; Royal Waggon Train; Field Train Department of the Ordnance; Royal Staff Corps; Commissariat Department; Ordnance Medical Department, and the following units of the King's German Legion: 1st and 2nd Light Dragoons; 1st, 2nd, and 3rd Hussars; 1st, 2nd, 3rd, 4th, 5th, and 8th Line Battalions, and 1st and 2nd Light Battalions: the total present being about 2,308 officers and 42,120 non-commissioned officers and men.

The medals of those regiments marked * are particularly sought after by collectors. The Dragoons lost heavily in their brilliant charge, the Scots Greys having 106 officers and men killed and 97 wounded, and the "Enniskilleners" 86 officers and men killed, their Colonel, Lieutenant-Colonel, 4 subordinate officers and 101 sergeants and rank and file wounded. The 27th went into battle with 698 men, and of these 480 were placed *hors de combat*, while the 28th was reduced to four companies. The 42nd lost at Quatre Bras and Waterloo 51 killed and 247 wounded, the former including their Colonel and a Major who was mortally wounded, and the latter a Lieutenant-colonel and a Major. The 92nd was reduced to less than 300 before 4 o'clock. The following regiments composing Sir Charles Colville's division, which was located at Halle owing to Wellington's belief that he might be attacked on his right flank, also received the medal, although not actually engaged in the conflict; 2nd Batt. 35th; 1st Batt. 54th; 2nd 59th and 1st Batt. 91st (Argyll and Sutherland Highlanders); likewise two Brigades of Artillery and the 6th Hanoverian Brigade.

The British Waterloo Medal.—When, on March 10th, 1816, it was decided to bestow upon every soldier, from Commander-in-Chief to drummer boy, the same type and quality of medal as a reward for services at Waterloo, the custom was established to grant a medal to all, irrespective of rank or occupation, who had given war service to his or her country. It was ten months after that auspicious day that there appeared in the *London Gazette* the notification that "a medal shall be conferred upon every officer, non-commissioned officer, and soldier present on that memorable occasion," but be it noted "the ribbon issued with the medal shall never be worn but with the medal suspended to it."

First Official War Medal.—The first general official medal issued at the suggestion of the Duke of Wellington—who, however, strenuously opposed the grant of one for the Peninsular—was of silver, $1\frac{2}{5}$ in. in diameter, bearing on the obverse the laureated head of the Prince Regent facing to the left, with the inscription GEORGE P. REGENT on either side of the head, and on the reverse a seated figure of Victory with wings outspread, holding in her right hand a palm branch, and in the left a sprig of olive; on a plain tablet is WATERLOO, and in the exergue JUNE 18. 1815, while above is the name of Wellington in Roman letters. As originally issued it had a large steel ring 1 in. wide, run through a steel clip, for suspension by a dark-crimson blue-edged ribbon $1\frac{1}{2}$ in. wide (sometimes found 2 in. wide), as illustrated facing page 56. The clip, however, was not always secure, and as a result the possessors of the medal frequently had silver suspenders made according to their own taste, or that of the local jobbing jeweller, who did not always do justice to Wyon's classic medal, slight differences in which may be observed upon comparison, as the medals were not all struck from one die, and the die-sinker's art was not then mechanical. The names and regiments of the recipients were indented in large capital letters on the edge of the medal, the tops and bottoms of the letters almost touching the rims of the medal.

Second General Award.—This was the first general award of medals for 166 years, when Simon's medal was given to those engaged at Dunbar in 1650. However, the official war medal for all ranks is now an established institution, and, it is said, sometimes given with too free a hand. But the soldier has a substantial record of his services on the field of battle—not even now, to my mind, satisfactorily complete—and but a small recognition of his country's appreciation; while those interested in the traditions of our race, those who recognise that in the grim and terrible business of war the better part of man is frequently called forth, will find the means of collecting an interesting record of great deeds, and incidentally materialising the history of the epoch-making nineteenth century, and these dawning days of the twentieth, when we had hoped that the dark spectre of war might have been replaced by the radiating figure of peace. But the war drums still throb, the battle flags are still unfurled; and it is good, when duty has called, in the nerve-wracking suspense and in the loud clamour of war, that fortitude, loyalty, honour, and the love of fatherland have shone above the horrors, and made us proud of the men who wear on their breasts the badges of the brave.

Continental Waterloo Medals.—Several other medals were issued for Waterloo, notably the Hanover, Nassau, Saxe-Gotha, Altenburg, and Brunswick medals, also the Belgian silver star.

Nassau Medal.—Fredrich Duke of Nassau was the first to issue a medal for Waterloo—that was on December 23rd, 1815. It is a small silver medal $1\frac{1}{10}$ in. in diameter, with a lug stamped with the medal, through which ribbon could be run for suspension. On the obverse is the bust of the Duke, and the inscription FRIEDRICH AUGUST HERZOG ZU NASSAU. Underneath the truncation are the letters I.L. On the reverse, to the left, is a standing figure of Victory, holding in her right hand a palm branch, while with the left she is represented placing a laurel wreath upon the brow of a Roman soldier; in the exergue is DEN 18 JUNI 1815, and round the figures DEN NASSAUISCHEN STREITERN BEY WATERLOO. The medal, issued unnamed, was suspended from a dark-blue ribbon with orange edges.

Hanover Medal.—The Hanover medal was struck by command of the Prince Regent of Great Britain. The order was given in December 1817, and was to the effect that surviving soldiers of his German dominions and the relatives of those who had fallen at Waterloo were eligible for the award. The medal is $1\frac{2}{5}$ in. in diameter, and bears on the obverse the laureated head of the Prince Regent facing to the right, the truncation draped, and surrounded by the inscription GEORG . PRINZ . REGENT 1815. The reverse bears in the centre WATERLOO JUN. XVIII, flanked by laurel wreaths, with a small trophy consisting of a cuirass and flags above, and the whole encircled by the inscription HANOVERISCHER TAPERKEIT. The medal was suspended from a steel clip and ring by means of a crimson and blue-edged ribbon, like that used for the British Waterloo.

Brunswick Medal.—The Prince Regent (as guardian of the minor Princes of Brunswick) was also responsible for the issue of a medal to the soldiers of Brunswick who were present in the engagements of June 16th, 17th and 18th, 1815. The order for the striking was given on June 11th, 1818, and the medals were struck from captured French cannon. On the obverse is the head of Duke Fredrich Wilhelm of Brunswick, who fell at Quatre Bras, and the inscription in German lettering FRIEDRICH WILHELM HERZOG; on the truncation of the bust is C. Häseler in script. On the reverse is a wreath of oak and laurel, encircling 1815, and the inscription in German lettering arranged round and divided by rosettes, Braunschweig Seinen Kriegern. Quatrebras und Waterloo (Brunswick to her Warriors. Quatre Bras and Waterloo). The medal, $1\frac{7}{20}$ in. in diameter, was suspended from a steel clip and ring by a $1\frac{1}{2}$ in. yellow ribbon, with broad blue stripes equal to one quarter the width near the edge. The officers' medals were gilt, with the names indented on the edge.

Saxe-Gotha-Altenburg Medal.—In 1816 Duke Emilius Leopold Augustus decided to have a medal struck for presentation to the members of the Foreign Legion of the Duchy of Saxe-Gotha-Altenburg who had taken part in the campaigns of 1814–15. The medals were of bronze; those awarded to the officers were $1\frac{3}{5}$ in. in diameter, gilt all over, and those given to the private soldiers $1\frac{13}{20}$ in. in diameter, parti-gilt, the Altenburg rose and ornamental border on the obverse being gilt, likewise the ducal crown and lettering on the reverse, which bears, in old German characters, IM KAMPFE FUER DAS RECHT (In the struggle for right), and round the edge, where the name of the recipient is usually placed, HERZOGTH GOTHA-VND-ALTENBVRG MDCCCXIV : MDCCCXV. The medal has a ball-shaped loop and ring for suspension from a dark-green ribbon 1 in, wide, with a border of black, through which gold is braided at intervals (see facing page 72).

Belgian Star.—A five-pointed star, with ring for suspension, was granted to the Belgian troops who were present at the battle of Waterloo. Like many continental medals of the period, it is very simple, having on the obverse 1813, and on the reverse 1815. It is not very coveted, owing to the uncertain conduct of the Belgian troops.

Hanoverian Jubilee Medal.—On the fiftieth anniversary of the battle of Waterloo the inhabitants of Hanover presented a bronze medal to the survivors of the Hanoverian troops. On the obverse is the arms of Hanover, encircled with the inscription STADT HANOVER DEN SIEGERN VON WATERLOO 18 JUNI 1815, and on the reverse, within a laurel wreath, ZUR 50 JAHRIGEN JUBELFEIER AM 18 JUNI 1865.

French or St. Helena Medal.—It was not until August 12th, 1857, that the survivors of that magnificent army of Frenchmen who followed *le petit Caporal* in his campaigns from 1792 to 1815 received a medal commemorative of their devotion and gallantry. It was due to the Emperor Napoleon III that the old military and naval warriors received, what their great General and Emperor would wish for them, " marks of merit." The St. Helena medal is oval and of bronze, $1\frac{1}{5}$ in.

across. On the obverse is a beaded circle within laurel leaves, which frame the whole medal (see facing page 64), the laureated head of Napoleon facing to the right with the legend NAPOLEON I EMPEREUR. On the reverse, within a beaded circle, is the inscription A SES COMPAGNONS DE GLOIRE SA DERNIERE PENSEE STE. HELENE 5 MAI 1821 (To his companions in glory his last thought St. Helena 5 May 1821), and around CAMPAGNES DE 1792 A 1815, with a small five-pointed star beneath. As part of the medal, but standing above the oval, is the French Imperial Crown, through the terminal of which a ring runs for suspension from a green ribbon, 1½ in. wide, with narrow stripes of red.

NEPAUL

While the attention of Britain was necessarily riveted upon the Peninsula, both East and West, as we have seen, compelled her to keep the torch of battle burning, and I must therefore mention the campaign in Nepaul, for which the Honourable East India Company awarded a special medal. The order dated Fort William, March 20th, 1816, stated that the Government had decided to present silver medals to every native officer who had served within the hills, and to as many non-commissioned officers and privates as might be recommended for distinguished zeal or gallantry. The campaign was brought about owing to the Goorka tribes having adopted an aggressive policy, and refusing to vacate the districts of Bootwal and Sheoraj belonging to the H.E.I. Co., that had to be retaken by an armed force, and by an attack upon the British outposts at Rourah, which had been left in possession when the rainy season commenced. The operations, which followed the declaration of war made on November 1st, 1814, covered a frontier stretching 600 miles. In the early part of the campaign General Gillespie was killed while leading a body of 100 dismounted men of the

8th Dragoons against the small but well-defended fort of Kalunga. The back of the war was broken when, in March 1815, General Sir David Ochterlony resisted a furious assault of the Gōrkālis for two hours and then, boldly charging, forced them to fly in confusion. For his brilliant services General Ochterlony was made G.C.B., the first officer in the Indian Army to receive the honour. Ummur Singh, Rajah of Nepaul, after the surrender of his officers and men was compelled to submit, but a year elapsed before the treaty was ratified at Fort William. The regiments occupied in the campaign were: H.M.'s 14th (the 14th did not, however, receive the medal, as they had not served in the field), 17th, 24th, 66th, a wing of the 67th, and 87th Royal Irish Fusiliers; a detachment of the 8th Light Dragoons, and the 20th and 25th Native Infantry.

The H.E.I. Co.'s Nepaul Medal.—The medal awarded by the Honourable East India Company was silver, $1\frac{10}{20}$ in. in diameter, and given to all ranks. On the obverse troops with fixed bayonets and colours flying are represented marching through a hilly country, the heights crowned with forts and stockades; in the foreground to the left is a field gun. On the reverse is a Persian inscription, "This medal is conferred by the Nawab Governor-General Bahadar in testimony of the energy, good service, skill, and intrepidity displayed during the campaign in the hills, in the years of the Hegira 1229 and 1230." It was worn suspended from a yellow silk cord, and, as it was only given to native soldiers for particularly distinguished conduct, is very uncommon.

When the "Army of India" medal was issued in 1851, those native soldiers who had not received the H.E.I. Co.'s medal were granted the new medal, with bar for NEPAUL (see facing page 56). The British regiments entitled to the medal with this bar were the 8th Light Dragoons; 17th, 24th, 26th, 66th, one wing of the 67th, 87th Regiments, and 20th and 25th Native Infantry. On November 5th, 1817, the Peishwa,

Bajee Rao, who had again given evidence of his treachery, was defeated at the village of Kirkee by Lieutenant-Colonel Burr of the 7th Bombay Native Infantry, with the 65th Foot and a native force consisting of the 2nd, 6th, 7th, 12th, 13th, and 23rd Bombay Native Infantry; Bombay Artillery, Field and Horse, and Pioneers; the Bombay Fusiliers (now 2nd Batt. Dublin Fusiliers); and Poona Horse. On November 17th Colonel Burr, together with Brigadier-General Lionel Smith commanding H.M.'s 65th British Regiment and the 2nd Grenadier Regiment of Native Infantry, captured the town of Poona. On the 26th of the same month a force under Lieutenant-Colonel Hopetoun Stratford Scott and Brigadier-General Doveton captured the camp of the Rajah of Berar, Apa Sahib, at Seetabuldee (£79 has been paid for a medal with the single bar for this engagement), outside the city of Nagpore, which, after a desperate defence by its garrison of 5,000 men, surrendered on New Year's day, 1818. Although eight companies of the Royal Scots were present at the battle of Seetabuldee, they were not awarded the bar for same when the Army of India medal was distributed. They were also present at the capture of Nagpore (400 medals were issued to officers and men, and 43 to Europeans in the H.E.I. Co.'s service) with the following native regiments: 1st, 2nd, 17th, 21st, 23rd, 26th, 28th, 29th Madras Infantry and 6th Bengal Light Cavalry and 6th Madras Cavalry.

Maheidpore.—At Maheidpore, on December 21st, 1817, Lieutenant-General Sir Thomas Hislop defeated the Pindarees under the Mahratta Rajah, Mulhar Rao Holkar. The British troops present were two flank companies of the 1st or Royal Scots (12 medals only with this bar were presented to officers and men of this regiment), who were specially mentioned in the Commander-in-Chief's orders, and one squadron of the 22nd Light Dragoons. Only 26 clasps for this battle were issued to survivors of the 22nd. The native

MEDAL FOR GHUZNEE, 1839.
Awarded to the Rt. Hon. the Earl of Auckland, G.C.B., P.C., Late Governor-General of India.

MEDAL FOR GHUZNEE, 1839.

troops comprised 3rd, 14th, 27th, 28th, and 31st Madras Infantry; 1st Madras Fusiliers; 3rd and 4th Madras Light Cavalry.

Corygaum.—On January 1st, 1818, the village of Corygaum was most gallantly defended against the whole force of the Peishwa, and when the "Army of India" medal was issued the surviving soldiers of the 2nd Bombay Native Infantry and 2nd Poona Horse were given the medal, with bar for the defence. Dr. Payne had in his collection a medal awarded to G. Bainbridge of the 65th Foot, with the bar for POONA AND CORYGAUM, which, according to the medal roll, was inaccurate, but it appeared that the man was actually entitled to this rare distinction.

Rare Medals.—Ten medals only were issued with bars for POONA and CORYGAUM. For the latter battle only 8 medals were issued. Only 20 with the single bar for SEETABULDEE AND NAGPORE were issued to Europeans and 194 to natives. For KIRKEE AND POONA and CORYGAUM 73 medals were issued. 16 medals with the bar for KIRKEE AND POONA were issued to officers and men of the 65th Regiment and 44 officers and men of the same regiment had the single bar for POONA. Medals with the bars for AVA and BHURTPOOR are rarely met with, or those for CHRYSTLER'S FARM and AVA to the 89th.

Ceylon, 1818.—In 1819 the Government of Ceylon issued a medal, 1½ in. in diameter, in connection with the Kandian rebellion of 1818. It bears on the obverse a wreath of bay and oak leaves, encircling CEYLON 1818, and on the reverse REWARD OF MERIT, the name of the recipient being engraved in the centre. A blue ribbon was used for suspension. Four men only of the 73rd Perthshire Regiment, two men of the 2nd and 39 of the 1st Ceylon Regiments were awarded the medal.

FIRST BURMESE WAR

In 1824 war was declared against Burmah, and the campaign opened by Brigadier-General Sir Archibald Campbell taking possession of Rangoon on May 12th, and the capture by storm of Cheduba on May 17th, 1824. A series of assaults was made by the British, and feeble attacks by the Burmese, before Maha Bandoola, a veteran of considerable force and ability—having provided himself with gold fetters with which to bind Lord Amherst when he was captured—took command of a force of about 50,000 infantry and horsemen, with 300 pieces of artillery, and began to entrench himself in front of the British position. On December 1st Major (afterwards Sir Robert) Sale—who gallantly defended Jellalabad and was killed at Moodkee, December 1845—making a daring and vigorous attack upon the left of his line with 450 men of the 13th Light Infantry and 18th Native Infantry, compelled the Burmese to fly. On December 5th another effort was made against the left wing, and 240 guns were captured; this success was followed, on December 7th, by a determined assault upon the trenches, which resulted in the rout of Bandoola's army with the loss of 5,000 men. With indomitable pluck the Burmese general rallied his army and entrenched himself again; but when the assault was made, upon the 15th, by the British, the Burmese were forced from their position and put to rout in fifteen minutes. In the meantime Brigadier General Morrison, with a force of about 11,000, had moved forward across the mountains into Ava, and on April 1st Arracan was captured, and the Burmese retreated to Donabu, where on the Irrawady, behind a strong teak stockade a mile long, backed by an old brick rampart upon which 150 guns were mounted, they defied a marine attack, and the British had to retire down the river, leaving their wounded to the fiendish treatment of their enemies. On the next day, however, Sir Archibald Campbell gained possession of Donabu, for Maha Bandoola having been killed the troops lost confidence in their other leaders, and they evacuated the place at night. After a period of inaction at Prome, necessitated by the season, hostilities were resumed in November, and a series of conflicts took place until after the evacuation of Meeaday, when the Burmese again made overtures for peace. On January 19th, 1825, hostilities were resumed, and the British army advanced upon the Burmese capital. In the advance the battle of Melloon was fought, and a new Burmese General took command of an army of 40,000 men. Nee Woon Breen, "King of Hell" or "Prince of Darkness" as he was called, was, however, defeated by the British near Pagahm Mew on February 9th, and fled to Ava, to meet a cruel death by order of his King. Following up this success, the British marched upon the old capital Amarapura, but when within four days' march of the city the King accepted the British terms, and the treaty of peace was signed at Yandaboo on February 24th, 1826. By this treaty Arracan, Mergui, Tavoy, Tenasserim, and Yé were ceded, and became part of the Indian Empire. 3,222 European soldiers and 1,766 sepoys fell in the war.

The following regiments were engaged in this campaign, and those marked * were mentioned in general orders with sentiments of unfeigned admiration: 1st,* 13th,* 38th,* 41st,* 44th, 45th,* 47th, 54th, 87th,* and 89th,* also 14th Madras Native Infantry, Madras Fusiliers (European, now 1st Royal Dublin Fusiliers), and 1st Madras Cavalry, Bengal and Madras Artillery (European).

The Ava Medal.—By a General Order, dated "Fort William, April 22nd, 1826," a medal was bestowed upon the native troops engaged in the war between 1824–26. The medal, 1½ in. in diameter, was struck in gold for the officers (Sir Archibald Campbell was the only British officer to receive one) and in silver for the men: it was attached to a large steel clip and ring for suspension from a crimson ribbon with blue edges, $1\frac{1}{2}$ in. wide. On the obverse is the White Elephant of Ava crouching before the British Lion; behind the Lion is the Union Jack unfurled, and behind the Elephant the Burmese colours submissively lowered; in the background are palm trees. The exergue is occupied by a Persian inscription, "The Elephant of Ava is obedient to the Lion of England, Year 1826." In the left upper corner of the exergue is the designer's name, W. DANIELL, RA. DEL. On the reverse is depicted a storming party advancing against the Great Pagoda of Rangoon, which is surrounded by stockades; to the left is the Irrawady flotilla of gunboats and the "Diana"; In the foreground is depicted Sir Archibald Campbell directing operations from the shade of a palm tree. In the exergue is the Persian inscription, "A medal for the victorious British soldiers on Ava" (on Ava territory). In the left upper corner of the exergue is W. WYON. Although 750 gold medals were struck for distribution among native officers and civil dignitaries, and over 24,200 silver medals were issued, they are very rare. It is noteworthy, as Dr. Payne points out, that in connection with the issue of this medal it was the first time a ribbon was mentioned for use with an Indian medal. It was ordered to be worn perfectly square upon the centre of the left breast, the upper edge of the ribbon being even with the button for ranks wearing sword-belts only, and even with the second button for ranks wearing cross-belts. A bar for AVA was issued with the "Army of India" medal in 1851.

Burmese Chiefs' Medal.—The H.E.I. Co. had a medal, $2\frac{3}{20}$ in. in diameter, struck in gold for presentation to six Talaing chiefs who served with the British army during 1825–26. On the obverse are the arms and motto of the East India College; above is a scroll with "Award of Merit," and underneath a scroll to take the name of recipient. On the reverse in the foreground are two steamers by a river-side, and two soldiers, one bearing a flag, addressing a group of Burmese; in the background are hills, a pagoda, and a palm tree, and behind all the setting sun. To the right are a cannon and bungalow. The medal has a loop for suspension, composed of two plates held together by means of a gold bezel.

Bhurtpore.[1]—Owing to the assumption of sovereign power by Doorjun Sal, the nephew of the deceased Rajah of Bhurtpore, Baldeo Singh, it became necessary for the Indian Government to take steps to ensure the succession of the dead ruler's young son, Bhulwunt Singh. An army under Lord Combermere was dispatched to lay siege to the city of Bhurtpore, which with its 5 miles of fortifications, as already described, was able to withstand the four assaults made by General Lake's army twenty years before. On December 10th, 1825, the British army of just over 25,000 men appeared before the city, and prepared to lay siege to the fortress, which was garrisoned by about 25,000. On December 24th the artillery began to fire on the city, but despite the continual bombardment, so well had the walls been built that breaches could not be made large enough to admit of stormers. Mining, therefore, was resorted to, and by the morning of January 18th, 1826, a determined assault was made, and within two hours the British had gained possession of the ramparts; by 4 o'clock the citadel surrendered, and the 14th, as a reward for their gallantry, placed to garrison it. The impregnable city, the capital of the Jants, had fallen after a twenty-six days' siege; Doorjun Sal was captured, spoil to the value of £500,000 taken, the young Rajah placed upon the throne, and the conquest of India confirmed by the success of Lord Combermere, who was rewarded with a Viscountcy. Hand grenades were last used in India at the siege of Bhurtpore. One hundred and eighty of the British troops were killed and 780 wounded. The following regiments took part in the siege: 14th and 59th Regiments,

[1] This is spelt BHURTPOOR on the bar of the "Army of India" medal.

11th Light Dragoons, and 16th Lancers (who had only just been armed with the lance). The Native Regiments included: 11th, 15th, 21st, 23rd, 31st, 32nd, 33rd, 37th, 41st, 63rd, and 66th Bengal Light Infantry; Simoor Rifles; 1st Bengal Cavalry; 6th Light Cavalry and one wing of the Bengal Fusiliers (Europeans), now 1st Royal Munster Fusiliers; Bengal Horse and Foot Artillery and Bengal Engineers. The 14th, "The Old and Bold," were specially mentioned in dispatches.

Coorg, 1837.—For suppressing the insurrection in Canara in 1837 the Coorg soldiers were, on the recommendation of the Commissioner, Lieutenant-Colonel Mark Cubbon, awarded a medal as a recognition of their noble conduct and courage. The medals of gold and silver were $1\frac{19}{20}$ in. in diameter, and of varying thicknesses according to value. They were worn suspended from the neck by chains by the superiors who received them; 20 gold medals and 200 silver medals given to junior officers and ryots were presented without chains. Two gold medals valued at 400 rupees, and 12 valued at 200 rupees, were bestowed with chains. On the obverse is a Coorg warrior in fighting attitude, his upraised right hand grasping a Coorg knife ready to strike, and in his left a matchlock; round the face of the medal is the inscription in Carnese, "For a memorial of Fidelity to the Government of the East India Company in suppressing rebellion in the months of April and May 1837"; on the reverse a trophy of Coorg arms and ornaments within a laurel wreath, surmounted by the inscription, FOR DISTINGUISHED CONDUCT AND LOYALTY TO THE BRITISH GOVERNMENT, and in the exergue COORG APRIL 1837.

FIRST AFGHAN WAR

The Shah of Persia, having laid claim to a considerable portion of Afghanistan after Shah Shoojah had been driven from his throne and his kingdom divided among a number of chieftains, sent an army to besiege Herat. The Indian Government then decided to effect the restoration of Shah Shoojah, and sent forward the "army of the Indus" under the command of Sir John (afterwards Lord) Keane. The Persians had meanwhile raised the siege of Herat, and retired toward their capital. Fighting their way against almost insuperable obstacles, short of food for man and beast, the divisions of the British army pushed their way through to Candahar, which they reached on April 27th, 1839. In the meantime Hyderabad had been captured, the Ameers of Scinde compelled to submit, the erstwhile rulers of Afghanistan forced from their thrones, and the deposed ruler restored, but considerable fighting had to be done ere he could settle down peaceably, and several important battles were fought.

Ghuznee, 1839.—The army of the Indus left Candahar at the end of June 1839, and arrived before Ghuznee, which Prince Mahomed Hyder Khan had strongly fortified; leaving only one gate unblocked by masonry. He had with him a garrison of 3,000 Afghans. As the British force possessed no means of breaching the walls, it was determined to take the place by storm, and in order to effect this the gate was blown in by the 13th (now Somerset Light Infantry), and the troops entered the fort, only 5 men being killed, and 6 officers (including Brigadier-General Sale, severely) and 63 men wounded out of Her Majesty's Regiments. On the fall of the fortress the Afghan cavalry outside fled in the direction of Cabul, the British forces following. At Ughundee Dost Mahomed Khan had resolved to try conclusions, but dissatisfied with the morale of his troops he decamped, and on August 7th Shah Shooja regained his throne, which, however, he did little to strengthen. The British losses were 18 men and 20 officers killed, and 153 men wounded. The force engaged in this comparatively bloodless campaign was composed of 8 companies of Her Majesty's 2nd (Queen's Royal); 13th Light Infantry; 17th (Leicestershire); two squadrons 4th Light Dragoons; 16th Lancers; and the following regiments of the H.E.I. Co.'s army: 1st Bengal Fusiliers (Europeans); Bengal Sappers and Miners; 1st Bombay Light Cavalry; 2nd and 3rd Bengal Light Cavalry; 2nd Skinner's Horse; 300 Poona Horse; Bengal Horse and Foot Artillery; two troops Bombay Horse Artillery; Bombay Foot Artillery; 16th, 19th, 35th, and 48th Native Infantry.

Order of Dooranée.—To reward the British officers who had been instrumental in restoring him to his throne, Shah Shooja instituted the Order of the Dooranée Empire, the first installation taking place at Cabul on September 13th, 1839, when Sir John Keane, the commander of the British army, received the First-Class Order from the Shah. There are three classes of the Order, which is not unlike the Guelphic Order of Hanover. Three of the first class, nineteen of the second, and thirty-six of the third were to be presented. The badge consists of a gold Maltese cross, the borders of which are raised, with eight points terminating in gold beads, resting upon two crossed swords. On a blue and green enamelled ground, which forms the centre, are two sentences in Persian characters; above "Authority is from God alone," and below "Every brave man recognises his sway." This translation, as Dr. Payne states, is very different from the usual one of Duri-i-Dauran, "Pearl of the Age." The enamelled centre is surrounded by a circle of pearls. The star is of cut silver shaped like a Maltese cross, overlaid with a smaller one of gold, in the centre of which is the same decoration as on the badge, except that a diamond is set in each arm of the gold cross, and the enamelled circle is surrounded by seventeen pearls. The star of the second class has rays between the arms, and the small gold cross which rests upon it has crossed swords between the arms; the centre is enamelled with a green flower on a blue ground, surrounded by eighteen pearls set in a gold band. The badge of the second class has twenty pearls set round the centre, and on

MEDAL FOR KELAT-I-GHILZI, 1842.

Awarded to Major Robert Leech, C.B., who had also the medal for Candahar, Ghuznee, and Cabul, 1842, and the Order of the Dooranee Empire (3rd Class).

FIRST, JELLALABAD MEDAL, 1842.

Awarded to Captain (later Lieutenant-Colonel) Peter R. Jennings, 13th Foot.

the reverse a red enamelled centre circled by a green enamelled border decorated with gold tracery. The third-class badge has only fourteen pearls. The badges of each class were suspended from a half red and green ribbon, like that used for the Ghuznee medal.

Ghuznee Medal, 1839.—Shah Shoojah had also determined to decorate the ordinary soldiers who had taken part in the campaign, but his assassination prevented the realisation of his desire, and subsequently—November 23rd, 1842—the Governor-General decided that the medals which had been struck at Calcutta in commemoration of the capture of Ghuznee should be given to the officers and men who had participated in the capture. This, the first Indian medal to be issued with a bar for suspension, is of silver, and, with the exception of the medal for Seringapatam, the first to be given to all the troops engaged, the previous medals having been bestowed upon native troops only. It is $1\frac{1}{2}$ in. in diameter, with broad flat edge, bearing on the obverse a view of the gateway of the citadel, with GHUZNEE on an ornamental panel underneath, running with the lines of the medal. On the reverse, within a laurel wreath, a mural crown above 1839, and between the terminals of the wreath the date of the capture of the fortress, 23D JULY, a space being left in the centre for the recipient's name, which had to be placed there at his expense. The medals are occasionally found with the names engraved or impressed upon the edge. Two dies were used for the obverse, and one has a much wider border. The medal was originally suspended from a half green and yellow ribbon, but it was changed to crimson and green $1\frac{2}{5}$ in. wide.

Governor-General's Medal.—Dr. Payne has kindly enabled me to reproduce the unique silver medal presented to the Right Honourable the Earl of Auckland, G.B.C., Governor-General of India and Governor of the Presidency of Fort William. The medal is $1\frac{3}{5}$ in. wide, has no bar for suspension, and is enamelled. On the obverse is represented the fortress of Ghuznee—

a very different rendering to that on the medal just referred to—with the victorious army marching through the gate. On the reverse is the inscription AFGHANISTAN-GHUZNEE—23rd JULY 1839. This striking is the only one known to exist. (See facing page 88.)

FIRST CHINESE WAR, 1842

The extension of British commerce caused the first Chinese war. The Government not only loathed the "Fan Kwei," or "foreign devils," but objected to the use of opium which the merchants sold. The destruction of about £2,000,000 worth of opium by Chinese officials, and the insult offered to Her Majesty's representative, Captain Elliot, R.N., led to the declaration of war against China, and a force under Brigadier-General Burrell was dispatched to China in June 1840, but following an attack on Ting-hae-hien the Chinese opened negotiations for peace, and after protracted *pourparlers* they agreed early in 1841 to pay an indemnity, and to cede the island of Hong Kong. In accordance with the treaty the island was occupied by the British, but this was resented by the Chinese, who had determined not to carry out the terms agreed upon. Hostilities were consequently recommenced, and an expedition sent up the Canton River, where it quickly dealt with the batteries at Wantong, landed a body of British and Indian infantry, and captured considerably over 1,000 Chinese soldiers without losing a man. Pursuing its way up the river, the expedition imposed severe punishment upon the different forts, and the Chinese again sued for peace, mainly, however, for the purpose of gaining time to bring up reinforcements. The forts at Canton were silenced, and on May 24th, 1841, the British troops landed and carried the fortified heights, the 18th Royal Irish and the 49th charging in brilliant style. A flag of truce was hoisted on May 26th, and as a result of *pourparlers*

£4,250,000 sterling was agreed to be paid for the evacuation of Canton, which was to be opened to trade.

The Emperor had no intention of carrying out any such agreement—on the contrary he issued a mandate for the extermination of the British. As a result of this flagrant breach of faith, an expedition was sent to the Island of Amoy, and appeared before the city of Amoy on August 25th. Two hours' bombardment sufficed to demoralise the garrison, despite their 500 pieces of cannon, and a force of troops was landed. Chusan was retaken, the military depot of Chinghai stormed, and the fortified and important city of Ningpo captured. A determined effort was made on March 10th, 1842, to retake it, but despite the surprise the enemy was beaten off. Later the British left Ningpo and again proceeded up-river, attacking Tsekee on March 15th, capturing the heights of Segon and forcing the Chankee pass ere returning to Ningpo on March 17th. In May Ningpo was evacuated, and the fortified city of Chapoo at the mouth of the Shanghai River attacked and a landing effected. In June Woosung and Poonshau, and the city of Shangee, were captured, and with the aid of the 98th Regiment under Lieutenant-Colonel Campbell (afterwards Lord Clyde), which now reinforced the expedition, the city of Chin-Kiang was attacked, and after a severe struggle taken on July 21st, 1842. Lieutenant-Colonel Sir Hugh (afterwards Viscount) Gough, whose decoration from Dr. Payne's collection is used to illustrate the medal awarded for this campaign, led his army to the ancient capital Nankin, which was reached on August 9th, 1842. The demonstration sufficed to bring the enemy to reason, and a treaty of peace was signed, the indemnity paid, the ports of Amoy, Ningpo, Foo-choo-foo, and Shanghai opened to British commerce, and Hong Kong ceded to us.

The army of about 3,000 men was composed of the 18th Royal Irish, 26th Cameronians (which lost nearly 600 men out of 900, by disease, before Canton fell),

49th, 55th (which particularly distinguished itself at the assault on the city of Chin-Kiang), 98th, 37th Madras Infantry, Royal and Indian Artillery, and a Naval Brigade, including marines, also took a prominent part.

The First China Medal, 1842.—In January 1843 the grant of a medal was made to commemorate "the signal successes of Her Majesty's Naval and Military forces," both European and Native, upon the coast and in the interior of the Empire of China. On the obverse is the diademed head of Queen Victoria facing to the left, with the legend VICTORIA REGINA on either side, and on the truncation of the head W. WYON, R.A. This was the first medal issued with the Queen's head. On the reverse is a trophy of naval and military weapons, very badly disposed, with an oval shield bearing the Royal Arms in the centre, the whole backed by a palm tree, and above ARMIS EXPOSCERE PACEM, with CHINA on a plain panel in the exergue, and underneath the date 1842. The medal is $1\frac{2}{5}$ in. in diameter, with a plain straight German-silver clasp soldered on for suspension from a ribbon of red with broad yellow edges, the same width as the diameter of the medal. The names of the recipients, and the number of the regiment or name of ship, were impressed upon the edge in bold Roman capital letters. The reverse of this medal has been used ever since for presentation to troops taking part in Chinese wars, the date 1842 being deleted, and fishtail bars recording the engagements and dates used to indicate the campaign.

SECOND AFGHAN WAR

Shah Shoojah made a very indifferent ruler, and, as I have stated, did little to strengthen his hold on the throne upon which the British had placed him—indeed, without British arms he possessed no authority The price we paid to keep this unpopular and bad ruler upon his throne—mainly, it is true, to ensure

Reverse.) (Obverse.)

SECOND JELLALABAD MEDAL, "FLYING VICTORY," 1842.

CHINA, 1842.
Awarded to Field-Marshal Hugh Viscount Gough, K.P., G.C.B., G.C.S.I

CABUL, 1842, WITH ORIGINAL STEEL SUSPENDER.

a barrier against Russia—was a heavy one. In October 1841 the Afghans openly rebelled at Cabul, murdered Sir Alexander Burnes and Sir William Macnaughton, and the troops were forced to retire upon Jellalabad, but were practically cut to pieces on the march, Dr. Brydon—who became one of the famous Lucknow garrison—being the only British officer to reach that city. It is estimated that of 20,000 who left Cabul, only a few hundreds escaped with their lives. The 44th (1st Essex Regiment) had 22 officers and 543 men killed during this disastrous retreat. Meanwhile Sir Robert Sale with the 13th Somerset Light Infantry, which had been detached from Cabul to deal with the insurgents in the Khoord-Cabul pass, had occupied Jellalabad.

Jellalabad.—The defence of Jellalabad by the little force under Sir Robert Sale, including Captain (afterwards the famous Sir Henry) Havelock, is the bright gem which sparkles in the dross of this unfortunate campaign. The city was in a very poor condition when the gallant defender seized the place and proceeded to strengthen it. On November 12th, 1841, the garrison had only one half-day's rations, but by a plucky sortie they managed to drive off the Afghans and obtain supplies. The place was again invested on the 27th, but by a successful sally on December 1st they were routed. On January 9th the leader of the rebellion called upon Sir Robert Sale to surrender the fortress, but he naturally refused. Not only was the indomitable spirit of the British tried by the repeated attacks of the enemy, but it almost appeared as if Nature herself was in league with the insurgents, for within a month a series of earthquake shocks demolished a third of the place and part of the defences, and necessitated continual repair of the works. Then came the blockade by Mahomed Akbar Khan, who after assassinating the British envoy at Cabul was responsible for the destruction of the army which left that city. On April 7th, however, the defenders made a determined attack upon the camp of the besiegers, and the 6,000 men under Mahomed Akbar were put to rout, and the siege abandoned. The regiments engaged in the defence of Jellalabad, under Sir Robert Sale, were H.M.'s 13th Light Infantry; a squadron of the 5th Bengal Light Cavalry; 35th Bengal Native Infantry; detachments of Broadfoot's Sappers and the 6th Shah's Infantry; half the Shah's Mountain Train; 2nd and 6th Batts. Bengal Artillery; a few native officers, and 682 armed followers.

First Jellalabad Medal.—The "Illustrious" garrison of Jellalabad, which had kept the flag flying for five months, was awarded a medal by General Order dated Allahabad, April 30th, 1842. It is a very simple medal, 1½ in. in diameter, bearing on the obverse JELLALABAD in capital block letters taking the line of the medal above a mural crown, and on the reverse the date APRIL boldly occupying the centre, with VII above and 1842 below. The suspenders were of two kinds—a steel clip and ring, or a silver wire loop attached to a ring through a hole in the broad flat rim. The medals were mostly issued without names; a few were indented, but the majority were engraved. The ribbon of watered silk is 1⅘ in. wide, and shaded rainbow fashion from crimson to yellow and blue. Two thousand five hundred and ninety-six medals were issued, and the relatives of those defenders who succumbed between April 7th, 1842, and the date of issue in December received the medal. The rarer medals are those awarded to the Shah's Mountain Train and the Shah's Cavalry, also the 5th Bengal Light Infantry.

Second Jellalabad Medal.—Lord Ellenborough, the Governor-General of India, being dissatisfied with the simple and somewhat crude character of the award, had a more decorative medal designed by W. Wyon and struck at the Royal Mint, that those who cared might exchange when they were issued in March 1845. The men, however, were loath to part with the original medal, and very few applied for the new one; it is stated that of the famous 13th—the only European regiment to receive the award—only five exchanged their medals. The obverse of the second Jellalabad, or the "Flying Victory," bears the head of Queen Victoria as on the China medal, with VICTORIA VINDEX above (a few were issued with VICTORIA REGINA), and on the reverse the fortress of Jellalabad, with a bold flying figure of Victory above bearing in her uplifted left hand the Union Jack, and in her right two wreaths; above is JELLALABAD VII APRIL, and in the exergue MDCCCXLII. The medal, $1\frac{2}{5}$ in. in diameter, was suspended from the crimson, yellow, and blue ribbon by a steel or plated bar $2\frac{2}{5}$ in. long. The recipient's name, etc., was impressed upon the edge in capital Roman letters, but some were issued engraved in script. In addition to the 13th (later called Prince Albert's Light Infantry, as one of the honours conferred for the occasion; they also had their facings changed from yellow to blue), the 35th Native Infantry, 2–6th Batt. Bengal Artillery, Anderson's Horse, one squadron 5th Bengal Light Native Cavalry, and Broadfoot's Sappers and Miners took part in the defence.

Cabul.—The prestige of Britain had been lowered as a result of the insurrections, and in order to relieve Jellalabad (but it is said to take vengeance upon the enemy), Major-General Pollock was dispatched with an army, with which he forced the Kyber Pass and relieved Jellalabad on April 16th, where he halted for some months to organise his transport. Meanwhile oppressive heat and pestilence played havoc with the troops encamped in the valley of Jellalabad, and it was determined to divide the force and advance on Cabul, General Pollock having a hard fight at the pass of Jugdulluck, where he beat off Mahomed Khan and the Ghilzie chiefs. Later the combined divisions of Pollock and Sale combated the enemy in the Tezeen pass, where, on September 13th, Akbar Khan with 20,000 men disputed passage. His force, however, was put to rout, and the leader escaped accompanied by a single soldier. The debacle was complete, and on September 15th the army encamped on the race-course at Cabul, and next day the British flag again flew over the city.

The following troops took part in the march and recapture of Cabul: 9th, 13th, 31st, 40th, and 41st Foot Regiments, and the 3rd Light Dragoons, who were awarded the medal for Cabul, also the following regiments in the H.E.I. Co.'s service: 4th, 5th, 6th, and 12th Bengal Infantry; 1st and 10th Bengal Light Cavalry; 3rd Irregular Cavalry; the 5th company of the Bengal Artillery, and Sappers and Miners.

Three thousand five hundred medals were issued to Europeans for Cabul, but very few were issued to the 40th Regiment.

Candahar.—Major-General's Knott's defence of Candahar, if not so famous as that of Jellalabad by Sir Robert Sale, exhibited the same tenacity and cool determination. The rebels in the vicinity of Candahar were led by Prince Sufter Jung, a son of Shah Shoojah, and Prince Timour; they and their men, however, were helpless in the face of the little British army that sallied forth on the morning of January 12th, 1842, and routed them. The success of the battle did but keep the enemy at a respectful distance, for they spent their time in plundering the villages in the neighbourhood. Bad weather militated against any further effort on the part of the British general until March 7th, when he advanced against the rebel Afghans, and again put them to rout, but while the detachment was engaged in combating one section of the insurgents, another made a daring attack upon the city; they were, however, repulsed with considerable loss. Another action was fought on March 25th, and, with the assistance of the brigade under Colonel Wymer, the investing enemy was driven in confusion across the Urghundaub. On April 28th Major-General (later Sir) Richard England, after attacking the enemy at Hykulzie, successfully reached Candahar, and enabled Major-General Knott to advance upon Cabul.

"The Fighting Fortieth" was the only British regiment present in the defence and in the actions outside. It is noteworthy that through disease the regiment had been considerably reduced before it took part in the defence of Candahar. Major Biddulph states that only 64 "Candahar" medals were struck for the 40th Foot, and of these 42 were for the relatives of deceased officers and soldiers, and 22 for sick, etc., who had returned to India. The 41st, the Welsh Regiment, was the only British unit that took part in the relief of Candahar. The 5th, 6th, and 12th Bengal Infantry; Poona Horse; 1st Bengal (Skinner's) and 3rd Bombay Cavalry, and Bombay and Bengal Artillery were engaged in the defence and relief.

Only 130 Europeans received the medal for Candahar.

Kelat-i-Ghilzie.—Colonel Wymer, C.B., with a detachment which included a few men of "The Fighting Fortieth," marched on Kelat-i-Ghilzie, a hill fort 84 miles from Candahar, in order to draw off the garrison, which had held the fort during the winter, and (on May 21st) five days before Colonel Wymer's arrival had succeeded in defeating 4,000 Ghilzees who had attacked the fort. The garrison of only 950, under the command of Captain Craigie, included about 100 Europeans. For this heroic defence a silver medal was issued to every man taking part; it is $1\frac{2}{5}$ in. in diameter, and depends from a steel clip and bar, as in the medal illustrated facing page 96, and was suspended from the same kind of ribbon. On the obverse is an ornamental shield inscribed KELAT-I-GHILZIE, surrounded by a laurel wreath with a mural crown above. On the reverse is a military trophy, with a breast-plate and helmet forming the central feature, and "INVICTA MDCCCXLII" on an ornamental tablet underneath. The names of the recipients were all engraved, generally in script. None of the Queen's Regiments took part in the defence, the majority (600) of the defenders being of the Shah's army. Three companies of the 43rd Bengal Infantry, 44 European Artillerymen including one officer, and 60 Sappers and Miners of the H.E.I. Co.'s service were also engaged. Only 55 Europeans received this medal.

Recapture of Ghuznee.—Ghuznee, with its garrison of exhausted and half-frozen men, had capitulated in March under the orders of Major-General Elphinstone, who had been taken prisoner at Cabul, and died shortly afterwards. Elphinstone's orders to surrender Jellalabad and Candahar had been ignored by Sir Robert Sale and Sir William Knott. They not only had the determination of Englishmen, but knew the treacherous character of their enemies. Colonel Palmer, unfortunately, forced by the weakness of his men and relying upon the promise of a safe escort for his sepoys to Hindoostan, agreed to surrender. The troops were unarmed, and with their womenfolk had hardly left the citadel when they were attacked and the majority massacred. Colonel Palmer was tortured, and with nine other officers thrown into a dungeon. When Sir William Knott appeared before Ghuznee, on September 5th, 1842, he found the city swarming with rebel troops, and great bodies of cavalry and infantry in the mountainous environs. He drove these off, erected breaching batteries, and prepared to attack the city, but on the morning of the 6th it was discovered that the fortress had been evacuated; and so with very few casualties Ghuznee was again taken, over 300 unfortunate sepoys released, and the sacred gates of Somnath, which Mahommedan invaders had taken from India centuries before, were taken down as mementoes of a campaign which did nothing to improve our prestige.

The 40th and 41st were the only British regiments to receive the medal with Ghuznee inscribed thereon. Members of these regiments who had been in the other actions were entitled to the medals bearing Candahar and Cabul also.

The Ghuznee, Candahar, and Cabul Medals.—For this war a silver medal was issued by the Indian Government to those who had taken part in it. Four types of medals were issued. One with the words CANDAHAR, GHUZNEE, CABUL, 1842, the names being above one another and the date below, all within a laurel wreath, the whole surmounted by the royal crown; another for GHUZNEE-CABUL, 1842, with the names in entwined wreaths and the date beneath them; and those for CANDAHAR, 1842, and CABUL, 1842, as indicated in the illustration facing page 100 of one with the original bright steel suspender attached to a hinged clip. The medals are $1\frac{2}{5}$ in. in diameter, the obverse bearing the diademed head of the young Queen Victoria with the legend "Victoria Vindex" above; some have the words "Victoria Regina," but these are scarce, while a few of the medals for Cabul are spelt thus, CABVL, but it is doubtful where they were actually issued to soldiers. It is noteworthy that on certain of the medals in this series the name of the executant, W. Wyon, R.A., is on the truncation of the Queen's head instead of beneath it. In this issue the recipients' names were engraved either in Italian script or in square Roman letters; a few were impressed, and some were issued unnamed. The "rainbow" or military ribbon for India is used for suspension.

One thousand four hundred Europeans received the medal with Candahar, Ghuznee, Cabul.

Meeanee.—In 1843 the "tail of the Afghan storm" arose. It followed the attempt on the part of several Ameers, with an army of 8,000 Belooches, to obtain possession of the British Presidency at Hyderabad; they were driven off, but owing to the weakness of the place it was decided to abandon it for the safety of the camp commanded by Sir Charles Napier, to whom the conduct of the campaign in Scinde had been entrusted, and who gained for himself the native sobriquet "Shai-tan-ke-bhaee," or "The devil's brother," by his determined action in dealing with the Ameers. He first destroyed the fort at Emaun Ghur on January 14th and 15th, a feat which drew forth the eulogiums of Wellington, and ruined the chance of the Ameers in the campaign. On February 17th the battle of Meeanee was fought, 2,800 men, with 12 guns, under Major-General Sir Charles Napier attacking, and in three hours defeating the Scindian army of 30,000 infantry and 5,000 cavalry, with 15 guns, but not before nearly all the European officers were killed or wounded. The Belooches, who fought with remarkable valour, lost 5,000 men, their camp and all its appurtenances falling into the hands of the British. Next day six of the Ameers surrendered. The only Queen's Regiment represented at Meeanee was the 22nd Foot, and only 65 medals were issued to the regiment, with two companies of Bombay Artillery; Grenadiers of the 1st Native Infantry; 12th and 25th Native Infantry; a detachment of Poona Irregular Horse, to which corps only 14 medals were issued; Scinde Irregular Horse; 9th Bengal Light Cavalry, and half company of Madras Sappers and Miners. The 22nd lost 1 officer and 23 men killed; its commanding officer, Lieutenant-Colonel Pennefeather, 6 officers, and 52 men wounded.

Hyderabad.—The Ameers, who still held out against the British army, made a stand with 20,000 men at the village of Duppa, near Hyderabad, under Meer Shere Mahomed, where on March 24th they were attacked by Sir Charles Napier with 5,000 troops, and defeated with great loss. The arch enemy of the British, Hoche Mahomed Seede, was killed, and Meer Shere Mahomed fled to the desert; the battle was decided by the 22nd Regiment and the troop of Bombay Horse Artillery. In this battle the 22nd, which numbered 560 rank and file, lost 23 men killed, 5 officers and 134 men wounded.

As a result of the campaign Scinde and Meerpore passed into British possession. The troops engaged at Hyderabad were: H.M.'s 22nd Foot; 1st Troop of Bombay Horse Artillery; 2nd-1st and 2nd-2nd

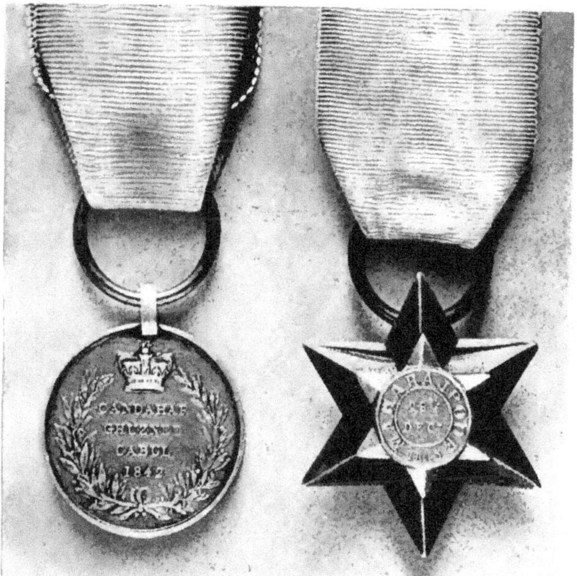

AFGHAN MEDAL FOR CANDAHAR, GHUZNEE, AND CABUL, 1842.

BRONZE STAR FOR MAHARAJPOOR.

Companies of the Bombay Artillery; 1st, 8th, 12th, 21st, and 25th Native Infantry; 3rd and 5th Companies of Golundaze; Madras Sappers and Miners; 3rd Bombay Cavalry; 9th Bengal Cavalry; Poona Horse and Scinde Horse; also men from the Indus Flotilla.

Medals for Meeanee and Hyderabad.—The medal for the war in Scinde in 1843 was the only war medal given by the Home Government to the troops employed in India during the period of the East India Company's rule. The medal is practically the same as that described and illustrated as given for Candahar, Ghuznee, and Cabul in the 1842 campaign, the name of the battle —MEEANEE or HYDERABAD—and date, 1843, being arranged inside a wreath surmounted by the royal crown. A third medal given to those who were present in both actions has MEEANEE-HYDERABAD within laurel wreaths tied together, the whole surmounted by the royal crown. A medal for the battle of Meeanee, awarded to the 22nd, has realised from £7 to £9. The steel suspender, as in the Cabul medal illustrated, was issued with the medal given to non-commissioned officers and privates, but those with the officers' medals were of silver; the Colonel, however, of the only English regiment present—the 22nd—had, at his own expense, silver suspenders made for his men in place of the steel ones. The 22nd were present at both battles, also detachments of royal and horse artillery. Some of these medals have the names of the recipients, etc., engraved in script; others are mostly impressed in Roman capitals. Those to the 22nd were generally engraved in block lettering. The ribbon used is the "rainbow" or Army of India type.

THE GWALIOR CAMPAIGN

The year 1843 also saw the Gwalior campaign, in which the two decisive battles of Maharajpoor and Punniar were fought on the same day, December 29th. "The Army of Exercise" had been assembled at Agra owing to troubles following the death of the Maharajah. Tyajee Rao Scindiah had been acknowledged by the British Government, and Mama Sahib was appointed Regent. He was, however, expatriated, and an attempt made to undermine the position of the young Maharajah. His authority was, however, re-established, and the British army prepared to leave the country under the command of Major-General Sir Hugh Gough.

Maharajpoor.—They crossed the Koharee River on the morning of December 29th, to find about 18,000 Mahrattas with 100 guns strongly entrenched at the village of Maharajpoor. The British army consisted of 14,000 men with 40 guns. A terrific conflict ensued owing to the superiority of the enemy in artillery, but attacked on all sides by the British force they were routed, despite the splendid determination of the Mahratta gunners, who were bayoneted at their guns, and the gallantry of the infantry, who discarding their matchlocks stubbornly fought hand to hand with the British soldiers.

The regiments engaged were the 39th and 40th Foot and the 16th Lancers. The H.E.I. Co.'s troops were the 2nd, 14th, 16th, 31st, 43rd, and 56th Bengal Native Infantry; 1st Bengal Light Cavalry; 4th, section of the 5th, 8th Bengal Light Cavalry; 10th Bengal Light Cavalry; 2nd Skinner's Horse; 4th Irregular Cavalry (in reserve); 2 companies of Bengal Foot Artillery (3 companies and 4 of natives in reserve); 3 troops of Bengal Horse Artillery and 3rd, 4th, and 7th companies Bengal Sappers and Miners.

Punniar.—On the afternoon of the 29th the left wing of the army, under Major-General Grey, discovered about 1,200 of the enemy in position on a range of hills near Punniar, and by a brilliant attack carried every position, and routed the enemy with considerable loss. These two victories which terminated the war were dearly bought, over 1,000 men being placed *hors de combat.*

The Queen's Regiments at Punniar were the 3rd

Buffs, 50th, Queen's Own, and 2 squadrons 9th Lancers; 5 companies of the 39th, 50th, and 51st Native Infantry; 2 squadrons 5th Bengal Light Cavalry; 8th Bengal and 8th Irregular Cavalry; 2 squadrons 11th Bengal Light Cavalry; 1st Company Bengal Sappers and Miners; 2 troops Bengal Horse Artillery and 1 company Bengal Foot Artillery.

Maharajpoor and Punniar Stars.—For these fights the victorious Britishers and native troops were awarded bronze six-pointed stars, $1\frac{7}{10}$ in. in diameter (as illustrated facing pages 108 and 112), made from the captured cannon, having in the centre a small silver star bearing the words PUNNIAR or MAHARAJPOOR, 1843, in a circle, and in the centre the date of the action, $\frac{29^{th}}{\text{DEC}^{r}}$ The star worn by Lord Gough had a silver elephant on the obverse instead of the silver star. These stars were originally fitted with broad brass hooks for attachment to a loop on the coat, but in some instances a steel ring was run through it, so as to allow for suspension by the rainbow ribbon used at the time with other Indian war medals, as illustrated in the pair for Candahar, Ghuznee, Cabul, and Maharajpoor, or a silver, or German-silver, suspender was attached, as shown in the Punniar star. The name, number, and regiment of the soldier were engraven on the back, at either side of the hook, in a light, slanting script.

THE SIKH WARS

The Sikh wars of 1845–6 and 1848–9, which concluded with the annexation of the Punjab, and the Koh-i-noor diamond being presented to Queen Victoria, were particularly hard-fought campaigns; and Lord Gough—one of the famous fighting Goughs—whom Wellington thought unworthy to continue the command in the second campaign, which, however, he successfully closed, is depicted on the medal for the Punjab campaign receiving from the defeated Sikhs the arms they were no longer allowed to bear against us. Some state that the mounted soldier depicted on the medal is Sir Walter Raleigh Gilbert.

Runjeet Singh, who died in 1839, left an army trained by Europeans and a young son, Dhuleep Singh. The former, numbering 60,000, inspired by the ambitious Ministers of the infant Maharajah, crossed the Sutlej and invested Ferozepore with the object of expelling the British from Hindoostan, but was encountered at Moodkee on December 18th by the "Army of the Sutlej," consisting of 3,850 Europeans and 8,500 natives, which with 42 guns had made a tiring march of 22 miles to relieve Ferozepore.

Moodkee.—It was 3 o'clock in the afternoon when the alarm was given that the Sikhs were to give battle. The fatigued and famished soldiers abandoned the meal they were preparing, and, many coatless, rushed to their arms and to meet the Sikh army, the strength of which was so great that it outflanked the British, which practically only formed a single line to oppose it. The infantry meeting them in front gave a brilliant account of itself with the bayonet. The second brigade of cavalry, with the 3rd Light Dragoons, swept the enemy's left, while the rest of the British cavalry turned their right, but the fight continued until darkness, aided by the great clouds of dust from the sandy desert, came to the aid of the Sikhs, who sullenly retired to their camp at Ferozeshuhur after the loss of 17 guns, while the British rested on the battlefield for a few hours and then returned to their camp. In this battle, Colonel James Robertson, C.B., relates, Captain (afterwards Lord) Napier had a remarkable escape with his life; when riding armed merely with a walking-stick several of the enemy rushed at him, but he escaped unscathed. He had ridden from Umballa alone and straight into the battle when it was at its height! In this battle, Colonel Robertson states, the Sikh gunners "were conspicuous for their reckless bravery and

MEDAL FOR SUTLEJ, 1845.

BRONZE STAR FOR PUNNIAR, 1843.

MEDAL FOR PUNJAB, 1849.

devotion to their guns." They "died rather than yield; and there were no white flags, and no quarter asked or given on either side, so we had just to fight it out," and these were the men with whom we had fought side by side in the advance on Cabul but three years before! In this encounter 84 officers and 800 men were placed *hors de combat*, and the hero of Jellalabad, Major-General Sir Robert Sale, mortally wounded, Sir John McGaskill killed, and 4 brigadiers wounded.

At Moodkee H.M.'s 9th, 31st, 50th, and 80th Regiments and the 3rd Light Dragoons were present, also the 2nd, 16th, 24th, 26th, 42nd, 45th, 47th, 48th, and 73rd Bengal Native Infantry; 4th and 5th Bengal Light Cavalry, 9th Irregular Cavalry; 4 troops of Horse Artillery and 2 companies of Foot Artillery. On the day following the battle H.M.'s 26th Foot, the 1st Bengal European Regiment, 11th and 41st Bengal Native Infantry, and 2 companies of Bengal Artillery joined the force.

Ferozeshuhur.—The Sikhs strengthened their position at "Ferushahr" and entrenched themselves with 120 guns; meanwhile the British army had been reinforced with heavy guns and a number of troops, while Major-General Sir John Littler had brought his division of 5,000 men from Ferozepore, making a total of about 5,674 Europeans and 12,053 native troops, with 65 guns, to be ranged against 25,000 Regulars and 10,000 Irregular Sikhs, with 83 guns, strengthened by an army of about 23,000, with 67 guns, under Tej Singh, encamped only ten miles away. The battle of Ferushahr, or Ferozeshuhur, as it is officially spelt, opened on December 21st, with a terrific cannonade by the Sikhs, under which the British infantry advanced, stormed the entrenchments, and took a number of the enemy's guns. So hardly did the Sikhs fight that they could not be forced from the whole of their entrenchments; indeed, so stubborn was the battle of Ferozeshuhur that it lasted for two days, while the intervening night was also kept lively by the Sikh artillery playing on to the ground gained by the British. The 80th Regiment was sent with the 1st European Light Infantry to silence the guns, and the 80th was successful in capturing three. "A more wonderful battle never was. Within 150 yards of one another were 8,000 British troops against an unknown number of enemies yet unbroken." All the Governor-General's staff were killed or wounded before the day of retribution arrived. The Commander-in-Chief and the Governor-General of India, Sir Henry Hardinge, placed themselves in front of the two wings of the army "to prevent the troops from firing" until they closed! Unchecked by the enemy's fire, the line advanced, dislodged the enemy and swept everything before it. Halting, as on the parade ground, the army vociferously cheered its leaders, for they were masters of the field. Sirdar Tej Singh, however, brought up his army of over 30,000 men, with a large field of artillery, and endeavoured to retrieve the day; but despite the fact that the British artillery had already expended its ammunition, he was compelled to retire, and the day was won. On the field, however, were 2,415 dead or wounded soldiers, and 115 officers, which the victorious army had lost. The men who fought and won this protracted and sanguinary battle had mostly been without food or water for forty-eight hours.

The Queen's Regiments present at Ferozeshuhur were the 9th, 29th, 31st, 50th, 62nd, and 80th; 3rd Light Dragoons; 1st and 3rd Brigades Bengal Horse Artillery, and the 4th, 6th, and 7th Batteries of Artillery; the 2nd, 12th, 14th, 16th, 24th, 26th, 33rd, 42nd, 44th, 45th, 47th, 48th, 54th, and 73rd Native Infantry; Royal Bengal Fusiliers; No. 6 Company Bengal Sappers and Miners; 4th, 5th, and 8th Bengal Light Cavalry; 3rd and 9th Bengal Irregular Cavalry; 2nd Skinner's Horse, and the Governor-General's Bodyguard. The 11th and 41st Bengal Native Infantry guarding the wounded at Moodkee and the men of the 27th and 63rd Bengal Native Infantry, Bengal Artillery, and

Sappers who garrisoned Ferozepore were also granted the medal or the bar for Ferozeshuhur.

Aliwal.—Within a month the battle of Aliwal was fought. The Sikh Sirdars having effected a passage of the Sutlej near Loodiana, likewise at Ferozeshuhur near Sobraon, and threatened the garrison at the former place, the first brigade under Sir Harry Smith—a Peninsular veteran who bore twelve clasps to his medal—was ordered to make a forced march to relieve it. At Budiwal the enemy had prepared to intercept the British force by strongly entrenching the range of sand-hills and the villages. Sir Harry Smith, however, refusing to take any risks with his army of footsore and fatigued men, decided to make a flank movement by the right, although it would mean running the gauntlet of a heavy fire from the forty guns which was opened upon his columns. The British troops passed through the ordeal with remarkable discipline, but with a loss (this was on January 21st, 1846) far greater than was sustained later at Aliwal.

Colonel Godby, who commanded the garrison, on observing the approach of the British force, moved out from Loodiana and joined the relieving force. General Wheeler's brigade also arrived, and on January 28th the battle of Aliwal was fought. About 10,000 men against 20,000 with 68 guns, Sir Harry Smith boldly advanced upon the Sikh position held by the flower of that fine army. The British cavalry drove the Sikh horsemen on their own infantry, and captured several guns, while the British infantry made terrible havoc with the bayonet. The battle commenced at 10 a.m., and by 1 o'clock the Sikh army was flying in disorder, the whole of their artillery being left in possession of the victors. In this action the British losses were 151 killed, 413 wounded, and 25 missing. The 16th Lancers, who had greatly distinguished themselves, lost 8 officers and 100 men. Colonel Robertson, who was present with the 31st, records a "magnificent charge of the 16th Lancers, who rode right through one of the enemy's squares. As they (the Lancers) came round our right flank a regiment of sepoys, our own men, opened fire upon them!" The gallant Colonel, however, at the risk of his life, ran along the line knocking up their muskets with his sword as he shouted in Hindustani, "Our men, our own men!"

The following regiments were engaged at Aliwal: H.M.'s 31st, 50th, and 53rd Foot and 16th Lancers. The H.E.I. Co.'s troops were the 1st and 2nd Goorkas, 30th, 36th, 47th, and 48th Bengal Native Infantry and a wing of the 24th; 1st, 3rd, and 5th Bengal Light Cavalry; 4th Irregular Cavalry and the Bodyguard; Bengal Horse and Foot Artillery; Bengal Sappers and Miners.

Sobraon.—The completeness of the victory at Aliwal did not utterly dispirit the Sikhs, for they made a bold stand at Sobraon, where, on February 10th, they desperately disputed the day with the army composed of the combined forces of Sir Hugh Gough and Sir Harry Smith. The Sikhs held their 3,500 yards of entrenchments, shielding 70 guns, with 34,000 men, while on the other bank of the Sutlej 20,000 more were ready for battle. Sir Hugh Gough's army of 16,224 men—6,533 Europeans and 9,691 native troops—prepared for battle at sunrise, when the British batteries opened fire, but with little effect upon the enemy's guns in the strong redoubts. At 9 o'clock the European infantry advanced to attack, and then, without firing a shot, under a galling fire rushed upon the Sikhs, while the cavalry, passing in single files through the gaps in the entrenchments made by the sappers, re-formed and sabred the stubborn Sikh artillery-men at their posts. Those who fled were drowned in thousands in the waters of the Sutlej, which was red in places with the blood of men and horses, and the fine army of Sikhs (Khalsas) melted away, after leaving about 8,000 on the field of battle. No longer was there any doubt as to the determination of the British to maintain their prestige and possessions, and the haughty chieftains sought peace from their conquerors. The youthful Maharajah Dhuleep Singh was, however, compelled to treat in person. As a result the victors occupied the rich district between the Sutlej and the Beeas, and he was allowed to nominally rule the Punjab until the war indemnity of £1,500,000 was paid. It was a dear victory for the British: the 10th lost 3 officers and 130 men; the 29th Regiment lost 13 officers and 135 men; the 31st, 7 officers and 147 men, two of the officers—Ensign Tritton and Jones—being killed while bearing the regimental colours, which were ultimately planted on the highest point of the enemy's entrenchments by Sergeant Bernard McCabe, who as Captain McCabe was mortally wounded when leading a fourth sortie from Lucknow on October 1st, 1857. The 50th lost 12 officers and 227 men. Generals McLaren and Cyril Taylor were among the dead, also Major-General Sir Robert Dick, who gained considerable fame as leader of the 42nd Highlanders during the Peninsular War and at Waterloo, and who fell at the moment of victory.

The regiments engaged at Sobraon were: H.M.'s 9th, 10th, 29th, 31st, 50th, 53rd, and 62nd Infantry Regiments; 3rd Light Dragoons; 9th and 16th Lancers; 1st Bengal Europeans; 1st and 2nd Goorkas; 1 company 12th, 16th, wing of 24th, 26th, 33rd, some of the 38th, 41st, 42nd, 43rd, 47th, 59th, 63rd, and 68th Bengal Native Infantry; with the 45th and 59th in reserve; 3rd, 4th, and 5th Bengal Light Cavalry; 2nd, 8th, and 9th Irregular Cavalry; 10 troops of Horse Artillery; 11 companies Foot Artillery. The 73rd Bengal Native Infantry stationed at Rhodawalla also received the medal. It should be noted that the record of regiments engaged does not necessarily represent all the units taking part. Details and drafts from various corps were often present through one cause or another, or by accident. This should be borne in mind whenever a medal comes under notice. The medal rolls will generally assist, but there have been instances, and very serious ones for the collector, where the want of accuracy in a medal roll has cost him very dearly. Dr. Payne gives an instance of this on page 74 of "British and Foreign Orders, War Medals and Decorations," a bar for Corygaum being on a medal awarded to a man of the 65th Foot, as well as the bar for Poona. The first-named bar made the medal one of the rarest of the series issued to a British soldier, but the medal roll led Dr. Payne to think it a fraud, which later evidence demonstrated was not the case.

The Sutlej Medal.—The first, or Sutlej medal, $1\frac{1}{4}$ in. in diameter, bearing the legend "ARMY OF THE SUTLEJ," was given with three bars for four battles; that is to say, that those who fought through all the battles had their record on the medal and three bars. A soldier taking part in the battle of MOODKEE would have the name impressed in raised letters, with the date, 1845, in the exergue of the medal, and the succeeding battles on the bars, as in the illustration facing page 112, of a medal awarded to a man of the 31st Royal East Surrey Regiment—this and the 50th Royal West Kent Regiment being the only two British regiments to receive the three bars, for FEROZESHUHUR, ALIWAL, and for SOBRAON, the battle which determined the first campaign. This was the first Indian medal issued with bars. The ribbon is dark blue, edged with dark crimson. The medal, with W. WYON, R.A., on the truncation of the Queen's head and W.W. above the left-hand corner of the exergue, is generally considered a good one, and certainly it does stand out among the many issued during the reign of Queen Victoria as symbolic and suited for its purpose. The names of the recipients are impressed in Roman capitals on the edges. The medal is attached to the ribbon by means of a scroll bar fixed to the medal by a claw clip acting on a swivel, so that the medal may be turned about when fixed on the breast.

SECOND PUNJAB CAMPAIGN

This originated in a murderous assault made upon the British officials accompanying the Sirdar Khan Singh, who was deputed to take over the charge of the fortress of Mooltan (Multan) and province from the native Governor Dewan Moolraj, who had resigned. On arriving at the gate of the fortress the officers and their escort were attacked and killed by some of the Governor's men. A punitive force, including a body of 1,400 Sikhs, was consequently sent to punish the city, but the Sikhs went over to the rebels, and it was found necessary to send a strong force to reduce the mutineers who had been excited by Moolraj. Mooltan, a city three miles in circumference, is of considerable antiquity; it was taken by Alexander the Great, and at the beginning of the eleventh century by Mahmud of Ghazni, and by Tamerlane at the end of the fourteenth century; besieged by Ranjit Singh in 1810, he was bought off by the Afghan Governor, but it fell in 1818, and was annexed to the Punjab.

General Whish marched upon the city from Lahore with a force which included H.M.'s 10th and 32nd Regiments, and concentrated at Mooltan in the middle of August; but he had realised, after the desertion of 5,000 Sikhs under Sheer Ali, that it was fruitless to maintain the siege, and therefore raised it on September 15th. Lieutenant Edwardes, however, had by superhuman energy and indomitable pluck raised a force of Sikhs and Mahommedans, which was of the greatest possible assistance in holding Moolraj and his army of about 8,800 infantry and 1,200 cavalry, with 54 guns, in check, twice defeating him in battle at Kineyree and Suddoosam.

Edwardes's Medal.—For his splendid services the H.E.I. Co. presented him with a gold medal designed by W. Wyon, R.A., bearing on the obverse the bust of Her Majesty Queen Victoria, and on the reverse, resting upon a lotus flower, the figures of Valour and Victory, crowning the arms of Major (afterwards Sir Herbert) Edwardes; beneath, in allusion to the youthfulness of the hero, the infant Hercules strangling serpents. Within the border is the inscription, FROM THE EAST INDIA COMPANY TO LIEUTENANT AND BREVET-MAJOR H. B. EDWARDES, C.B., FOR HIS SERVICES IN THE PUNJAB A.D. MDCCCXLVIII.

Where "El Chico Blanco" fell.—While Whish and Edwardes had been holding the ground, Lord Gough assembled an army at Ferozepore, and marching into the Punjab met the rebels at Ramnuggur on the banks of the Chenab on November 28th, 1848, where after a determined battle the enemy drew off beaten. It was here Lieutenant-Colonel William Havelock, K.H., "El Chico Blanco," the fair-haired and gallant boy-leader of the Peninsular War, was killed while leading the 14th Dragoons in the cavalry charge. Colonel Havelock was brother to Sir Henry Havelock, and his six-bar Peninsular medal is in the great collection of my friend Dr. A. A. Payne, of Sheffield. Brigadier-General Cureton, C.B., the commander of the cavalry division, also fell. The brilliant charges made by the 13th and 14th Light Dragoons, and the 5th and 8th Light Cavalry, against the great bodies of hostile horsemen numbering nearly 4,000, called forth special mention in general orders. Various minor engagements were fought while Major-General Whish renewed the siege of Mooltan, which lasted twenty-five days, during which time a shell from one of the British mortar batteries struck the "Jumma Musjid," or Great Mosque, while another blew up a powder-magazine with 400,000 lb. of powder, when, states Colonel Maude, "as if by mutual consent the firing on both sides ceased a while, every one gazing upwards with silent awe and wonder! Then from the British camp arose one long burst of triumph, which was speedily answered by a furious cannonade from our courageous and still unsubdued enemy." Two breaches had been made in the walls

FIRST INDIA GENERAL SERVICE MEDAL.
Issued for Second Burmese War, 1854.

MEDAL FOR SOUTH AFRICA, 1853.
Awarded to Lieutenant-General A. C. Cavendish Bentinck.

(Obverse.)
INDIA GENERAL SERVICE MEDAL.
With bar for Persia, 1856–7.

by January 1st, 1849, one at the Delhi Gate and one near the "Khonee Boorj," or Bloody Bastion; this the Bombay European Fusiliers, the 4th Rifles, and the 19th Native Infantry were told off to storm, but the "Old Toughs," as the Fusiliers were called, had to make three desperate attempts before they and their comrades could gain the summit. Meanwhile the 32nd, with two Bengal regiments in support, had been ordered to storm the Delhi Gate; but their efforts were unsuccessful owing to the imperfect breaching, and they entered the city by the "Khonee Boorj." So Mooltan was won, but no rest was obtained until the troops reached the Soharee Gate, after fighting through the streets of the blood-stained city, in the siege of which the British force lost 1,200 in killed and wounded.

The British regiments which took part in the siege were the 10th, 32nd, and 1st Batt. 60th Rifles (only 100 of this regiment received the single bar for Mooltan, and fakers frequently remove the Goojerat bar from a medal in order to enhance its value); likewise a Naval Brigade (Indus Flotilla) of about 100 men; 1st (Royal) Bombay Fusiliers; 3rd, 4th, 9th, and 19th Bombay Native Infantry; 8th, 49th, 51st, 52nd, and 72nd Native Infantry; Bengal and Bombay Sappers and Miners; 1st and 2nd Scinde; Irregular Horse and 1st and 11th Bengal Light Cavalry; and 7th, 11th Irregular Cavalry with a detachment of the 14th. Likewise a detachment of Guides and Horse and Foot Artillery.

Chilianwala.—After the fall of Mooltan, Major-General Whish, with the 10th, 32nd, and the 60th Rifles, proceeded to join Gough's army, and was in time to take part in the final battle of Goojerat. The sanguinary battle of Chilianwala was fought on January 13th, 1849, and although described by Sir Henry Havelock as "the most sanguinary, and the nearest approximation to a defeat," it was not at first decided to issue a bar for the engagement. It was not an officers' battle; all the more reason, therefore, that the bar should be given to the soldiers "whose dauntless valour rectifies the errors of its commanders," for, as the *Calcutta Review* rightly affirmed, the soldiery "redeemed their errors with its blood," the unfortunate 24th, which advanced on the Sikh guns with unloaded muskets, losing no less than 14 officers and 241 men killed in the action, and 10 officers and 253 men wounded, besides losing one of their colours which was left on the field, but four others shared the same fate when after this confused battle Lord Gough drew his men off a mile in order to find water. The total loss of the British was 38 officers and 564 men killed, 94 officers and 1,557 men wounded, and 104 missing. The Sikhs, who "fought like devils," lost about 3,000 killed and 4,000 wounded.

The following British regiments took part: 24th, 29th (whose "undaunted bravery" was placed on record), 61st Foot; 3rd, 9th, and 14th Light Dragoons; Bengal Horse and Foot Artillery; the 2nd Bengal Fusiliers; 15th, 20th, 25th, 30th, 31st, 36th, 45th, 46th, 56th, and 69th Bengal Native Infantry; 1st, 5th, 6th, and 8th Bengal Light Cavalry; 3rd and 9th Irregular Cavalry and Bengal Pioneers.

Goojerat.—The last battle of the campaign was fought at Goojerat (Gujerat) on February 21st, 1849. Hitherto the Sikhs had proudly attacked; in this instance they acted upon the defensive under Sheer Singh, who commanded a force of 34,000 Sikhs, with 59 guns, and about 1,500 Afghans under the command of a son of the Ameer. In this battle the British General was superior in artillery, having 97 guns, but only 24,000 men. In advancing upon the Sikh position, Gough, as usual, intended sending his infantry forward before the artillery had done its work. Colonel Robertson alleges that, "Sir Hugh Gough's simple strategy was to put the strongest regiment into the hottest place, and no attempt was made to outflank or turn a position, which must account for the heavy list of casualties." However, on this occasion, Malleson states his staff induced the veteran to mount the top story of an isolated building which commanded a complete view of the battlefield, and then quietly removed the ladder, and only replaced it when the artillery had done its work. Then this indomitable army went for the well-served Sikh guns, and put the enemy to rout, leaving the cavalry to pursue them for twelve miles. The power of the Sikhs was broken, the Punjab became a possession of the British Crown, and the Maharajah Dhuleep Singh granted a pension. The Sikh losses were terrific, but only 29 officers and 671 men of the British force were placed *hors de combat.*

The following regiments were represented at Goojerat: 10th, 24th, 29th, 32nd, 53rd (which arrived in the evening), 60th, and 1st Batt. 61st Foot; 3rd Light Dragoons, 9th Lancers, and 14th Light Dragoons; 1st Bombay Europeans; 1st and 2nd Bengal Europeans; 3rd, 8th, 13th, 15th, 19th, 20th, 25th, 31st, 36th, 45th, 46th, 51st, 52nd, 56th, 69th, 70th, 72nd Bengal Native Infantry; 19th Bombay Infantry; 1st, 5th, 6th, and 8th Bengal Light Cavalry; 3rd, 9th, 11th, and 14th Irregular Cavalry, and Scinde Horse; 4 field batteries of Bombay Artillery, and 9 troops of Horse Artillery; 2 companies of Bengal Sappers and Miners, 6 companies of Bengal Pioneers, and 1st Company of Bombay Pioneers. A Brigade of Seamen from the Indian Navy which accompanied the Bombay column were awarded the medal.

The Punjab Medal.—The obverse bears the diademed head of Queen Victoria by Wyon, and the legend VICTORIA REGINA, and the reverse Sikh chiefs laying down their arms before a victorious British General seated on horseback in the front of his troops, who are drawn up in line with colours flying. In the background are palm trees, and above TO THE ARMY OF THE PUNJAB. In the exergue is the date MDCCCXLIX. The medal is $1\frac{2}{5}$ in. in diameter, and has the names of the recipients impressed on the edge in Roman capital letters. The clasps for suspension, and the bars, are similar to those on the Sutlej medal; the ribbon is

$1\frac{3}{4}$ in. wide, and of dark blue with bright-yellow stripes at the side.

FIRST NEW ZEALAND WAR

During this time trouble had arisen in the distant colony of New Zealand, the sovereignty of which had been ceded to Britain in 1840, a year after the first settlers had established themselves. The trouble arose mainly owing to the occupation of tribal land, although other causes were alleged. The first outbreak took place in July 1844, when a police magistrate and a number of settlers were killed, and a powerful chief, Hone Heke, tore down the British flag at Kororareka—in the north islands—took possession of the township and plundered it. H.M.S. "Hazard," with a detachment of the 95th, was sent to quell the disturbance and guard the place, but so persistent and successful was Heke in his attacks that it was deemed advisable, in March 1845, to abandon the settlement and embark the inhabitants for Auckland. On the arrival of reinforcements from Sydney, the Union Jack was again hoisted, and martial law proclaimed. A force consisting of the 58th Regiment, a detachment of the 96th, Seamen and Marines from the "Hazard" and "North Star," and a number of natives under Colonel Hulme of the 96th Regiment, then proceeded to Okaihau, where Heke had built a Pah or fortification. Although only 18 miles inland, the troops were four days in reaching their goal, and then the place was found impregnable, and as no artillery accompanied the expedition, it was forced to return, having lost 14 killed and 39 wounded. The expedition being strengthened by reinforcements, another attempt was made to try conclusions with Heke, who had entrenched himself at Oheawai. The 58th Regiment, with detachments of the 96th and 99th, a number of volunteers from Auckland, and 30 men from the "Hazard," in all 630 men with 250 natives and 4 guns, under the command of Colonel Despard, appeared before Oheawai on June 23rd; the guns, however, proved useless for the work, but breaches were made with a 32-pounder brought up by the commander of the "Hazard." On July 1st an assault was made by 200 soldiers and seamen, but half the stormers were killed or wounded, and they were compelled to retire. They consequently sat down in front of the stockades, and when preparing to make another assault, on July 10th, they discovered that the enemy had evacuated their Pah. In November 1845 more active measures were taken by the newly appointed Governor, Captain (afterwards Sir George) Grey. A force of about 1,170 soldiers, volunteers, and seamen, with 450 natives, 4 guns, and rocket tubes, were sent against the chiefs Heke and Kawiti. While a friendly Maori chief, Macquarrie, kept Heke engaged at Ikorangi, the main force attacked the Pah at Ruapekapeka held by Kawiti, which after withstanding a siege from December 31st until January 10th, 1846, was practically abandoned by the Maoris, and then taken with a loss of 13 men killed and 30 wounded. The chiefs were pardoned, and the war in the North Island ended. Meanwhile trouble had been brewing in the South Island, which was brought to a head by an attack in May 1847 on a party of the 58th Regiment. A reprisal was made by the British attacking the stockade of Rangihaeata, and dispersing the natives, 600 of whom, however, on May 19th made an attack on the Wanganui settlement, and maintained it for several hours until night fell, when they retired. The place was held by 170 of the 58th, assisted by a gun-boat. Early in June an attack was made on a detachment of the 65th; but following a sortie from the block-houses, held by the 58th, the Maori chiefs sued for peace.

The regiments engaged were the 58th, 65th, 96th, 99th; Royal Artillery; Royal Engineers; H.E.I. Co.'s Artillery; Royal Marines, and the crews of nine of Her Majesty's ships, the names of which are recorded in the naval section. The medal for the war was not authorised until March 1869.

Medal for Peninsular War.—It would appear strange that although a medal was so readily awarded for Waterloo, no recognition was given to the veterans of the Peninsular, and that they should have to wait until 1847 before a medal was issued as a record of, and reward for, their services against the veteran troops of France. That they ultimately received a decoration was due to the efforts of the veteran Duke of Richmond, who, on July 21st, 1854, presented a petition in the House of Lords, "requesting their Lordships' recommendation of them to their Sovereign, that they may receive a decoration for the services they had performed." The gallant Duke urged the claims of the junior officers, non-commissioned officers, and men in simple but powerful terms, while the Duke of Wellington, who had previously refused to present a petition to King George, opposed the grant to the survivors of over sixty battalions of the army who had, as he admitted, served under him in six campaigns.

He appeared, according to the record of the debate, to have thought only of the senior officers. "Have no rewards been bestowed upon these officers?" he asked. "New modes were discovered and adopted of distinguishing and rewarding the officers." Medals had been struck and distributed to 1,300 officers, special brevets were issued, and 7 officers raised to the peerage, as the Duke of Wellington pointed out; but he had no word of remembrance, no word of support for the thousands of officers and men whose only rewards had been hardship and scars. No word of commendation for the men who kept "grim Busaco's bloody ridge," and ultimately hurled the veterans of Marengo and Austerlitz into the ravine below; or for those who added laurels to Britain's fame at Fuentes d'Onor (Fountain of Honour); the Fusiliers and their brave companions in arms who fought so splendidly and so fiercely on the heights of Albuera, where the 57th—the first battalion of the Middlesex Regiment—gained their coveted *nom de guerre*, "Die Hards," for, as Napier states, "nothing could stop that astonishing infantry," which turned the day in the bloodiest battle of the war. It was indeed a soldiers' battle, but the senior officers, as usual, received the gold medal. What a case the Duke of Richmond had if only by stating the result of this one battle, after which "1,800 unwounded men, the remnant of 6,000 unconquerable British soldiers, stood triumphant on the fatal hill." The Iron Duke must have forgotten the gallant storming of the breaches at Ciudad Rodrigo, the success of which gained for him his viscountcy; the storming of Badajoz, which caused him to weep at the cost of the victory, and Orthes, where Picton's men climbed the rough, rocky bank, and Colborne's 52nd undauntedly pressed forward knee-deep in the marshy swamp and up the hill to the victory which practically determined the Peninsular campaign.

In his efforts, however, to oppose and belittle the claims of the petitioners for the Peninsular medal he incidentally and unconsciously brought forward the claims of those who had fought in Egypt and on the high seas, and at the battle of the Nile and Trafalgar, from the "Glorious First of June" to the blockade of France and the Bay of Biscay during the Peninsular War. As a result, we may assume, of his arguments against the issue of what is known as the "Military General Service Medal," when the claims came to be considered it was felt that justice must also be done to the sister service, and the "Naval General Service Medal" was likewise instituted on June 1st, 1847, when by a General Order it was commanded that medals be struck and conferred upon every officer and soldier who was present at any battle or siege for which gold medals had been awarded. The medal was given to those who had taken part in the Peninsular Campaign, Egypt, Italy, North America, and in the East and West Indies; and although thirty-four years had elapsed since the last battle, Toulouse, in 1814, over 30,000 applicants made good their claims to the medal. Only six applied for fifteen bars, and of these two made good their claims, Private Talbot of the 45th, and Private Loochstädt, formerly of the King's German Legion. This medal is in Lord Cheylesmore's collection.

The Military General Service Medal.—The medal, $1\frac{2}{5}$ in. in diameter, bears on the obverse the diademed head of Queen Victoria, with VICTORIA REGINA on either side of the head, and the date of issue, 1848, underneath. On the reverse is depicted Queen Victoria, robed and crowned, standing on a daïs in the act of placing a laurel wreath on the head of the Duke of Wellington, who kneels on his left knee and holds his field-marshal's baton in his right hand. By the side of the daïs is the British Lion dormant; above is the inscription TO THE BRITISH ARMY, and in the exergue 1793–1814. In the left-hand corner of the exergue is the letter "W," the initial of W. Wyon, R.A., the medallist. The medal depends from a straight clasp, which is attached to the medal by a swivelled four-claw clip. The ribbon is $1\frac{1}{4}$ in. wide, of dark crimson with blue edges. The recipient's name, rank, and regiment are impressed in Roman capitals. It is noteworthy that the bars, which are $1\frac{3}{10}$ in. by $\frac{1}{10}$ in., are arranged in sets of three, where the number warrants, the earliest date being nearest the medal. Those awarded to the infantry are arranged $\frac{3}{40}$ in. apart until the number exceeds six, when they are arranged, like those issued to the cavalry, closely together.

The following regiments received the medal with bars—the bar for EGYPT being granted by General Order dated February 11th, 1850; and as an indication of the depletion of the ranks of those who fought in 1801, I might mention that only three officers—Hill, Beaven, and Deane—of the 18th Royal Irish Regiment were alive to receive the Military General Service Medal with the bar for Egypt.

EGYPT, 1801.—The 2nd and 3rd Foot Guards; 1st, 2nd, 8th, 10th, 13th, 18th, 20th, 23rd, 24th, 25th, 26th, 27th, 28th, 30th, 40th, 42nd, 44th, 50th, 54th, 58th,

English Crimea, Turkish Crimea.

MEDALS FOR CRIMEAN WAR.

61st, 79th, 80th, 86th, 88th, 89th, 90th, 92nd, and 96th Foot; 11th, 12th, and 26th Light Dragoons; Hompesch's Hussars; De Rolls and Dillon's Regiments; Corsican Rangers; and the 2nd and 13th Bombay Infantry.

MAIDA, July 4th, 1806.—20th, 27th, 35th, 58th, 61st, 78th, and 81st Foot; 20th Light Dragoons.

ROLEIA, August 17th, 1808.—5th, 6th, 9th, 29th, 32nd, 36th, 38th, 40th, 45th, 50th, 60th, 71st, 82nd, 91st, 95th, and 97th Foot; 20th Light Dragoons.

VIMIERA, August 21st, 1808.—2nd, 5th, 6th, 9th, 20th, 29th, 32nd, 36th, 38th, 40th, 43rd, 45th, 50th, 52nd, 60th, 71st, 82nd, 91st, 95th, and 97th Foot; 20th Light Dragoons.

SAHAGUN, December 20th, 1808.—7th, 10th, 15th, and 18th Hussars; 3rd Hussars, K.G.L., and two batteries of Horse Artillery.

CORUNNA, January 16th, 1809.—1st, 2nd, 4th, 5th, 6th, 9th, 14th, 20th, 23rd, 26th, 28th, 32nd, 36th, 38th, 42nd, 43rd, 50th, 51st, 52nd, 59th, 60th, 71st, 76th, 79th, 81st, 82nd, 91st, 92nd, 95th; 1st and 3rd Batts. 1st Foot Guards; 7th, 10th, 15th, and 18th Light Dragoons; 3rd Dragoons, K.G.L.; three batteries Royal Horse Artillery; five batteries Royal Artillery; 1st and 2nd Light Battalions, K.G.L.

MARTINIQUE, February 24th, 1809.—7th, 8th, 13th, 15th, 23rd, 25th, 46th, 60th, 63rd, and 90th Foot; 1st West India Regiment, and the Royal York Rangers.

TALAVERA, July 27th–28th, 1809.—2nd and 3rd Foot Guards; 2nd, 3rd, 4th, 5th, 6th, 7th, 20th, 24th, 28th, 29th, 31st, 32nd, 36th, 38th, 40th, 42nd, 43rd, 45th, 48th, 50th, 52nd, 53rd, 60th, 61st, 66th, 71st, 79th, 82nd, 83rd, 87th, 88th, 91st, 92nd, 95th, and 97th Foot; 3rd Dragoon Guards; 4th, 14th, 16th, and 23rd Light Dragoons; 1st Light Dragoons, K.G.L.; 1st and 2nd Light Battalions, K.G.L.; 1st, 2nd, 4th, 5th, and 7th Line Battalions, K.G.L.; 1st, 2nd, 4th, 5th, and 7th Line Battalions, K.G.L.; 1st Hussars; 2nd and 4th Batteries Foot Artillery, K.G.L.

GUADALOUPE, January to February 6th, 1810.—1st,

13th, 15th, 25th, 46th, 60th, 63rd, 70th, 90th, and 96th Foot; Marine Artillery and Naval Brigade; the Light Companies of the 1st, 2nd, 3rd, 4th, and 6th West India Regiments; the 8th West India Regiment; York Light Infantry Volunteers; Royal York Rangers; Royal Artillery and Military Artificers.

BUSACO, September 27th, 1810.—1st, 3rd, 5th, 7th, 9th, 11th, 24th, 27th, 29th, 31st, 34th, 38th, 39th, 40th, 42nd, 43rd, 45th, 48th, 50th, 52nd, 53rd, 57th, 60th, 61st, 66th, 74th, 79th, 83rd, 88th, 95th, and 97th Foot; three batteries of Royal Artillery; and the 1st and 2nd Light Batts. and the 1st, 2nd, 5th, and 7th Line Batts., K.G.L.; 4th, 14th, and 16th Light Dragoons; 2nd and 3rd Foot Guards.

BARROSA, March 5th, 1811.—1st, 2nd, and 3rd Foot Guards; 9th, 28th, 47th, 67th, 82nd, 87th, and 95th Foot.

FUENTES D'ONOR, May 5th, 1811.—1st, 4th, 5th, 9th, 24th, 30th, 38th, 42nd, 43rd, 44th, 45th, 50th, 51st, 52nd, 60th, 71st, 74th, 79th, 83rd, 85th, 88th, 92nd, 94th, and 95th Foot; 1st Dragoons; 14th and 16th Light Dragoons; 1st Hussars, K.G.L.; 1st and 2nd Light Batts., K.G.L.; 1st, 2nd, 5th, and 7th Line Battalions, K.G.L.; 2nd and 3rd Foot Guards.

ALBUERA,[1] May 11th, 1811.—3rd, 7th, 23rd, 27th, 28th, 29th, 31st, 34th, 39th, 40th, 48th, 57th, 60th, 66th, and 97th Foot; 3rd Dragoon Guards; 4th and 13th Light Dragoons.

JAVA, August to September 18th, 1811.—14th, 59th, 69th, 78th, and 89th Foot; 22nd Light Dragoons; Indian Native Cavalry, Infantry, and Bengal Volunteers.

CIUDAD RODRIGO, January 19th, 1812.—2nd and 3rd Foot Guards; 5th, 7th, 23rd, 24th, 30th, 40th, 42nd, 43rd, 45th, 48th, 52nd, 60th, 74th, 77th, 83rd, 88th, 94th, and 95th Foot.

BADAJOZ, March 17th to April 6th, 1812.—2nd Foot Guards; 1st, 4th, 5th, 7th, 23rd, 27th, 30th, 38th,

[1] Spelt ALBUHERA on bars.

40th, 43rd, 44th, 45th, 48th, 52nd, 60th, 74th, 77th, 83rd, 85th, 88th, 94th, and 95th Foot; 13th and 14th Light Dragoons.

Salamanca, July 22nd, 1812.—2nd and 3rd Foot Guards; 1st, 2nd, 4th, 5th, 7th, 9th, 11th, 23rd, 24th, 27th, 30th, 32nd, 36th, 38th, 40th, 42nd, 43rd, 44th, 45th, 48th, 51st, 52nd, 53rd, 58th, 60th, 61st, 68th, 74th, 79th, 83rd, 88th, 94th, and 95th Foot; 5th Dragoon Guards; 3rd, 4th, 11th, 12th, 14th, 15th, and 16th Light Dragoons; and the following regiments of the K.G.L.: 1st Hussars; 1st and 2nd Light Battalions, 1st, 2nd, and 5th Line Battalions, and Artillery.

Vittoria, June 21st, 1813.—2nd and 3rd Foot Guards; 1st, 2nd, 4th, 5th, 6th, 7th, 9th, 20th, 23rd, 24th, 27th, 28th, 31st, 34th, 38th, 39th, 40th, 43rd, 45th, 47th, 48th, 50th, 51st, 52nd, 53rd, 57th, 58th, 59th, 60th, 66th, 68th, 71st, 74th, 82nd, 83rd, 87th, 88th, 92nd, 94th, and 95th Foot; two squadrons 1st and 2nd Life Guards and Horse Guards; 1st Dragoons; 3rd and 5th Dragoon Guards; 3rd, 4th, 10th, 11th, 12th, 13th, 14th, 16th, and 18th Light Dragoons; 15th Hussars.

Pyrenees, July 28th to August 2nd, 1813.—2nd, 3rd, 5th, 6th, 7th, 11th, 20th, 23rd, 24th, 27th, 28th, 31st, 32nd, 34th, 36th, 39th, 40th, 42nd, 43rd, 45th, 48th, 50th, 51st, 52nd, 53rd, 57th, 58th, 60th, 61st, 66th, 68th, 71st, 74th, 79th, 82nd, 83rd, 87th, 88th, 91st, 92nd, 94th, and 95th Foot; 13th and 14th Light Dragoons.

Fort Detroit, August 16th, 1812.—Three hundred men of the 41st Foot; Newfoundland Regiment; Canadian Militia, and 600 Indians; Royal Artillery (30 men).

San Sebastian, August and September 9th, 1813.—200 of the Foot Guards; 1st, 2nd, 4th, 7th, 9th, 11th, 20th, 23rd, 24th, 27th, 36th, 38th, 40th, 43rd, 47th, 48th, 51st, 52nd, 53rd, 59th, 68th, 82nd, 85th, 87th, 88th, and 95th Foot, and a Naval Brigade.

Chateauguay, October 26th, 1812.—Royal Artillery;

Indians; Canadian Fencibles; Canadian Militia; Voltigeurs; and Chasseurs.

NIVELLE, November 10th, 1813.—1st, 2nd, and 3rd Foot Guards; 1st, 2nd, 3rd, 4th, 5th, 6th, 7th, 9th, 11th, 20th, 23rd, 24th, 27th, 28th, 31st, 32nd, 34th 36th, 38th, 39th, 40th, 42nd, 43rd, 45th, 47th, 48th, 50th, 51st, 52nd, 53rd, 57th, 58th, 59th, 60th, 61st, 62nd, 66th, 68th, 71st, 74th, 76th, 79th, 82nd, 83rd, 84th, 85th, 87th, 88th, 91st, 92nd, 94th and 95th Foot; and the following regiments of the K.G.L.: 1st and 2nd Light Battalions; 1st, 2nd, and 5th Line Battalions; 12th, 13th, 14th, and 18th Light Dragoons.

CHRYSTLER'S FARM, November 11th, 1813.—49th and 89th Foot; Canadian Fencibles; Canadian Militia; Indians; Voltigeurs; and Royal Artillery.

NIVE, December 9th to 13th, 1813.—1st, 2nd, and 3rd Foot Guards; 1st, 2nd, 3rd, 4th, 5th, 7th, 9th, 11th, 20th, 23rd, 27th, 28th, 31st, 32nd, 34th, 36th, 38th, 39th, 40th, 42nd, 43rd, 45th, 47th, 48th, 50th, 52nd, 53rd, 57th, 59th, 60th, 61st, 62nd, 66th, 71st, 74th, 76th, 79th, 83rd, 84th, 85th, 87th, 88th, 91st, 92nd, 94th, and 95th Foot; 7th, 12th, 13th, 14th, 16th, and 18th Light Dragoons.

ORTHES, February 17th, 1814.—2nd, 5th, 6th, 7th, 11th, 20th, 23rd, 24th, 27th, 28th, 31st, 32nd, 34th, 36th, 37th, 39th, 40th, 42nd, 45th, 48th, 50th, 51st, 52nd, 57th, 58th, 60th, 61st, 66th, 68th, 71st, 74th, 82nd, 83rd, 87th, 88th, 91st, 92nd, 94th, and 95th Foot; 3rd, 7th, 10th, 13th, 14th, 15th, and 18th Light Dragoons.

TOULOUSE, April 10th, 1814.—1st, 2nd, 3rd, 5th, 7th, 11th, 20th, 23rd, 27th, 28th, 31st, 32nd, 34th, 36th, 39th, 40th, 42nd, 43rd, 45th, 48th, 50th, 52nd, 53rd, 57th, 60th, 61st, 66th, 71st, 74th, 79th, 83rd, 87th, 88th, 91st, 92nd, 94th, and 95th Foot; 2 squadrons 1st and 2nd Life Guards and Horse Guards; 1st Dragoons; 3rd and 5th Dragoon Guards; 3rd, 4th, 7th, 10th, 13th, 14th, 15th, and 18th Light Dragoons.

Twenty-eight bars were authorised. Twenty-one for

FIRST FRENCH MÉDAILLE MILITAIRE.

SARDINIAN MEDAL.

MEDAL FOR SECOND CHINESE WAR.

the Peninsular, three for North America, two for services in the West Indies, one for Italy, and one for Egypt, 1801, "to those who were still alive." The bars, it should be noted, vary in their arrangement, some being farther apart than others, those awarded to the cavalry being very close together, and those to the infantry $\frac{3}{10}$ in. apart. If more than six are attached they are placed quite close like those for the cavalry. The bars are attached in sets of three, so that no rivets were used in the fixing of two or three bars to a medal, the set being struck in one piece; six, nine, or more bars were riveted together with very neat, small-headed rivets.

The India Medal, 1799–1826.—The European veterans of the Indian wars had been as badly treated in the way of war decorations as the heroes of the Peninsular, but an effort to reward the survivors was made public by the announcement on April 14th, 1851—forty-eight years after the first battle, for which a bar was issued, was fought—that the Queen had assented to the measure proposed by the Court of Directors of the East India Company to grant a medal at their expense to the surviving officers and soldiers of the Crown who were engaged in the services in India from 1799 to 1826. On the obverse is the diademed head of Queen Victoria with VICTORIA REGINA similar to that on the obverse of the medals already described, and on the reverse a seated figure of Victory, holding in her left hand, which falls by her side, a chaplet of laurel, and in her right hand an olive branch. In the background is a palm tree, and in front a trophy of oriental arms. Above all is inscribed TO THE ARMY OF INDIA, and in the exergue 1790–1826 and W.W., the initials of the medallist, in small capitals. The medal is $1\frac{2}{5}$ in. in diameter, attached to a scroll suspendor as used with the Punjab and Sutlej medals; the bars are also of the same design. The ribbon of pale-blue corded silk is $1\frac{1}{4}$ in. wide. The names, etc., of the European recipients were impressed in capital Roman letters, but most of those awarded to native officers and soldiers were officially engraved in a light, slanting script.

Bars Awarded.—Twenty-one bars were issued with the medal: ALLIGHUR, DELHI, ASSYE, ASSEERGHUR, LASWARREE, ARGAUM, GAWILGHUR, DEFENCE OF DELHI, DEIG, CAPTURE OF DEIG, NEPAUL, KIRKEE, POONA, KIRKEE AND POONA, SEETABULDEE, NAGPORE, SEETABULDEE AND NAGPORE, MAHEIDPOOR, CORYGAUM, AVA, BHURTPOOR. Only four medals were issued with the single bar for Seetabuldee and Nagpore, and one of these with edge impressed and verification of award to a man of the 39th Native Infantry realised £74 at auction. Nineteen only were issued for Assye, thirteen for Gawilghur, twenty-six for Maheidpoor, forty-eight for Laswarree, seventy-nine for Corygaum, but only a few of any of these to Europeans. The Duke of Wellington's medal had three clasps; for ASSYE, ARGAUM, and GAWILGHUR. Only thirteen with this combination were issued. A single bar for Capture of Deig awarded to a Lieutenant-Colonel has realised £50. An officer's four bar has realised £100, and a five bar £150. (See sales prices.)

FIRST KAFFIR WARS

In 1834–5 the Kaffirs in South Africa became very troublesome and necessitated the employment of armed force to subdue them, but the first serious Kaffir war broke out in 1846 owing to the outrages perpetrated by the Gaikas, and considerable fighting had to be done before the Kaffirs, by this time armed with firearms, were subdued, and Sandilli and his brother surrendered. In this campaign the following regiments were represented: 6th, 27th, 45th, 73rd, 90th, 91st, and 1st Battalion Rifle Brigade; 7th Dragoon Guards; Royal Artillery; Engineers, Sappers and Miners, also Cape Mounted Riflemen.

At the end of the year 1850 the Kaffirs were again in a turbulent condition, and Sir Harry Smith, a Peninsular veteran whose record in India I have indicated, summoned the chiefs to meet him, but Sandilli ignored the invitation and was outlawed. Then began another protracted war with the natives, during which several disasters befell the British troops. In the Keiskamma defile misfortune overtook Colonel Mackinnon, where with a force of 600 men, which included detachments of the 6th and 73rd Regiments, he was ambuscaded. He, however, reached Fort Cox, where the Governor, Sir Harry Smith, was surrounded by the dusky hordes. He managed to get away with a flying escort, and safely reached King William's Town. In June 1851 operations were conducted in the Amatola and Wolf Valley with comparatively severe losses to the British troops. In December the passage of the Kei was effected, despite the enemy's ingenious attempt—a new method in warfare to them—at constructing breastworks. Ultimately the chiefs of the Gaikas and Seyolo requested peace upon terms which could not be conceded, and the war was continued. In the spring of 1852 a determined advance was made by the British against Sandilli's stronghold in the Amatola mountains, from whence the 74th Highland Regiment, the "tortoises" as the Kaffirs called them in allusion to the markings of their kilt, after much hard fighting cleared them out. Sir Harry Smith was relieved of his command at the Cape by Sir George Cathcart, and ultimately the Water Kloof was cleared, and the Basutos under Moshesh defeated at Berea; Sandilli again surrendered, and the recalcitrant natives were expelled from the territory they had previously occupied.

The regiments represented in the campaign were the 2nd, 6th, 12th, 43rd, 45th; 2nd Batt. 60th Rifles; 73rd, 74th, 91st; 1st Batt. Rifle Brigade; 12th Lancers; Royal Artillery; Engineers, Sappers and Miners; Cape Mounted Rifles, Seamen and Marines.

Loss of the "Birkenhead."—The South African Wars of 1834–53 recall the loss of the troopship "Birkenhead" on February 26th, 1852, when 9 officers and 349 men, drawn up in parade order on the deck, went to a watery grave after placing the women and children, and sick, in the boats. They were on their way out as drafts for the regiments engaged in the third Kaffir war, which began at the end of 1850, and was not terminated until March 1853. The German Emperor was so impressed with the heroism of these British soldiers, who so calmly went to their death, that he caused an account of their gallant conduct to be read to every regiment of his army.

Those who took part in these wars were awarded a medal in a General Order issued in November 1854, and although many of the recipients had fought their battles twenty years previously, the medal, $1\frac{2}{5}$ in. in diameter, bore the date 1853 in the exergue, above which is an admirably modelled lion, crouching under a mimosa bush; above all the record SOUTH AFRICA. No bars were given with the medal, which was only awarded to survivors, and the only way to distinguish the campaign for which it was granted is by the name of the regiment indented on the edge of the medal; but this means of identification is not possible in medals issued to the Naval Brigade, for the rank only is frequently given, as for instance, J. SHORE, STOKER. The diademed head of Queen Victoria, as issued with the medal previously described, was used on the obverse of this medal, which was suspended by a scroll clasp. The ribbon is orange and watered, with two thick and thin stripes of dark blue, leaving a narrow orange margin at the edge; the names, etc., were indented on the edge in square Roman capitals. The illustration of the medal is from a photograph of the one awarded to Captain (afterwards Lieutenant-General) A. C. Bentinck of the 7th Dragoon Guards—father of the present Duke of Portland.

The regiments engaged in the three Kaffir wars were: 1834–5: 27th, 72nd, 75th. 1846–7: 6th, 27th, 45th, 73rd, 90th, 91st; Rifle Brigade and 7th Dragoon Guards. 1850–3: 2nd, 6th, 12th, 43rd; 2nd Batt. 60th; 73rd,

VICTORIA CROSS.

INDIAN ORDER OF MERIT.

MERITORIOUS SERVICE MEDAL.

74th, 91st; 1st Batt. Rifle Brigade; Royal Marines; a Naval Brigade and Cape Mounted Rifles.

SECOND BURMESE WAR

As a result of the violation of the treaty of Yandaboo, which was really never kept, and the refusal of the King of Ava to redress the grievances of the Europeans in Rangoon, who had been compelled to seek safety on board the "Proserpine," a force of 5,767 men, composed of the 18th Royal Irish, 51st and 80th, together with Artillery, Sappers and Miners, Gun Lascars, and three Infantry Regiments of the H.E.I. Co., under the command of General Godwin, occupied, after very little resistance, Martaban and Rangoon. In the reduction of the latter place the naval contingent and the ships of the fleet participated. The key to the position, the Golden Pagoda, Shwe Dagon, was carried by the Royal Irish and the 80th after a very trying time caused by the excessive heat, five of the senior officers being struck down by solar apoplexy—Major Oakes, who commanded the artillery, and another officer fatally. In the capture of Rangoon 2 officers were killed, and 14 wounded; 15 non-commissioned officers and men killed, and 114 wounded, of whom one-third belonged to the Royal Irish, which lost its adjutant at the foot of the Shwe Dagon.

Pegu.—Expeditions were sent out to the north and west, Bassein being captured by one force and Pegu taken on November 21st, 1852, by the other. The province of Pegu was then annexed; but the Burmese continued to harass the garrison which held the city, and it was not until after the arrival of reinforcements from Rangoon that steps could be taken to successfully deal with the enemy. The war was concluded in June 1853, but not before the army had been considerably reduced by disease, cholera claiming a large number of victims, and 22 men had been killed, and 14 officers and 94 men wounded in the taking of the village of

Donobyu, held by Myat Toon, a robber chief who had about 8,000 guerillas at his call. It was in the attack on this village that Viscount, then Ensign, Wolseley of the 80th, in leading a small body of men from the 18th, 51st, and 80th, was struck down at the moment of victory. On June 30th, 1853, the war was declared at an end without any formal treaty, and a few months later the troops left a country which had claimed a terrible toll of brave men's lives in the swamps and jungles, the 18th Royal Irish having alone lost 365 officers and men mostly by disease.

First India General Service Medal.—On January 23rd, 1854, Queen Victoria sanctioned the issue of a silver medal with bar for PEGU to those who had taken part in the Burmese War of 1852–3. This, later called the India General Service Medal, 1854, has been given for all the various campaigns up to the fighting in the Kachin Hills in 1892–3. Twenty-three bars were given, representing the different "little wars" in which the recipients were engaged during forty-one years. This medal, illustrated facing page 120, has the same kind of suspender and bars as the Punjab medal, likewise the same head of Victoria, but on the reverse a figure of Victory crowning with laurel a seated warrior in classic pose, holding in his right hand a Roman sword, and in his left a sheath. In the exergue is a lotus flower and leaves, and L. C. WYON beneath it. The medals, $1\frac{2}{5}$ in. in diameter, have the rank, name, and the regiment or ship, impressed in Roman capitals for Pegu, Persia, North-West Frontier, Umbeyla, Bhootan, but for Perak and the rest of the series, except Jowaki, the lightly engraved running hand or Roman capitals were used for naming. The ribbon, $1\frac{3}{4}$ in. wide, is dark crimson with two dark-blue stripes.

The regiments entitled to the medal with bar for Pegu were the 18th, 51st, and 80th; Artillery; Sappers and Miners; 1st Bengal Fusiliers; 1st Madras Fusiliers; 5th Madras Native Infantry; and the Naval Brigade from thirteen of Her Majesty's ships.

The V.C. and Distinguished Conduct Medal.—These were comparatively small but significant campaigns, although the period between Waterloo and the breaking out of the Crimean War has been termed the long peace. The war in the Crimea was of a very different kind. Once again it was the meeting of Europeans face to face, and a long struggle resulted. An important outcome of the war from one point of view was the institution on January 29th, 1856, of that coveted decoration, the Victoria Cross, preceded by the institution of the Distinguished Conduct Medal on December 4th, 1854, and the Conspicuous Gallantry Medal for naval men on August 13th, 1855. The ribbon for the V.C. is dark red—dark blue for the Navy—that for the Distinguished Conduct Medal (illustrated facing page 140) dark blue with broad, red stripe down the centre, and the ribbon blue with a white stripe down the centre for the Conspicuous Gallantry medal. In the first issues the names were impressed on the edge, but in the later issues they are engraved, and the date of the action is also included. It is particularly noteworthy that the first medals of this class issued were struck from the Meritorious Service Medal, the two words being erased and Conspicuous Gallantry engraved in place thereof (see facing page 140).

THE CRIMEAN WAR

In October 1853 the Czar of Russia declared war against the Sultan of Turkey, and to defend "the sick man" Great Britain and France landed troops at Varna, a Bulgarian port in the Black Sea. At the end of February 1854 Queen Victoria bade farewell to the Guards at Buckingham Palace, and these with other regiments sailed for the East to take part in a war which, declared on March 28th, 1854, was to cost us the lives of thousands of brave men: 21,815 were killed, or died of wounds or disease, and 11,876 were wounded—a total of 32,691 officers and men. Disease accounted for the deaths of 16,041, battle only 4,774! The old soldiers, as Sir Evelyn Wood states, died without a murmur, and it "is impossible to overpraise the disciplined silence of men under privations which in a few weeks reduced one battalion from nearly 1,000 effectives to a strength of 30 rank and file." At the end of February 1855, although the strength of the British army was on paper 44,000, only 18,000 were "present and fit for duty"; and yet the medals of the brave and patient survivors of this war are selling at a few shillings apiece unless they happen to have the bar for Balaklava or Inkermann—Sebastopol, that long-drawn agony, counts for nothing!

On September 7th, 1854, the Anglo-French army put to sea, its objective being the great naval station of Sebastopol in the Crimea. There were, in addition to the crews of the warships, about 27,000 British, including the Light Brigade of 1,000, with 60 guns, under Lord Raglan, who had served on Wellington's staff in the Peninsular; 28,000 French under Marshal St. Arnaud, and 7,000 Turks under Omar Pasha. The Sardinian contingent under General La Marmora did not reach the Crimea until May 1855. The troops were landed, between September 14th and 18th, on the beach at Kalamita Bay, about twenty-five miles from Sebastopol, and two days after the disembarkation had been effected they had to force the Alma, when they defeated the Russians under Prince Mentschikoff so effectively that the allies were allowed to march southwards unmolested, and to form bases within a few miles of Sebastopol. The British established themselves at the little port of Balaklava, eight miles to the south of Sebastopol, and the French at Kamiesch Bay, six miles north of the place.

The Alma.—The Russians had taken up a position commanding the road to Sebastopol, with a front of two miles, mainly on the range of hills varying in height from 120 ft. to 200 ft. on the left bank of the River Alma, and on the morning of September 20th the allies pre-

(Reverse.)
DISTINGUISHED CONDUCT MEDAL.

(First Type.)
CONSPICUOUS GALLANTRY MEDAL.

(Obverse.)
DISTINGUISHED CONDUCT MEDAL.

pared to attack them. The French and Turks began the battle, Marshal St. Arnaud attacking the left of the Russian position, and the centre with the aid of one British division, and Lord Raglan the right. The French General Bosquet made a brilliant advance, and drove the enemy from some high ground, but the French centre did not support him, and he was isolated; Raglan therefore decided to order the second and light divisions to advance, and this they did under a murderous cannonade. "Opposed to the English were at least two-thirds of the Russians," but the young and untried soldiers of Britain once again demonstrated the splendid fighting material of which they were made. Through the entangling vineyards and into the rapid waters of the Alma they steadily marched; emerging, they were ordered "to fix bayonets, get up the bank, and advance to the attack." They quickly climbed the steep and rugged hill-sides, but received a check, and then some one calling "Cease fire and retire," they became somewhat confused; they soon rallied, and the light division stormed into the great redoubt with its 14 guns of heavy calibre, carried the Kourgane hill, defended by 17,000 men with two redoubts and 14 guns, and then, with their depleted and broken ranks swelled by the Guards and the Highlanders, the Black Watch to the fore, who had pressed into the fight and up the Kourgane hill, charged the heights and at the point of the bayonet drove the Russians off in confusion. Thus the battle of the Alma was won, not by the brilliant tactics of the officers, but by the courage and tenacity of the men, and so by 6 o'clock in the evening Marshal St. Arnaud's tent was pitched on the spot where Mentschikoff had his head-quarters in the morning. After the battle, a staff officer relates, Lord Raglan's eyes filled with tears when he shook hands with Sir Colin Campbell, and "he could not speak when the brave old veteran said to him, pointing to the dead Highlanders, 'Sir, it was they who did it,' and then the cheer went up when he asked to wear a bonnet! With its horrors, war has its romance." The same officer, writing a little later, said, "Glory when looked at close, and while it is being earned, is rather an ugly thing." The heroes of the Crimea certainly learned that.

The late Major-General (then Colour-Sergeant) Luke O'Connor, gained his Victoria Cross in this battle. Ensign Anstruther of the Royal Welsh had with boyish impetuosity gallantly rushed forward and, outpacing all, proudly planted the Queen's colour on the parapet of the Redoubt; riddled with Russian bullets he fell, but O'Connor, taking the colour from his dying grasp again planted them on the Redoubt, and, although wounded in the breast, carried them through the rest of the fight. Captain (afterwards Major-General) Bell, C.B., also won his V.C. at the heights of Alma for his daring conduct on that day, when, all the senior officers of the regiment being placed *hors de combat*, he found himself in command, and successfully brought the 23rd out of action. The gallant Welsh Fusiliers, who were the first to land in the Crimea, lost their colonel and 8 officers, and the Light Division, of which they formed part, lost 47 officers and 850 men. The Division was composed of the 7th, 19th, 23rd, 33rd, 77th, and 88th. The total British losses were 2,000, and those of the French 3 officers, including Generals Canrobert and Thomas wounded and 560 men. The Russians had 5,000 placed *hors de combat*, including 45 officers, among whom were Generals Karganoff and Shokanoff.

The regiments present at the Alma were the Grenadier, Coldstream, and Scots Guards; 1st, 4th, 7th, 19th, 20th, 21st, 23rd, 28th, 30th, 33rd, 38th, 41st, 42nd, 44th; two companies of the 46th; 47th, 49th, 50th, 55th, 63rd, 68th, 77th, 79th, 88th, 93rd, 95th Foot; 4th and 13th Light Dragoons; 8th and 11th Hussars and 17th Lancers. The cavalry were never called into action. Lord Raglan had said he preferred to keep his "cavalry in a band-box."

Siege of Sebastopol.—On September 23rd, 1854, the army began its slow and "most remarkable flank march in military history." Balaklava was ultimately reached, "the result of blind accident rather than knowledge of how war should be conducted," and the armies of Britain and France settled down on the Chersonese upland, and the siege of Sebastopol commenced. From September 28th, 1854, it ran its long course until September 11th, 1855, when the Russians, who had fought so well under Todleben, retired and the allies entered it unopposed. This long and complex siege cannot here be dealt with in detail. How the unfortunate troops fared in the trenches has passed into the history of this country as a blot upon the administration which allowed brave men to die like flies; "between November 1854 and February 1855 there were 9,000 deaths in hospital, and by the end of February there were no less than 13,600 officers and men in hospital," and no wonder when men, emaciated by disease and tried by the severe weather, "carrying great-coat and blanket, marched through a sea of mud to the trenches," where they remained in a cramped position all night until day broke, and, after a few hours' rest, "often in a puddle, which chilled their bones under a worn-out tent through which the rain beat"; they who had the heart of a lion were compelled to do asses' work. The hospitals were frequently merely bell tents, and "the men lay, often in mud, on the ground, and in many cases their diet was only meat and biscuit. They were, moreover, so crowded together that the doctors could scarcely pass between the patients." The heroic Florence Nightingale, as every one knows, with her band of nurses helped to lighten the burden of pain and suffering which the men endured, but the lasting disgrace which attaches to the callous administrators of the period is only accentuated when we consider the dire necessity for her action.

Capture of the Mamelon.—Sebastopol was bombarded, with the co-operation of the fleet, on October 17th, and the Russians made their first sortie on the 26th; in March 1855 they made another, and on April 9th the place was ineffectively bombarded; but on June 6th a bombardment was followed by an assault, and the Mamelon, an outwork of the Malakoff, was taken. Then the allies essayed the capture of the Malakoff and the Redan, the French to take the former and the British the latter; but despite the strenuous efforts of the men, the forts remained in the possession of the enemy. In these abortive attempts on June 17th and 18th the French lost 3,500 men, and the British 1,500; and of the six British and French Generals and Commanders who led the attacks on the 18th, 4 were killed and 1 wounded. The Royal Irish lost 259 officers and men by death and wounds in the attack on the Redan, and Captain Thomas Esmond of that regiment gained the V.C. and Sergeant John Grant the Meritorious Service medal. The Russians lost 4,500 officers and men killed and wounded. The siege was, following these repulses, resumed in the ordinary way, and the stubborn Russians, who were losing 250 men each day, began to realise that their magnificent courage could not stand much longer against the determination of the allies; and at the end of August, when the brave garrison was being depleted by the loss of over 900 each day, they determined upon the evacuation of the fortress they had held so well. So they built a bridge across the outer harbour, and made preparations to cover their retreat.

Capture of the Malakoff.—Then the allies again determined to reduce the fortress, and bombarded it for practically three days and nights, at the end of which, on September 8th, the French again assailed the Malakoff, and carried and held it against the enemy, but the British failed to capture the Redan. The British losses were 2,271 officers and men placed *hors de combat,* including three generals wounded, and the French 7,567, including 5 generals killed and 4 wounded. The Russian losses were 12,915, including 9 generals. During the night the Russians exploded their mines, and Prince Gortschakoff successfully took his garrison, with most

(Obverse.)
MEDAL FOR INDIAN MUTINY,
1857–8.

(Reverse.)
MEDAL FOR NEW ZEALAND WARS,
1861–6.

(Reverse.)
MEDAL FOR INDIAN MUTINY,
1857–8.

of the wounded, across the bridge, which they destroyed, and took up a position on the heights to the north of Sebastopol, from whence they cannonaded the fortress which the allies had occupied. The war was thus virtually at an end, although hostilities continued until February 1856, and the treaty of peace was not signed in Paris until March 29th.

Meanwhile the battles of Balaklava and Inkermann had been fought, both brought about by the efforts of the Russians to raise the siege, and both battles remarkable for the courage and fearlessness of the British troops, and alas! for the lack of simple military genius on the part of many of the officers. As a military authority has stated, "The army ought to have been beaten according to all the canons of war, but it wasn't!" Tommy Atkins saved the face of his superiors.

Balaklava.—The charge of the Light Brigade in this battle was an example of what appears to have been either muddle-headedness or overweening belief in the importance of our cavalry, but it gave to the Light Cavalry of the British Army undying fame. Their "death-ride" gave them eternal life! The battle of Balaklava was fought on October 25th, 1854, the anniversary of Agincourt. The Russian relieving force, numbering 22,000 infantry and 3,400 cavalry, with 78 guns, advanced from the Tchernaya by Kamara across the Woronzoff Road to attack the front of the position at Balaklava, and to meet them Lord Raglan took down the first and fourth divisions to the plain, which General Canrobert—St. Arnaud had died—reinforced with the first division of French Infantry and the Chasseurs d'Afrique.

"The Thin Red Line."—The Russians obtained possession of three forts, and managed to retain two of them with 7 guns; then "their artillery advanced with a large mass of cavalry, and their guns ranged to the 93rd Highlanders, which, with 100 invalids, under Lieutenant-Colonel Daveney, in support, occupied very insufficiently, from the smallness of their numbers, the slightly rising ground in front of No. 4 Battery." Then a body of the enemy's cavalry, "amounting to about 400, turned to their left, separating themselves from those who attacked Lord Lucan's Division, and charged the 93rd Highlanders, who immediately advanced to the crest of the hill, on which they stood and opened their fire, which forced the Russian cavalry to give way and turn to their left; after which they made an attempt to turn the flank of the 93rd . . . upon which the Grenadiers of the 93rd under Captain Ross were wheeled up to their right and fired on the enemy, which manœuvre completely discomforted them." Thus the gallant Sir Colin Campbell officially described the deed of the "thin red line" at Balaklava; that line of imperturbable Highlanders armed only with muzzle loaders, that, despite the rounds of shot and shell which harried them, dared to accept the charge of cavalry and drove them off.

Charge of the Heavy Brigade.—Then the main body of the Russian horsemen, about 3,000, advanced against the British Heavy Brigade, 900 sabres strong. The light-blue jackets, with embroidery of silver lace, the grey coats of the Dragoons, and the glitter of the lances, made a brave picture as the enemy's cavalry cantered toward their objective. They "nearly halted. Their first line was at least double the length of ours—it was three times as deep. Behind them was a similar line, equally strong and compact. They evidently despised their insignificant-looking enemy," but the trumpets rang out again in the valley; then Brigadier-General Scarlett, with about 300 Scots Greys and Enniskilleners, went right at the centre of the Russian cavalry, and then, as W. H. Russell, the famous *Times* correspondent, graphically described, "as lightning flashes through a cloud the Greys and Enniskilleners pierced through the dark masses of the Russians. The shock was but of a moment. There was a clash of steel, and a light play of sword blades in the air, and then the Greys and the redcoats disappeared in the midst of the shaken and quivering columns. In another moment we saw them emerging with diminished numbers, and in broken order, charging against the second line. It was a terrible moment. . . . With unbated fire the noble hearts dashed at the enemy. It was a fight of heroes. . . . By sheer steel and sheer courage, Enniskillener and Scot were winning their desperate way through the enemy's squadrons, and already grey horses and red coats had appeared right at the rear of the second mass, when, with visible force, like one bolt from a bow, the 4th Dragoon Guards riding straight at the right flank of the Russians, and the 5th Dragoon Guards, following close after the Enniskilleners, rushed at the remnants of the first line of the enemy, went through it as though it were made of pasteboard, and put them to utter rout." Then followed that " superb incident—at once a blunder and a miracle."

Charge of the Light Brigade.—Lord Raglan, being desirous that the withdrawal of the Russians should be taken advantage of to regain the heights, and fearing that the enemy might attempt to remove the guns from the redoubts captured from the Turks, gave orders for the Light Cavalry to be moved forward. Captain Nolan misconstrued the order, and indicated to Lord Lucan that the heavy battery of guns a mile away, supported by masses of cavalry and infantry, with other batteries on either flank, was to be charged. Lucan hesitated to carry the mad order into effect, but ultimately decided that it was his duty to do so, and gave the order to the Earl of Cardigan to take the terrible odds which he saw arrayed against him. Cardigan, like Scarlett, was a brave and daring cavalry leader, and into the " valley of death " he rode with his Light Brigade of 621 men. Encircled by fire they charged down the valley, dashed through the guns, sabred the gunners where they stood, cut their way through a brigade of Russian cavalry and a company of infantry, turned about, what was left of them, when four squadrons of Lancers were hurled at them, but they met the charge gallantly, and then the Russian artillerymen, recovering from their shock, but without regaining their senses, turned their guns upon the daring British cavalry and their own men who were struggling with them! The Greys and Enniskilleners, and the Chasseurs d'Afrique created a diversion, otherwise it is doubtful whether a single man of that brave little band, which rode so fearlessly and fought so uselessly, would have survived to receive the plaudits of his countrymen, and—shame of shames—as was the case with several of them, to end his days in the workhouse or eke them out with the aid of charity. It were almost idle for the poet to sing:

> "Honour the charge they made!
> Honour the Light Brigade,
> Noble six hundred!"

Four hundred and twenty-six of the brigade were placed *hors de combat,* 13 officers and 162 men were killed or taken prisoners, the former including Captain Nolan, who was struck early in the charge by a piece of shell and killed; 27 officers and 224 men were wounded. There struggled back to camp, in scattered groups, a remnant of 195 mounted men. In this charge Quartermaster Charles Wooden, whose group of medals forms the frontispiece to this volume, as Sergeant-Major in the 17th Lancers, gained the Victoria Cross by assisting Surgeon Mouat in saving the life of Lieutenant-Colonel Morris of his own regiment when lying exposed to a heavy fire of shot and shell, and carrying him to a place of safety. The Sergeant-Major had a horse shot under him.

The regiments engaged at Balaklava were the Heavy Brigade, comprising the 1st, 2nd, and 6th Dragoons; 4th and 5th Dragoon Guards; the Light Brigade, consisting of the 13th Light Dragoons and 17th Lancers in the first line, 4th Light Dragoons and 8th and 11th Hussars in the second line, and the 93rd Highlanders; but men of the Rifle and Naval Brigades, and of the

ABYSSINIAN WAR MEDAL, 1867–8.

CANADA MEDAL, 1866 AND 1870.

ASHANTEE WAR MEDAL, 1873–4.

ZULU WAR MEDAL, 1877–9.

artillery and various line regiments, including the 4th, 19th, 21st, 30th, 33rd, 44th, 47th, 50th, 53rd, 68th, and 77th, were also present and received the medal with the bar for Balaklava.

Inkermann.—On Sunday, November 5th, 1854, the battle of Inkermann was fought. The Russians had received considerable reinforcements; it is estimated that by that time 120,000 troops were under Prince Mentschikoff's command, and the Czar's soldiers were in high fettle, owing to the presence in camp of the Czar's sons, Michael and Nicholas. A general advance was made by the Russians, Mount Inkermann being the objective, and the battle commenced by an assault thereon by General Danneburg with about 40,000 men. The Mount of Inkermann and the district thereof was held by the Second Division under General Pennefeather. At daybreak masses of Russians were discerned marching to the attack, and soon afterwards the armies were in conflict. Then began the "soldiers' battle," which was to add lustre to the record of the several British regiments whose gallantry gained the day. How can one adequately paint the picture of 200 men of the 30th charging a couple of Russian battalions and putting them to rout, or of the 49th defeating a strong column and chasing their flying foemen? How can one do justice to the 525 men of the 41st who fought and dispersed five battalions of the enemy, and the 260 men of the 77th who put to rout 1,500 of the famous Tomsk Regiment? In the first stage of the battle 4,000 British soldiers beat back 15,000 Russians from the slopes of Inkermann, where, in the detached fighting that had taken place, many a V.C. was well earned. In the second stage the terrific struggle ranged around the worthless sand-bag battery; seven times did the Russians capture it, and as often was it retaken by the British, each time with awful slaughter, until the dead lay around the battery in heaps and filled the entrenchment. In this useless and stubborn fight Generals Adams, Cathcart, and Torrens fell. There young Captain Stanley of the "Die Hards" fell as he called to his men to "remember Albuera" and follow him. There a few hundred Coldstreams, who had 8 officers killed, fought back to back against 6,000 Russians, and sustained "the bloodiest struggle ever witnessed since war cursed the earth." Then two 18-pounders were ordered up by Lord Raglan, and, as happened at the Alma, helped to determine the events of the day. An artillery duel resulted, many a British gunner falling by his gun, in the discomfiture of the enemy ere the Zouaves, led by their Vivandière, pushed into the fight and drove the Russians toward the sand-bag battery, where the dead had formed a wall. The French Zouaves and British soldiers then made short work of the contest, and the Zouaves' standard was firmly planted on the sand-bag battery. By 1 o'clock the battle was decided, but such was the straggling and intermittent nature of the fighting that it was not until 3 o'clock that Mount Inkermann was again entirely in the possession of the allies. In this great battle 10 British Generals were killed or wounded—Lord Raglan alone being unharmed—39 officers and 558 men killed, and 1,760 officers and men wounded (rather more than one-third of the total strength of the army engaged). The French lost 130 killed and wounded, and the Russians over 11,000, including 256 officers killed.

The following regiments were present at Inkermann: Grenadiers, Coldstreams, and Scots Fusilier Guards; 1st, 4th, 7th, 19th, 20th, 21st, 23rd, 28th, 30th, 33rd, 38th, 41st, 2 companies 46th, 47th, 49th, 50th, 55th, 57th, 63rd, 68th, 77th, 88th, 95th; 4th, 8th, 11th, and 13th Hussars; 17th Lancers and 2 batteries of the Royal Horse Artillery and Royal Field Artillery.

Medals with bars for the Alma or Inkermann, or both, are rare to the 46th Foot, while the medals awarded the Coldstream Guards and to participants in the charge of the Light Brigade realise good prices when offered for sale.

The Crimean Medals.—In December 1854 Queen

Victoria commanded that a medal bearing the word "CRIMEA" should be struck, and that bars for the ALMA and INKERMANN ("the soldiers' battle") should be awarded to those who had taken part in the battles; but the world had wondered at the "death-ride" of the Light Brigade—on October 29th, 1854—and so no one was surprised when, in February 1855, a clasp for BALAKLAVA was granted to those who rode so bravely into the jaws of death, and, wonder of wonders, came out again; likewise to the Heavy Brigade and to the regiments engaged in the vicinity. The bar for SEBASTOPOL was added to the list in October 1855, so that the four-bar medal illustrated is a complete record of one of the greatest of modern wars, significant because of the fact that we fought shoulder to shoulder with our old adversaries the French, and side by side with the Turks and Sardinians against the Russian Army. The medal, designed by W. Wyon, represents on the reverse a Roman warrior with a flying figure of Victory crowning him with a laurel wreath; to the right of the figure is the word CRIMEA arranged perpendicularly. The obverse is the same as that on the Peninsular and India General Service medals. The bars are the most ornate of the whole series given with British medals. A special bar for AZOFF was given to the Navy. The suspender is of a very appropriate character, suggesting a palm wreath issuing from a conventional cusp. The ribbon is of pale blue with yellow edges, and, be it noted, the ribbon for the Baltic medal—illustrated facing page 296—granted to sailors and a few marines, is yellow, with pale-blue edges. The Baltic medals were all issued unnamed, as were also the Crimean; but some recipients of the latter had their name and regiment engraved privately. Others were officially named later, with the same stamps as were used for the Army General Service and early Kaffir War medals, in square Roman capitals.

Five bars in all were issued, but four is the greatest

number awarded with any medal. The bars should read upward from the medal as follows: Alma, Balaklava, Inkermann, Sebastopol, but a number of the medals were issued without the bars being fixed, and the order is consequently sometimes found to be inaccurate; care should therefore be taken to verify the record of the person named on the medal. Many medals were issued before the grant of the bar for Sebastopol at the end of October 1855, and many exist without this bar, although all who took part in the battles of Balaklava and Inkermann were entitled to it. The troops who landed in the Crimea after September 9th, 1855, the day Sebastopol fell, were not entitled to the medal unless they had been engaged against the enemy after that date.

The Turkish Medals.—The Turkish Government, in whose cause we took up arms, gave to the allies a silver medal suspended by a bright-red ribbon with green edges $\frac{1}{2}$ in. wide running through a small ring. This is illustrated beside the British Crimean medal. These medals vary; they were intended for the British, French, and Sardinian soldiers, and had among the trophy of flags that of the country the recipient belonged to in the front, beside the Turkish flag, and in the exergue either CRIMEA 1855, LA CRIMEA 1855, or CRIMEE 1855, but as the ship which was bringing the medals to England foundered, many men received the French or Sardinian variety. These medals, $1\frac{2}{5}$ in. in diameter, bear on the obverse the Sultan's cypher encircled by a laurel wreath, with the date in Arabic Hegira "1271."

They were generally issued unnamed, but I have several impressed.

The French Medal.—The Emperor of the French awarded crosses of the Legion of Honour to officers and men who had been conspicuous during the war, and the Médaille Militaire to about 500 non-commissioned officers and men who had distinguished themselves. The Duke of Wellington and Sir William Cordington, who was presented by Marshal Pellisier with his own medal, were the only two British officers to receive it. The medal is silver, the eagle and centre being gilt, and the band surrounding the head of Louis Napoleon enamelled, likewise that on the reverse, encircling VALEUR ET DISCIPLINE. The medal is suspended from an orange-coloured ribbon with green edges.

The Sardinian Medal.—The King of Sardinia awarded the Sardinian war medal to 400 officers, non-commissioned officers, and men of the Army, sailors, and marines. The obverse has the arms of Savoy within a wreath of palm or laurel encircled by the legend AL VALORE MILITARE. On the reverse is a laurel wreath and the inscription SPEDIZIONE D'ORIENTE 1855–1856. The medal of silver is suspended by blue watered silk ribbon through a broad loop with a flat top, much the same as the handle of a flat-iron; medals of gunmetal with similar loops were given by the German States for certain campaigns.

Turkish General Service Medal.—The Turkish General Service medal, incorrectly called the "Danube Medal," was awarded by the Turkish Government in 1855 to the officers and 30 men comprising the crew of a British gunboat, and to a Colonel and 16 men of the Royal Engineers, for services rendered on the Danube in 1854. On the obverse is the cypher of the Sultan Abdil Mageed Khan II within a beaded circle, with flags and laurel branches, and above all a crescent and star. On the reverse is an elliptical star of twelve points, with a smaller one of six in the centre; underneath is a scroll bearing an inscription in Persian characters, reading "Mischani Iftikar" (Medal for Glory), and under all a small star between laurel wreaths. The medal is $1\frac{1}{5}$ in. in diameter, and was suspended from a silver scroll bar by the same ribbon as used with the Turkish Crimean medal. The General Service medals were issued in gold and silver, gold to the officers and silver to the men.

The Silistria Medal.—Sir John Lintorn Arabin Simmonds, who was one of the officers to receive the medal referred to above, was awarded another in gold for his services in the defence of Silistria in 1854. Six other British officers also received the medal, but in silver. The obverse is the same as the General Service medal, but on the reverse is, within a beaded border, the fortress of Silistria, over which the Turkish flag is flying, with the Danube in the foreground. In the exergue on a scroll is SILISTRIA and the year of the Hegira "1271" in Arabic characters. The medal is $1\frac{9}{20}$ in. in diameter, and was suspended from a steel ring run through it by the same ribbon as used with the Turkish Crimean medal.

Kars Medal.—The Sultan of Turkey granted a silver medal for his brilliant defence of Kars to Sir William Fenwick Williams, and to several other British officers who had served with him, and also to his artillery servant. It has the same obverse as the Turkish General Service medal, but on the reverse is depicted the citadel and town of Kars, with the Turkish flag flying over the citadel, and underneath KARS and the date Hegira "1272" (1856). The medal, $1\frac{9}{20}$ in. in diameter, has a milled edge and was suspended from a straight silver bar and clip by $1\frac{1}{2}$ in. crimson ribbon with green stripes at the side.

The Order of the Medjidie.—The Sultan of Turkey rather profusely awarded the newly established Order of the Medjidie to officers in the British Army and Navy for services rendered in the Crimean War. Field Marshal Sir Evelyn Wood was awarded the fifth class. He was then a midshipman of seventeen. Five classes were given, and the recipients exceeded 1,000. The size of the decorations gradually decreased according to grade. The centre of the Badge of the first four classes is gold, and that of the fifth silver; the first three classes wear the Badge round the neck, suspended by a crimson ribbon with green edges, whilst the other two classes suspend the Badge from the left breast by a similar ribbon. The Badge or

Cross is of silver, with seven triple points or rays, between which are disposed seven small crescents and stars of five points. On a red enamelled band in Arabic characters are the words "Loyalty—Patriotism—Zeal" and the year of the Hegira "1268" (1852, when the order was founded) around the cypher of the Sultan. The badge is suspended from a red enamelled crescent and star.

Persia.—The Persian War followed in 1856–7; it was brought about by the intimate relations of the Persian Government with Russia. A rebellion broke out in Herat, "the Gate of Persia," and the place was besieged by the aid of Russia; as this was contrary to the provisions of a convention made between the Shah and the British Minister in 1853, an expedition under General Outram was dispatched. It landed near Bushire, and on December 9th, 1836, a battle was fought at Reshire, and the next day another at Bushire. Conflicts also took place at Kooshab—where the 3rd Bombay Cavalry broke a Persian square—Barajoom, Mohummerah and Ahaz, before the "King of Kings" sued for peace.

The following regiments were engaged: 64th and 78th Foot, and 14th Light Dragoons; and the regiments in the H.E.I. Co.'s service: 2nd Bombay European Fusiliers; Bombay Sappers and Miners; 2nd, 3rd, 4th, 5th, 8th, 11th, 15th, 20th, 22nd, 23rd, 25th, 26th, 28th, and 29th Bombay Infantry; 1st Scinde Horse; 3rd Bombay Cavalry and Poona Horse.

The India General Service medal, illustrated facing page 120, was awarded with the clasp for PERSIA, or the clasp was added to the medal of the soldier who had taken part in the second Burmese War; indeed, the medal was instituted to obviate the necessity for designing medals for each campaign, and multiplying the number which might adorn the soldier's breast.

THE INDIAN MUTINY

On Sunday, May 10th, 1857, the native soldiers in Meerut openly mutinied. It was not the fat used in the making of cartridges which caused the outbreak; that was merely the ostensible reason, for there can be little doubt that a section of the natives, failing to understand European institutions and principles, were easily led by political agitators to take the desperate step of rebellion. The Kings of Oude and Delhi had been practically deposed by the Indian Góvernment, and it was the effort to restore the Mogul dynasty, engineered by those who for political reasons wished to see the English expelled from India, that had much to do with the mutiny—for mutiny it certainly was, bred among men who had gathered extravagant notions of their own importance and their ruler's weaknesses. The common people were hardly affected—indeed, the princes were also, with the exception of the Ranee of Jhansi and the Náná Sáhib, loyal to the British. The arch-fiend of the mutiny was the Náná Sáhib, who had been quietly waiting his opportunity for revenge upon the British for refusing to continue the pension granted to his foster-father. The cartridge incident merely helped him in his nefarious designs, and forwarded the efforts of our external enemies. It was not, however, expected that the mass of the people of India, with their princes, would remain loyal as they did. At the time the mutiny broke out, in a population of 18,000,000, 200,000 of whom were sepoys, there were only 38,000 British soldiers, so that it is quite evident that the outbreak was not national, but military, otherwise they would have been swallowed up in the tempest. Disaffection had occurred at two or three stations, and the 19th and 34th Native Infantry had to be disbanded, the latter after Mangul Pandy had shot two officers at Barrackpore. Then the more determined outbreak occurred at Oude and Meerut; the English officers were murdered, and the European civilians slaughtered

MEDAL FOR AFGHAN WAR.

ROBERTS STAR FOR MARCH
TO KANDAHAR.

before the two regiments of mutineers marched off to join others in Delhi, which became the centre of mutinous activity. They proclaimed as king a descendant of the great Mogul, and attempted to capture the magazine, but, as we shall see hereafter, ten Britishers gallantly defended it, and ultimately blew it up. At this time the 8th were at Jallundur, the 32nd at Lucknow, the 60th and 6th Carabineers, with a troop of horse artillery and details of other regiments, at Meerut. The 61st at Ferozepore, the 75th at Umballah, the 81st at Meeau Meer, the 37th were called up from Ceylon, and the Madras Fusiliers from Madras. The 64th and 78th hurried from Persia, and the 84th were recalled from Burmah. The troops, which included the 93rd Highlanders, *en route* for China were ordered to India, and Sir Colin Campbell hurried off from England to take command of the operations. Meanwhile the insurrection spread, until the whole region between the Punjab and Lower Bengal was in revolt, a district as great as Austria, France, and Prussia combined. Within six weeks of the murder of Colonel Finnis at Meerut, of the 120 sepoy regiments only 25 had not mutinied, and of these only 5 could be relied upon.

Cawnpore.—Náná Sáhib took his opportunity directly the news of the rebellion at Meerut and Delhi became known. Placing himself at the head of the rebels, he proclaimed himself Peshwa of the Mahrattas, and on June 6th informed General Sir Hugh Wheeler that he was about to attack the position held by him, a miserably slight fortification within which he had gathered about 1,000 souls, of whom only 465 were males of all ages and classes. How bravely they fought, despite hunger, thirst, and disease, for twenty days, and only surrendered when a safe passage to Allahabad had been guaranteed by the bloodthirsty Náná, and how by his treachery they were butchered by gunshot and sword, or burnt to death, is one of those heroic episodes which will live for ever; while the butchery of the women and children also by order of the Náná Sáhib will blot the pages of Indian history for all time. On July 16th General Havelock, who had hastened to the relief, defeated the Náná with 7,000 men and powerful artillery, a few miles south of the city, and he fled during the night. Next morning the British soldiers entered Cawnpore, too late to save the unfortunate captives. Havelock's little army, which had marched 126 miles in eight days and captured 24 guns, consisted of 435 men of the 64th, about 300 78th Highlanders (Ross-shire Buffs), 190 of the 84th, 400 of the 1st Madras Fusiliers, 20 Volunteer Cavalry, 76 Royal Artillery, 450 Sikhs, and 50 Irregular Native Horse. "Better soldiers have never trod the earth," states Archibald Forbes; and when we consider the sweltering heat of the Indian summer, the men wading "in a sea of slush . . . while the flood of tropical rain beat down," and fighting against overwhelming odds, we may heartily endorse Dr. Fitchett when he says, "In the whole history of war, men have seldom dared, and endured, and achieved more than did Havelock's column in the gallant but vain struggle to relieve Cawnpore"; and of the "Ironsides" who were to press on to relieve and reinforce Lucknow, only 250 remained to participate in the glorious march.

Delhi.—On May 11, 1857, the 3rd Native Cavalry, who had outstripped their infantry companions in revolt on the long march of 38 miles from Meerut, appeared before the King's palace at Delhi, and declared they had "slain all the English at Meerut, and had come to fight for the faith." The old King had no desire to become embroiled in the tumult that had begun to rage around him, but his guards, sympathising with the mutineers, opened the gates and stood by while the miscreants murdered the English officers and officials, and their families, and any white person they could find. Lieutenant Willoughby, with 2 other officers and 7 British soldiers, determined not only to sell their lives dearly, but to do as much hurt to the rebels as possible, before they accepted their fate. Their daring exploit of keeping nearly 2,000 mutineers at bay as long as they could serve the 10 guns, and coolly taking their chance of life and death as they blew the magazine and the surging masses of sepoys into the air, is one of the most brilliant in history. Brave Conductor Scully, who fired the train, and four of his companions heroically met their death, but the three officers, Willoughby (who was captured later and killed), Raynor, and Forrest, and Conductor Buckley, miraculously escaped for the time being.

Meanwhile the Commander-in-Chief, General Anson, had collected all the available troops and marched upon Delhi, but dying of cholera at Kurnaul, the command devolved upon Major-General Reed. Major-General Sir Henry Barnard had collected a number of men at Aleepore, and Brigadier-General Archdale Wilson had quitted Meerut with about 700 men in order to join forces with the Commander-in-Chief. On May 30th he was attacked by the mutineers at Ghazee-ood-deen-nuggur, but drove off his assailants numbering over 4,000. They returned to the attack on the following day (Whit-Sunday), but the 6th Carabineers and the Rifles drove them off with considerable slaughter, and captured 26 guns.

Then commenced the siege of Delhi, but for three and a half months the ancient city withstood the assaults of the besiegers. First Badlee-ke-Serai was taken, the 75th particularly distinguishing itself in the valiant fight which the English troops made in acquiring it, and then, after the Delhi cantonments had been taken and burned, the Goorkas giving splendid evidence of their loyalty and gallantry, the army sat down before the city, and the British lines were slowly but deliberately pushed forward. On June 17th one of the enemy's positions near the Ajmeer gate was taken; and on June 23rd, the fatal day of prophecy when British rule was to cease, the mutineers made a determined advance from the city, but after a hard day's fighting, in which they had charged the Rifles, the Guides, and the Goorkas again and again, they were compelled to retire.

In June cholera broke out in the camp, and early in July Sir Henry Barnard was among its victims, while the 8th and 61st regiments suffered considerably. Again and again the rebels sallied forth, only to be beaten back, and the fighting proceeded gallantly week after week, during which time Major Tombs, Lieutenant Hill, and a number of men gained the V.C. Then Brigadier-General John Nicholson, having gained a brilliant victory at Nujjuffhur, and prevented an attack upon the rear of the British camp, rode in on August 7th. Shortly after his little army marched in; it comprised 680 men of the 52nd, a wing of the 61st, the 2nd Punjab Infantry, and a field battery with detachments of Beloochees and military police.

The besiegers were waiting for the siege train with which to breach the walls, and the mutineers having learned that it was *en route*, sallied forth in considerable numbers to intercept it; but Nicholson, with 1,600 infantry, 400 cavalry, and a battery of field guns, set out on August 25th to save the train. The troops had to cross the swamps, made more difficult by the ceaseless rain; but after wading through them, the water sometimes over the horses' backs, and along the rain-drenched roads, they came up with the enemy at Nujutgurh after a march of twelve hours. The mutineers numbered 6,000, but the tired and drenched soldiers of Britain, led by the daring Nicholson, charged the enemy, placed 800 of them *hors de combat*, captured 13 guns, and sent the sepoys flying back to Delhi. Retracing their steps, the little force, having lost 60 men, arrived back in camp, having in forty hours, during twenty-four of which they were foodless, marched 35 miles and beaten and dispersed an army of well-trained men.

The siege train arrived on September 4th, together with a wing of the 8th Detachment of the 9th and

HONG-KONG PLAGUE MEDAL.
Awarded to about 350 of the Shropshire Light Infantry, Engineers, and men of the R.N.

(Obverse.)
EGYPTIAN MEDAL FOR 1882.

(Obverse.)
KHEDIVE'S BRONZE STAR FOR 1882.

60th, and a battalion of Beloochees. The guns were mounted, and by the 11th ready for action, when fire was opened and incessantly sustained against the Cashmere and Watergates until breaches had been made, and then it was determined to assault the city.

Blowing in the Cashmere Gate.—It was found necessary, however, to blow up the Cashmere Gate, and a party of twenty men under Lieutenants Home and Salkeld of the Engineers were detailed for the purpose. They effected their object, but Salkeld was mortally wounded and many of the men killed and wounded in their gallant effort. Bugler Robert Hawthorne of the 52nd, having thrice sounded the regimental call of the 52nd for the column to advance, took charge of the wounded lieutenant, and having bound up his wounds, removed him to a place of comparative safety. The bugler was decorated with the V.C. for "as noble a deed as any that has ever graced the annals of war." In the assault on September 14th the gallant Irishman John Nicholson fell, mortally wounded while leading his men near the Lahore Gate, and 1,169 officers and men were killed, wounded, or missing, but the whole of the outer parts of the city were in possession of the British. On the 16th the magazine which Willoughby had, it was found, only partially blown up was captured by the 61st, but desultory fighting was continued until the 20th, when the Lahore Gate and the palatial Jumma Musjid were captured. Then the gates of the palace were blown in, and the 60th Rifles led the way into the ancient home of the Mogul Kings, and Delhi was in the hands of the British Army. In the grand assault upon Delhi 8 European officers and 162 men were killed, 52 officers and 510 men wounded, and 103 sepoys killed and 310 wounded. It is estimated that over 5,000 mutineers perished in the defence of the city.

Hodson's Daring Feat.—But the king and his family had taken refuge in Humayon's Tomb, about 7 miles from the city, and Hodson, the daring Captain of the

Light Horse bearing his name, determined to capture him. Taking only fifty men, he essayed one of the most daring feats on record. With his little band he rode along the rebel-infested road to the tomb where, in the gigantic marble dome, the King and his two sons had concealed themselves; for two hours he parleyed with the intermediaries of the decrepit King, undismayed by the thousands of retainers who guarded him. At last the old man, on promise of his life, surrendered, and Hodson marched off before the wondering natives with his royal prisoner. There yet remained the two sons of the King, whose conduct in torturing English prisoners had made them notorious. They had barbarously slaughtered innocent women and hapless children, and Hodson determined that they should also become his prisoners. On September 21st he and his second in command, Macdowell, with 100 men rode off to effect the capture of the princes, who had 6,000 or 7,000 armed followers at their command. Reaching Humayon's Tomb, he demanded their surrender, but they at first refused to submit unless their lives were promised. Hodson calmly refused, and then they came forth with 3,000 armed men. But the daring Hodson interposed his troopers between the bullock-cart in which the princes were riding and the armed men, sent forward the princes with an escort of troopers, and then—calmly ordered the retainers to lay down their arms! Having collected them, Hodson quickly said to Macdowell "We'll go now," and then rode off with his troop. Overtaking the princes, he found that a crowd appeared to be threatening the troopers in charge of the cart, and fearing that he might lose them and that the ends of justice would be defeated, he ordered the princes to strip, and, after stating the nature of the crimes they had committed, shot them with his own hand. Brave, gallant, and daring Hodson was prepared to take the consequences which he fully appreciated, and, convinced that he was right, did not flinch from moral censure any more than he had from physical consequences when he sallied forth with a handful of men to capture the King and his bloodthirsty sons. With the fall of Delhi the back of the mutiny was broken, but it had cost the besiegers a loss of 3,854 killed, wounded, and missing.

Defence of Lucknow.—The neighbourhood of Lucknow, however, remained in possession of the mutineers, although over the Residency

"Ever upon the topmost wall our banner of England flew."

For several weeks Lucknow had been in a state of unrest, and then at 9 o'clock on the night of May 30th, 1857, the smouldering fire broke into flame. Sir Henry Lawrence, that stalwart, cool northern Irishman, as just and firm as he was unselfish, had but 700 Europeans in a city of 700,000, of whom 7,000 were sepoys. He did not lose heart, but quickly and steadily made preparations for the defence of the place; he knew he could rely upon the Sikhs and a small number of the sepoys—700 actually remained true to their salt during the siege; he made no error in his calculations, and took few chances. He turned the Residency into a fortress, and generally prepared for the worst, while his sense of humour and his smiling face gave no sign of the stern practical heart within him. He was, as he said, "virtually besieging four regiments—in a quiet way—with 300 Europeans," while he resided "in cantonments guarded by the gentlemen" he was besieging!

On June 30th he decided to "blood" the native troops, and he accordingly sent half a dozen guns with sepoy artillerymen in the little force of just over 800, of whom only 336 were Europeans, to meet the mutinous regiments which were marching upon the city from Eastern Oude. The estimated force was 5,000; it turned out to be 15,000, and when they were encountered at Chinhut the Sikh horsemen bolted, and the artillerymen disabled and deserted their guns. The fates generally went against the bold step that Lawrence had taken. The remnants of the little band had to retire in face of the great moving mass of mutineers; "regiment after regiment of sepoys steadily pursued towards us," and the 32nd who had gone into battle —300 foodless and badly armed men—were reduced to a skeleton, 5 of their officers and 112 men being killed. Lawrence returned with his straggling men to Lucknow; it was one of the few mistakes which the hero of Lucknow made, but he saved the survivors of the desperate fight by a masterly stroke in placing empty guns upon the iron bridge (the ammunition was exhausted), and with gunners standing beside them with lighted port-fires stayed the advance of the victorious sepoys. The time for desperate action had arrived. On July 1st he blew up the Mutchee Bhawan with its 1,000,000 cartridges and 250 barrels of gunpowder, and concentrated upon the Residency, and there for eighty-eight days, with a force of 927 Europeans and 700 sepoys, made one of the most famous defences in history. And there, despite the supreme efforts of the mutineers to shoot it down, the British flag was only temporarily out of position when the staff was shot away, and so

"Ever aloft on the palace roof the old banner of England blew."

On July 2nd Sir Henry Lawrence was mortally wounded, and died on the 4th; then Brigadier Inglis took command of the troops. During the defence the populace of 3,000 and the troops were harassed by cholera, smallpox, and an indefinable disease, but the spirit of the troops remained undaunted until Jessie Brown's keen Scots ears heard the far-off skirl of the bagpipes of the 78th, Outram's Highlanders—"the saviours of India"—and Havelock, marching into Lucknow on September 25th, reinforced, as well as relieved, the brave garrison which still had, for another six weeks, to hold the Residency against 60,000 mutineers.

The original defenders of the Residency were 535 men of the 32nd, 50 of the 84th, with 89 artillerymen,

EGYPTIAN MEDAL FOR THREE CAMPAIGNS. (Reverse.)

INDIA MEDAL FOR FOUR CAMPAIGNS. (Reverse.)

EAST AND WEST AFRICA MEDAL FOR FIVE EXPEDITIONS. (Reverse.

and 100 British officers whose native regiments had mutinied, and 153 civilians who took up arms to assist the regulars; and these, with 700 sepoys, undeterred by "the terrific and incessant fire by day and night," had for eighty-eight days defied not less than "8,000 men firing at one time into the position."

Relief of Lucknow.—Havelock, having rested his men after the advance on Cawnpore, followed up the Náná Sáhib, destroyed his palace and stronghold at Bithoor, and then with his tiny force, which was daily lessening through wounds and disease, marched towards Lucknow. He attacked Oonas *en route*, and passed through it, despite the 15 rebel guns which guarded the only road; pushed past the opposition at Busserut Gunge, but with a loss of 88 officers and men killed and wounded; then in despair, owing to his enfeebled force, Havelock started to return to Cawnpore. On reaching the Ganges the mutineers made a determined attack upon the little force, but the 78th, Ross-shire Buffs, dashed at the enemy's guns, and, as Havelock said, saved themselves and their comrades. Retracing his footsteps, Havelock was not aware that he had unconsciously helped the besieged in Lucknow by drawing off the rebel force to meet him, thus giving the garrison breathing-space in which to strengthen the fortifications and increase its stock of provisions. After a four-days rest the undaunted Havelock again set out for Lucknow with 1,300 men, but again meeting with opposition at Bithoor, and although the enemy was defeated, he decided to return to Cawnpore and await reinforcements. The 5th and 90th Regiments arrived early in September, five companies of men came in to make up for the terrible losses of the 78th, and then Sir James Outram, the "Bayard of India," arrived to take command of the Cawnpore and Dinapore divisions, but chivalrously delegated the command to Havelock "in gratitude for the brilliant deeds of arms achieved by General Havelock and his gallant troops." The relieving force consisted of the 1st Brigade—5th Fusiliers, 84th, and 100 men of the 64th under Brigadier-General Neill; 2nd Brigade—78th Highlanders, 90th (Perthshire) Light Infantry, and Brasyer's Sikhs under Colonel Walter Hamilton of the 78th; 3 batteries of artillery under the dauntless Maude, "Hell-fire Jack" Olpherts, and brave Vincent Eyre; 109 volunteer cavalry and 59 native cavalry under the dashing Barrow.

Crossing the Ganges on September 19th and 20th, the relieving force of 2,500 men attacked the enemy at Mungulwah on the 21st, and, although an obstinate opposition was met with, defeated the enemy and captured a couple of guns. Pushing forward, drenched by heavy rains for three days and through quagmires of mud, the badly fed little force came upon the enemy, 12,000 men, entrenched on the outskirts of Lucknow at the Alambagh, but after the artillery had opened fire the 78th Highlanders and the Fusiliers rushed the position, and in ten minutes the mutineers were flying in all directions. "The petticoated devils" were too much for them. After a day's rest, leaving 300 sick men to hold the place, the army advanced towards the Residency, and through a storm of shot gallantly rushed the Charbagh bridge, leaving the 78th as a rear-guard to hold it, which they bravely did, although stormed at by rifles and field pieces. The column then pushed on to the Kaiser Bagh, or King's palace, taking in reverse the battery which had been firing on the main body; and then the Highlanders, followed by the Sikhs and Fusiliers, made a desperate effort to reach the Residency. For three-quarters of a mile, with desperate and dauntless courage, the men pressed forward through a street in which from every housetop, door, and window a relentless hail of bullets poured upon them. In this grand advance Brigadier Neill fell, shot through the head, but the troops pushed on, and then through the embrasure by the side of the battered archway of the Baillie Guard they pressed. The picture has often been painted of the big rough-bearded soldiers—those stern but soft-hearted Highlanders—seizing the little children out of their mothers' arms and "kissing them with tears running down their cheeks." The relieving column lost over 700 men by death or wounds, or nearly one in four of its total complement of 3,000.

Defence Continued.—As I have stated, the relief of Lucknow was also a reinforcement, and those who relieved the garrison, continued its defence for nearly fifty days, and survived the war, were awarded the clasp for the Defence of Lucknow.

Havelock, having relieved Lucknow, did not deem it advisable to take the risk of escorting the women and children and non-combatants through the thousands of sepoys who still encircled the Residency, so for six weeks the garrison, under Outram, kept the enemy at bay, until Sir Colin Campbell, the cool and daring Scots veteran of the Peninsular War, fresh from the Crimea, who was ready at the age of sixty-five to take up the arduous duties of Commander-in-Chief in India, prepared for the final relief. He landed at Calcutta on August 13th, but through delays, largely occasioned through want of transport, he could not start out on the march to Lucknow until November 9th. Three days later he had under his command 4,700 men, a number of whom had been engaged in the siege of Delhi; detachments of the 4th, 5th, and 23rd Fusiliers; a wing of the 53rd, a number of the 82nd, and the 93rd Highlanders 1,000 strong, 700 wearing the Crimean medal; remnants of the war-worn 8th King's (Liverpool) Regiment; the 75th (now 1st Gordons); a heavy battery Royal Artillery; Bengal Horse and Field Artillery; two squadrons of the 9th Lancers; Hodson's Horse; a squadron each of the 1st, 2nd, and 5th Punjab Cavalry; 2nd and 4th Punjab Infantry; detachments of Bengal and Punjab Sappers and Miners; and the Naval Brigade with 8 guns under Captain Peel.

Sir Colin Campbell arrived at the Alambagh, where Sir James Hope Grant had resolved to await reinforcements, on November 12th, and having determined upon his plan of advance forced his way round the north of the city, pushed the enemy through the Dilkoosha Park, and after taking the Secundrabagh and the Shah Nujeef, entered the Residency, where on the afternoon of November 17th he was shaking hands with Outram and Havelock in front of the Mess House, with bullets and shells pouring around them, for although the relief force had reached the Residency they stood in fear of being themselves besieged. Sir Colin Campbell therefore determined to take the 1,000 sick and wounded men, and 600 women and children, out of the place that had so long protected them from the "devil's work" of the mutineers. On the night of November 19th the work of removal commenced, and by the morning of the 22nd the Residency had been completely evacuated. As Captain R. H. Burgoyne, who was present, states, "Thus was accomplished one of the most difficult and daring achievements ever attempted, for such it must be acknowledged it was, when we consider that with a force scarcely exceeding 4,000 Sir Colin Campbell, opposed by upwards of 40,000 regularly trained soldiers supplied with munitions of war far exceeding ours, and holding one of the strongest positions imaginable, penetrated into their midst, carried one fortified position after another, and finally brought away in safety every living man, woman, and child shut up in the Residency, together with their baggage, treasure, etc.," and "the guns it was thought worth while to keep."

The Secundrabagh.—The storming of the Secundrabagh is full of heroic deeds. "There never was a bolder feat of arms," stated Sir Colin Campbell in his dispatch. There seven companies of the 93rd, as the 4th Punjabis halted for a moment when their British officers were shot down, raced ahead of their Sikh comrades, and Lance-Corporal Dunlay gained his V.C. for being one of the first through the breaches and supporting Captain Burroughs, who had been

JUMMOO AND KASHMIR MEDAL, 1895. (Obverse.)

BRONZE HUNZA NAGAR BADGE, 1891.

JUMMOO AND KASHMIR MEDAL, 1895. (Reverse.)

BRONZE DECORATIONS GIVEN BY THE MAHARAJAH OF JUMMOO AND KASHMIR.

wounded, against superior numbers of the enemy. There also Lieutenant Kirke Ffrench and Private Irwin of the 53rd, and Private J. Smith of the 1st Madras Fusiliers, gained the V.C. for being among the first to enter the Secundrabagh by the gateway. When the roll of the 93rd was called after the storming, it was found that 8 officers had been wounded, 28 non-commissioned officers and men killed, and 71 wounded. At the Shah Nujeef, a domed mosque, the enemy withstood a heavy cannonade by Captain Peel's naval siege train, and the field battery with some mortars for three hours, when "it was stormed in the boldest possible manner by the 93rd Highlanders under Brigadier Hope," while Captain Peel took his guns forward with such daring that had it not been for the withering fire of the Highlanders the naval brigade would have suffered considerable loss. As Sir Colin Campbell stated in his dispatch, it should not be forgotten that the heroes of the relief of Lucknow had made the longest forced marches, some from Agra, some from Allahabad, and had undergone great fatigues and privations in pressing forward for the attainment of this great object.

No Bar for Cawnpore.—On the afternoon of November 20th Sir Colin Campbell, with his great convoy, arrived at the Alambagh, and encamped on the open space his force had occupied before advancing on Lucknow. Three days after the march was resumed, and as he advanced towards Cawnpore the sounds of firing were heard, and he learned that the Gwalior contingent, 10,000 strong, under Tantia Topi—the only leader the mutiny produced—had joined with the force of the Náná Sáhib in attacking Major-General Windham (Redan Windham). Leaving the convoy in charge of the rear-guard, Sir Colin hurried forward with the main column, which, despite its fatigued condition, footsore and hungry, pressed onward to the assistance of their comrades at Cawnpore, fighting their way forward until on December 6th the city was taken, the rebel armies split in twain, and the portion under Tantia Topi driven across the Jumna, and that with the Náná across the Ganges. This was a general's battle. With a force of about 5,000 men, Sir Colin Campbell had not only defeated an army of 25,000 well-trained men, but had captured all their baggage and 32 guns. It is noteworthy that in order to take part in this battle the 42nd, Black Watch, marched 80 miles in fifty-six hours, no mean feat in a tropical climate. Actions were fought at Kâla Muddee, and Futteghur—where the 53rd spontaneously charged the enemy and captured several guns—and having occupied the latter place Sir Colin made arrangements for the retaking of Lucknow.

Lucknow.—To retake Lucknow, which Sir Colin Campbell had evacuated on relieving the city, a splendidly equipped army was organised to march upon the place under the Commander-in-Chief. On March 2nd, 1858, he advanced with about 18,700 men, to be later strengthened by Brigadier Frank's column and the Nepaulese under Jung Bahadoor. Approaching Lucknow by the Dilkoosha Park, the rebel pickets retired before the 42nd and 93rd Highlanders, who swept the rebels from their works in front of the Martinière, and the British troops took possession of the palace and the Mahomed Bagh. On the morning of March 9th the Martinière was assaulted. The Punjabis, with the 42nd Highlanders, took the rebels in flank, the 93rd Highlanders in skirmishing order advanced at the double, supported by the 90th, and as they approached the Martinière the mutineers bolted, and took refuge in their entrenchments across the canal.

On the 11th the Begum's palace (Begum Kotee), sheltering 5,000 sepoys, was stormed, the regiments rushing forward under "a perfect storm of musketry," but "not a man wavered." The rebels had cunningly devised "every obstacle that could be opposed to the stormers," but the 93rd and the 4th Punjab Rifles were not deterred, and they fired their muskets and plied their bayonets for two hours, until, with the aid of a party of the 42nd, the enemy was forced to disperse, and the Begum Kotee, the key of the position, was won after the most sanguinary fight in the siege of Lucknow, nearly 1,000 rebels being killed. In this fight Adjutant (later Lieutenant-Colonel) McBean, an Inverness ploughman who rose from the ranks to the command of the 93rd, gained the V.C. for his intrepid conduct in forcing his way through the breach and killing eleven of the enemy with his own hands. Here the 93rd lost a captain, a lieutenant, and 13 men killed, and 45 men wounded, several dying afterwards of their wounds. Here, also, the dauntless Major Hodson, of Hodson's Horse, met his death. Some of the flying sepoys had taken refuge in the rooms abutting on a narrow lane, and Sergeant Forbes Mitchell, of the 93rd Highlanders, records that Hodson, sabre in hand, demanded of him "Where are the rebels?" The sergeant pointed to the door of a room, and Hodson, shouting "Come on!" heedless of Mitchell's entreaties to wait, made a step forward and was shot through the body.

The Secundrabagh (Alexander's Garden) was again taken, this time with comparatively little opposition, two companies of Highlanders being conspicuous by the alacrity with which, on being faced by a wall, they obeyed Sir Colin Campbell's command to "Tear off the tiles, and in at the roof!" The Imambarrah was next stormed, and Brasyer's Sikhs rushed the Kaiser Bagh, or King's palace; the different smaller points of defence were carried, the mutineers scattered in all directions, and the outworks of the city were soon in the possession of the besiegers. It only remained to take the city itself, and this was effected by a combined effort on the part of Sir James Outram, Sir James Hope Grant and Brigadier Campbell, and after a stiff fight on March 21st Lucknow was again under the British flag. The defeat of the retreating mutineers by Sir Hope Grant about 12 miles out of the city removed all fears for its safety.

Central India.—Between January and June 1858 a number of engagements were fought, for which the clasp inscribed Central India was awarded. The troops under Sir Hugh Rose (afterwards Lord Straithnairn) had cannonaded the fort of Rathghur for a couple of days, and then on January 28th, after a section of the mutineers had made an attack on the rear of the British camp, took the place by storm. Three days after they thrashed a force near Baroda, and then pressed on to the relief of Saugor, where a number of Europeans with their wives and families had been besieged for six months. Sir Hugh's force relieved the fort on February 3rd, 1858. Major-General Whitlock, with the Madras column, had started on the same errand, and in pushing his way onward to the goal had cleared the Jubbulpore district. On March 17th Brigadier Stuart took the fort of Chandairee with the 86th Queen's and 26th Bombay Native Infantry, and then with Sir Hugh Rose proceeded to the investment of Jhansi, where the mutineers had shot down some of their officers in cold blood, and treacherously murdered others. On March 21st the army appeared before the place, but had hardly done so when it was discovered that the remnants of the Gwalior contingent, under Tantia Topi, which had retreated from Cawnpore, and had gathered strength in its march, were advancing upon Jhansi from Kalpee. Sir Hugh Rose gave the enemy no time to think, but turning his troops about charged the rebel hordes with such vigour that it is estimated 2,000 of them were killed. The effect of this punishment so impressed the Ranee, and the garrison of 12,000 men in Jhansi, that the bulk of them fled during the night, and next day, after a considerable amount of opposition, the British troops occupied the place. In the pursuit of the rebels about 1,500 were killed, their guns, ammunition, and baggage falling into the victors' hands. In this affair, in which Lieutenant Leith won the V.C., were engaged the 14th Dragoons (who lost, out of 243, 5 killed and 24 wounded);

MEDAL FOR MATABELELAND, 1893, WITH BAR FOR RHODESIA, 1896.

CAPE OF GOOD HOPE GENERAL SERVICE MEDAL, WITH BAR FOR BECHUANALAND, 1896.

(Reverse.)
QUEEN VICTORIA'S SUDAN MEDAL.

(Obverse.)
KHEDIVE'S SUDAN MEDAL.

207 Native Cavalry; 208 of the 86th Queen's; 226 3rd Bombay Fusiliers; 298 of the 14th, and 400 of the 25th Bombay Native Infantry.

Kotah.—General Roberts, in order to deal with the mutineers who had murdered the Resident, Major Burton, and his two sons, pushed past all opposition in Rajpootana, and in spite of great hardships advanced against Kotah, and on March 30th, under the leadership of the 72nd the troops poured through the Kittenpole Gate, which had been blown up by the Engineers; and after terrific fighting, in which Lieutenant Cameron of the 72nd gained his V.C., the mutineers gave way, and by the evening Kotah, with its 70 guns, was denuded of mutineers. The force engaged included 250 of the 72nd Highlanders; 500 of the 83rd; 250 of the 95th; a like number of the 10th and 13th Bengal Native Infantry; a number of Sappers, and a detachment of the 8th Hussars.

The Highland brigade, after the capture of Lucknow, was engaged in assisting to stamp out the rebellion in the province of Rohilcund, and to that end were employed in Brigadier Walpole's force, which marched from Lucknow with the object of advancing upon the district from one point while Brigadier John Jones advanced from Roorkee. In Walpole's advance he called upon the Rajah in possession of the jungle fort of Rhooyah to surrender the place, and on his refusal ordered four companies of the "Black Watch" to advance without protection or cover against the strongest face of the fort, from which, after gallantly remaining in an exposed position for six hours, they ultimately retired, but not before Brigadier Adrian Jones of the 93rd, Lieutenants Douglas and Bramley of the 42nd, and Lieutenant Willoughby of the Punjabis had been killed, also a sergeant and 6 privates of the 42nd, a lieutenant, 3 sergeants, and 34 privates wounded, in an affair that, but for the want of spirit on the part of the commander, might have been settled in a few minutes by a bayonet charge into the mud fort! Next day the place was found to be deserted, and after burying their dead the British marched in pursuit of the mutineers, whom they caught up and defeated at Allahgunge.

Three V.C.'s won by 42nd.—In this unfortunate affair Quartermaster-Sergeant Simpson of the 42nd gained the V.C. for bravely going back and rescuing an officer and a private who lay dangerously wounded. Private Davis of the 42nd was also awarded the coveted distinction for going up to the very walls of the fort and rescuing the body of Lieutenant Bramley. Lance-Corporal Alex Thompson of the 42nd also received the Cross for assisting Major-General W. M. Cafe to bear away, under a heavy fire, the body of Lieutenant Willoughby, and going to the rescue of Private Spence, who had been mortally wounded in the same effort.

The Occupation of Bareilly.—The advance on Bareilly was begun on April 28th, and on May 5th the army came up with the rebels on the plains to the east of the fort, where after a fight lasting four hours, the mutineers were compelled to retreat, and the city of Bareilly was occupied with little opposition. In the contest on the plains the "Black Watch" particularly distinguished itself when a body of Ghazis, Mussulman fanatics, charging down with mad fury upon the column forced the 4th Punjabis back on the 42nd. The charge of the Ghazis was described by Sir Colin Campbell as "the most determined effort he had seen during the war," but it was coolly met by the Scots, and in the hand-to-hand fight which ensued between them and the Ghazis the latter were speedily exterminated. During this bloody episode Sergeant Gardiner of the 42nd gained the V.C. by killing two fanatics with his bayonet, and saving Colonel Cameron, who was dragged from his horse by four of them, and was in danger of being killed. In this struggle the famous *Times* correspondent, W. H. Russell, would have been killed but for the timely intervention of Sergeant Forbes-Mitchell of the 93rd, who shot the attacking rebel. The back of the mutiny being broken the Highland brigade was ordered to encamp at Bareilly, where the last serious fight had been made by the mutineers. Parties of them, however, continued to give trouble for some time, and in a sharp encounter at Maylah Ghaut, on the banks of the Sarda, where on January 15th, 1859, Captain Lawson and 37 men of the "Black Watch" kept at bay 2,000 rebels from sunrise to sunset, and still further added to the roll of honour of the 42nd by Privates W. Cook and D. Millar earning the V.C. by going to the front and directing their company when their officers had been shot down.

The mutiny was now crushed, thanks to the bravery and endurance of the European soldiers, and the staunch conduct of those Indian troops who, with their Princes, remained true to Britain. The gallant Sir Colin Campbell at last received something like due reward for his services to his country by being raised to the peerage as Lord Clyde. The Honourable East India Company ceasing to exist, the Government of India was transferred to the Crown by the India Act of 1858, and the government of the country vested in a Viceroy and Council, the last Governor-General under the old order becoming the first Viceroy of our Indian Empire.

The following British regiments took part in the war: 5th, 8th, 10th, 13th, 23rd, 32nd, 34th, 38th, 42nd, 52nd, 53rd, 60th, 61st, 64th, 71st, 72nd, 75th, 78th, 79th, 80th, 82nd, 83rd, 84th, 86th, 88th, 90th, 93rd, 95th, 97th, 101st Royal Bengal Fusiliers (now 1st Royal Munster Fusiliers), 102nd Royal Madras (now 1st Royal Dublin Fusiliers), 104th Bengal Fusiliers (now 2nd Royal Dublin Fusiliers), 108th Madras (now 2nd Royal Inniskilling Fusiliers; Rifle Brigade; 7th, 8th, and 14th Hussars; 9th and 12th Lancers.

The medal was also granted to a naval brigade from the "Pearl" and "Shannon" (see Naval Section).

The Mutiny Medal.—This medal ($1\frac{2}{5}$ in. in diameter) was granted by a General Order of the Indian Government to those in the service of Her Majesty, and of the H.E.I. Co., the men who composed the naval brigade, and to all persons not in the military service who were employed in the suppression of the mutiny. By a General Order, May 19th, 1859, a clasp for DELHI was granted to those employed in the operations against and at the assault of Delhi from May 30th to September 14th, 1857. A clasp for the DEFENCE OF LUCKNOW was granted to all the original garrison under Major-General Sir John Inglis (who had succeeded to the command after the death of Sir Henry Lawrence) and to those who succoured them and continued the defence under Major-Generals Havelock and Outram, until relieved by Sir Colin Campbell, from June 29th to September 25th, 1857. A clasp for the RELIEF OF LUCKNOW was awarded to those who, under Havelock, had for nine weeks striven to succour the beleaguered garrison. Outram joined him on September 16th, and on the 25th they clambered through the battered gun embrasure beside the Baillie Guard at Lucknow. The members of the force under Generals Havelock and Outram were awarded this clasp. A clasp for LUCKNOW was granted to all the troops engaged in the several operations against Lucknow under the immediate command of Sir Colin Campbell in November 1857 and March 1858. A clasp for CENTRAL INDIA was granted to the troops of the column under Major-General Sir Hugh Rose engaged in the operations against Jhansi, Kalpee, and Gwalior, and also to the troops under the command of Major-Generals Roberts and Whitlock. Subsequently, by a general order, May 3rd, 1860, the troops employed at and prior to the battle of Banda and siege of Kotah were held entitled to the clasp for Central India, but those of either of the columns commanded by Major-Generals Roberts and Whitlock, who were employed in the subsequent minor engagements, were considered to have a claim to the medal only. On August 21st, 1860, it was notified that the few officers and soldiers who

KHEDIVE'S SUDAN MEDAL.
(Reverse.)

QUEEN VICTORIA'S SUDAN
MEDAL. (Obverse.)

QUEEN'S MEDAL FOR BOER
WAR. (Obverse.)

KING'S MEDAL FOR BOER
WAR. (Reverse.)

may have been present in the actions in which the Malwa Force was engaged under Brigadier Stuart, but who, from the consequence of wounds received in action or other causes, were prevented from accompanying the force on its junction with Sir Hugh Rose, were entitled to receive the bar for Central India. A General Order, dated January 10th, 1860, stated " that a clasp for the Relief of Lucknow was granted to the troops engaged in the operations against that city under the immediate command of Lord Clyde in November, 1857."

It is noteworthy that the infantry did not receive more than two bars, and that it was impossible for any participant in the war to obtain more than four, and very few received as many. This is a very striking medal, a very dignified figure of Britannia being represented on the reverse holding a wreath in the outstretched right hand, and an oval shield with the Union Jack thereon on the left arm, which also holds a laurel wreath; INDIA is arranged above following the line of the medal, and 1857–58 in the exergue. The obverse bears the same head of Queen Victoria as the Military General Service medal. The names were impressed on the edge in capital Roman letters, and the suspender and fish-tail bars are of the same kind as those used with the China medal of 1857. The ribbon, which is 1¼ in. wide, consists of scarlet and white stripes arranged alternately, three of white and two of red. This was the last medal issued by the H.E.I. Co., and it might be said almost the last of the classic medals (L. C. Wyon modelled the reverse).

Reissues of Earlier Medals.—On January 21st, 1859, the Government of India sanctioned the reissue of medals gratis to all officers and men who had lost their decorations through the mutiny. This is an important fact to remember when considering war medals issued prior to the mutiny, as reissues are not generally considered as valuable as originals.

SECOND CHINESE WAR

Fatshan, 1857.—The seizure by the Chinese of a small trading-vessel, the "Arrow," led to the second Chinese War, the first stages of which were confined to operations by seamen and marines under Admiral Sir Michael Seymour, who, attacking the Chinese fleet in Fatshan Creek with 11 gunboats and about 50 ships' boats from the fleet, defeated the enemy's fleet of 80 junks armed with 800 guns, and manned by 6,000 seamen; only 3 of the junks escaped destruction.

Canton, 1857.—Having defeated the fleet, it was considered that there was little chance of obtaining satisfaction without resort to arms, and a force of 5,000 men was dispatched to the Far East, but the mutiny breaking out they were diverted to India. The 59th Regiment, however, was on the China station, and with a body of marines took part in the bombardment and storm of Canton on December 29th, 1857, and the blockade of the fort until some semblance of satisfaction had been obtained. With the deportation of the late Imperial Commissioner Yeh to India the possibility of a satisfactory treaty was imminent, and when the treaty of Tientsin was signed, on June 26th, 1858, it was hoped that peace was assured.

Taku Forts, 1860.—The signature of a treaty was one thing, the ratification another; and when the British Envoy was proceeding up the Peiho River, in order to obtain the ratification, the forts at the mouth of the river opened fire upon the ships accompanying him, and they had to retire. It was therefore determined to achieve by force of arms what diplomacy had failed to do. An army of 10,000 British troops, and a force of 7,000 French, were organised to take the offensive against the Celestials. The British were under the command of Lieutenant-General Sir Hope Grant, and the French under General de Montaubon, a "beau sabreur." At the end of July 1860 the British and French fleets were ready, and on August 1st the allied troops were disembarked a couple of thousand yards from the Pehtang Forts in the Gulf of Pecheli, and they marched towards Sin-ho, where a battle was fought in which the new Armstrong guns were used in warfare for the first time. In this battle the allies outnumbered the enemy by two to one, and their steadfastness in the face of the most modern artillery and seasoned troops called forth the praise of General Sir Robert Napier, who stated that the enemy "had behaved with courageous endurance." After another engagement on the following day, and the rout of the enemy, the way was clear to attack the Taku Forts. The northern forts were assailed on land and bombarded from the sea on August 21st; the assault was made difficult by an arrangement of spiked bamboo stakes that had been planted over a space 20 feet wide, and while the allies were gingerly creeping over it missiles of all kinds were hurled at them. Ultimately the walls were reached, and a daring French soldier planted the tricolour on the parapet, only to fall dead with the cry of victory on his lips. Lieutenant Rogers of the 44th and Ensign Chaplin of the 67th were more fortunate, and the Queen's colours were firmly fixed over the fort. In their retreat from the first northern fort the unfortunate Chinamen had to cross the ground they had so ingeniously staked out as an obstacle for their enemy, and were spitted on their own stakes. The fall of the first fort advised the commander of the second fort that opposition was useless, and he hoisted a white flag; later the troops took possession of it, and then the southern forts were likewise surrendered, and the capture of the Taku Forts effected with the loss to the British force of 67 rank and file killed, and 22 officers and 161 men wounded; and to the French of 30 killed, and about 100 wounded. The Chinese are said to have had at least 2,000 killed.

Pekin, 1860.—The Buffs were left to garrison Taku, and the 44th Shanghai, which was threatened by the Taiping rebels, while the army marched on to Tientsin, from which on September 8th the 1st Division and about half the French force set out on the advance to Pekin. At Changdia-wan there was a sharp skirmish, and the enemy abandoned their entrenchments with 74 guns, and offered to sign a convention. Mr. (afterwards Sir Harry) Parkes, Lord Elgin's secretary, together with four officers and an escort, set out to arrange preliminaries, but were treacherously taken prisoners, and all the officials, with the exception of Mr. Parkes, barbarously murdered. Sir Hope Grant then prepared to meet the Chinese attack, and in the action which was fought at Chow-ho the Chinese were put to rout, and a great number cut down in the pursuit by the Dragoons and Probyn's Horse. The allies then advanced on Pekin, and the French army occupied the Summer Palace of Yuan-ming-Yuan. At the Bridge of Palichaio a last desperate effort was made by the Chinese Imperial Guard, but the vigorous French attack drove them back with the loss of 25 guns, and gained for General Montauban the title of Comte de Paliko.

On October 7th the enemy was informed that unless the prisoners were restored and one of the gates of the city given up to the allies, the city would be stormed. The Chinese agreed to this, and the Au-ting gate of the city was occupied without opposition by 200 British and French troops on October 13th. It was found, however, that twenty of Mr. Parkes' party had succumbed to the terrible treatment of their Chinese captors, and Lord Elgin having given his word to spare the city if it were surrendered, ordered the Summer Palace, where the atrocities had been committed, to be razed to the ground. The Summer Palace consisted of thirty buildings in grounds extending for about 6 miles, and although the troops set to work with a will to burn—and to plunder—it took two days to destroy the buildings by fire. Of the loot of the Summer Palace at Pekin many fabulous stories have been told, but there can be no doubt that in this wilful destruction

BOER WAR MEDALS FOR COMPARISON. THE BARS ON CENTRE MEDAL ARE UNOFFICIAL.

of the palace, and the indiscriminate dispersal of its valued contents, gems of art and wonderful pieces of jewellery and goldsmith's work were annexed by men who had not the faintest notion of their exceeding value. It is said that a French officer found a string of beautiful pearls, each about the size of a small marble, and sold it for £3,000! The loot taken by the British troops, such as was given up, was sold at public auction by order of Sir Hope Grant, and it is recorded the share of each private soldier was not less than £4 sterling. The city of Pekin was then occupied by the allies, and, the Chinese having agreed to cede the island of Kowloon, opposite Hong Kong, to Britain, to make Tientsin an open port, and to pay an indemnity of £100,000 to the relatives of the murdered men, the war was concluded.

The following regiments were engaged: 1st, 2nd, 1st Batt. 3rd, 31st, 44th, 59th, 60th, 67th, 99th; Royal Artillery; Sappers and Miners, and two squadrons 1st Dragoon Guards; 11th and 19th Bengal Lancers; 20th and 23rd Bengal Cavalry.

The China Medal, 1857–60.—This medal was granted on February 28th, 1861, to commemorate the successes of Her Majesty's British and Indian forces, both naval and military, employed in the operations in China which terminated in the capture of Canton on December 29th, 1857, and in the operations which terminated in the capture of the city of Pekin. The bars issued with the medal were, CHINA 1842 to those who had been in the first war; CANTON 1857, TAKU FORTS 1858, TAKU FORTS 1860, PEKIN 1860, and FATSHAN 1857 to seamen and marines only. The medal, which is the same size and design as that issued for 1842, but with the date left out on the reverse, is affixed to an ornamental clasp for suspension; the ribbon is also the same, and the names of the recipients, their regiments and rank, are impressed on the edge in Roman capitals. The medals given to the navy were mostly issued unnamed, with the exception of those awarded to the

Indian navy, and sometimes those to the marines. The suspender and bars of this medal are of the same pattern as those used with the "Mutiny Medal." A bar, or bars only, was granted for the second China War, 1857–60, to be affixed to the medal of those who had also served in the first China War, 1842, but such specimens are of the greatest rarity.

The medal given with the two bars TAKU FORTS 1860 and PEKIN 1860 to the 1st Dragoon Guards is rare, as only two squadrons of the regiment were present, no other British cavalry being engaged, and because of the brilliant action of the men on September 21st in riding at "a very ugly place," and scattering the Tartar cavalry. Lord Cheylesmore has in his collection the only five-bar medal issued for the war; it was awarded to a gunner in the Royal Marine Artillery, and has the bars for FATSHAN 1857, CANTON 1857, TAKU FORTS 1858, TAKU FORTS 1860, and PEKIN 1860.

NEW ZEALAND, 1860–6

War again broke out in New Zealand in 1860, and so serious did it become that at one time as many as 10,000 regulars and 15,000 colonials were under arms. At Taranaki, on June 13th, the Maoris were defeated, and again at Mahoetaki on November 6th. In order to protect New Plymouth, or the Taranaki district, a chain of redoubts was built, and from one of these, on the Waitara River, a little party of men of the 40th Regiment moved out to locate the Maories known to be in the bush. Directly they entered it a number were killed and wounded by the fire of the hidden enemy, and in a very short time hardly a man was unhurt. In this unfortunate affair, on March 18th, 1861, Sergeant Lucas gained the V.C. for his heroic conduct in assisting Lieutenant Rees, who had been wounded, into cover, and, although himself wounded, remaining at his post under a galling fire. Next day the Maoris surrendered, and for a short spell peace was restored.

In 1863 war again broke out. The Maoris in the North Island had elected a young, characterless man as King, and established a capital at Ngaruawahia, near the junction of the Waikoto and Waipa Rivers. British law was defied, and when the supporters of the weak and pliant "King" had interfered in a dispute affecting Europeans and Maoris, it became necessary to take strong measures. Lieutenant-General Sir Duncan A. Cameron, learning that the Waikoto tribe had planned to invade Auckland, advanced into the wild country north of the Waikoto River, where at Kohera, above the Mangatawari Creek, the rebels had taken up a strongly fortified position. Attacking them on July 12th, 1863, Cameron drove them out and into the swamps, but owing to difficulties in transport was unable to follow up his success. An instance of the chivalry which characterised the brave Maori warriors throughout the war may here be mentioned; the Maoris at Meri-Meri, hearing a rumour that Cameron and his men were short of food, sent a little fleet of canoes under a flag of truce with potatoes and milch goats for their enemy. When he was able to move, Cameron set out to attack Meri-Meri, but on his arrival he discovered that the Maoris had retired southwards across country, whither, owing to the condition of the land caused by the heavy rains, they could not be followed. Cameron pushed on up the river, and at Rangireri attacked the strongly fortified Pah on November 20th; but owing to the assault being made before the artillery had completed its work, the troops, after their repeated assaults, were only partially successful, and during which 6 officers were killed and 9 wounded, and 120 men killed and wounded. At night a number of the enemy escaped from the Pah, and the remnant of 183 surrendered at daybreak.

Four V.C.'s in 1863.—In this fight about 50 artillerymen, armed with swords and revolvers, led by Colonel

Mercer, made a brave effort to enter the Pah, but the Colonel was mortally wounded and the gunners had to retire: brave efforts were made by some of his men to rescue him, and Surgeon Temple of the artillery well earned his V.C. for dressing the dying man's wound while the bullets showered around him. Prior to this, Sergeant McKenna of the 65th had gained the V.C. for his coolness and excellent handling of a small detachment of men after the officers had been mortally wounded near Cameron Town on September 8th, and Ensign Donn and Drummer Stagpoole of the 57th had won theirs by a very gallant action in bringing in a wounded man under a heavy fire at Pontoko, on October 2nd. On December 9th General Cameron occupied Ngaruawahia and the Maori "King" surrendered.

The "Gate Pah."—On February 21st, 1864, Rangiawahia was taken, and on March 31st Orakau was invested by a force under Major-General Carey, and captured on April 2nd. Sixteen officers and men were killed and 52 wounded in this engagement. The Maoris then fortified Tauranga, which, on April 29th, after the artillery had played on the place for several hours, a party of the 43rd and about 150 seamen and marines stormed, and thinking the enemy had deserted the Pah, took little precautions for their own safety. The Maoris, however, breaking from the inner entrenchments, mortally wounded the officers, and the storming party, which had entered the "Gate Pah" with ringing cheers, poured out again in confusion. The Pah was evacuated during the night. The "Fighting 43rd" lost 7 officers killed and mortally wounded, and 97 men killed or wounded, the naval brigade losing its Commander, also 4 officers and 40 seamen and marines killed or wounded. In the assault on the Gate Pah Samuel Mitchell, Captain of the Foretop of H.M.S. "Harrier," gained the Victoria Cross for refusing to leave Commander Hay when ordered by him to do so, and, under a shower of bullets, carrying him in his arms out of the Pah. Surgeon

Distinguished Service Order. | Albert Medal (3rd Class). | Jubilee Medal. | Zulu War Medal. | Queen's South Africa. | Royal Humane Society's Medal.

DECORATIONS AWARDED TO THE LATE MAJOR EDWARD DENMAN THORNBURGH CROOPER, D.S.O.

Mauley of the R.A. also earned the V.C. for his gallantry in succouring the dying Commander and the wounded generally under the most dangerous circumstances.

On June 21st the 43rd and a detachment of the 68th marched against the Maoris at a Pah they had started to fortify at Te Ranga. After a short but stubborn fight the enemy was worsted and routed, among the killed being the chief, Rawhiwi, who had led the Maoris at the Gate Pah. In this affair Sergeant J. Murray of the 68th gained the V.C. for his bravery in charging one of the enemy's rifle pits and, single-handed, killing or wounding the eight or ten men who occupied it.

V.C. and three D.C.M.'S.—On January 25th, 1865, a daring attack was made on the British camp at Nukumaru, but was repulsed with great loss. At this place, on the 24th, Captain Shaw of the 18th Royal Irish gallantly won his V.C., and Privates John Brandon, George Clampitt, and James Kearns Distinguished Conduct Medals for the daring rescue of a comrade who had been wounded within about 30 yards of the Maori position.

After the repulse at Nukumaru the Maoris retired upon Wereros, a Pah considered by them to be impregnable and apparently also by General Cameron, for he moved slowly up the coast; and it was not until July 20th, 1865, that the place was attacked, and then under orders from Sir George Grey, the Governor of the Colony. One hundred of the 14th and a like number of the 18th Royal Irish were deputed to threaten an attack on the front of the position, while the 470 colonials and friendly Maoris worked round the rear. This strategy disconcerted the enemy and they abandoned their fortress without firing a shot. During the rest of 1865 and 1866 the fighting was restricted to the West Coast.

In January 1866 an expedition was organised against the Hau-Haus, a body of fanatics whose religion was a strange mixture of cannibalism, paganism, and Christianity. The force consisted of 139 of the 14th, 100 men of the 18th Royal Irish and a like number of the 50th, 45 Forest Rangers, and 300 native auxiliaries under Major Rocke of the Royal Irish. This little army captured the palisaded village of Otahuhu and the Putahi Pah, a fort in the top of a hill 500 feet high covered with dense and primeval jungle. On the night of October 17th Papoia, a native village in the depths of the primitive forest, was captured, and the Hau-Haus no longer offered serious resistance. This was the last serious engagement in which the regular troops took part, and for the rest of the campaign, which terminated in 1869, the Colonials were engaged under the direction of Colonel (afterwards Sir George) Whitmore.

The New Zealand Medal.—This, the second in the chronological order of the series facing page 188, was granted to the troops and seamen who had taken part in the Maori Wars of 1845–6–7 and 1860–1–2–3–4–5–6. The medal, however, was not authorised by a general order until 1869, and there are twenty-two different dates for the army and ten for the navy in raised letters in the centre of a laurel wreath, above which, following the outline of the medal, is "NEW ZEALAND" and below "VIRTUTIS HONOR." On the obverse is the bust of Queen Victoria with a coronet, in which fleur-de-lis stand out prominently, holding a veil which falls upon the shoulders, and the legend VICTORIA D. G. BRITT. REG. F. D. The suspender is fashioned after the manner of a fern frond. This type is only seen with this medal, which is $1\frac{2}{5}$ in. in diameter. No bars were issued, but the dates of service were struck as stated above, for example: 1845–7, 1860, or 1860–6; the rank, name, and ship or regiment being impressed on the edge. The medallists were Joseph S. and Alfred B. Wyon, but their names are not always found on truncation of the bust on the obverse of the medals. The ribbon, $1\frac{1}{4}$ in. wide, is dark blue with dark-red centre of equal width to the blue which borders it.

North-West Frontier.—Between the end of 1849

and the beginning of 1868 no less than seventeen expeditions had been sent to deal with the turbulent inhabitants on the North-West Frontier of India, but the troops which took part in the operations were not granted a medal until 1869, when by a general order dated July 1st it was decided to issue a medal for the services rendered, and then the India General Service Medal, with a bar inscribed NORTH WEST FRONTIER, was given to those who had taken part in fifteen of the expeditions enumerated below; to those who already possessed the medal the bar was added. The names and the regiments were impressed on the edge of the medal in Roman capitals.

The following regiments took part in the various expeditions. Regiments present, 1849: 60th and 61st. 1850: 60th, 61st, and 99th. 1852: 32nd, 53rd, and a battery of Royal Horse Artillery. 1853: 22nd, Guides, 66th Goorkas. 1857: 81st and 98th. 1863: bar for UMBEYLA—7th and 101st Fusiliers; 71st and 93rd Highlanders; half a battery of Royal Artillery and a Mountain Mule battery; the Guides, both Foot and Horse; a squadron of Probyn's Horse; 4th and 5th Goorkas; 3rd and 14th Sikhs; 1st, 3rd, 5th, 6th, and 20th Punjab Infantry; 23rd and 32nd Punjab Pioneers and a company of Native Sappers. End of December 1863 to January 1864: 79th Highlanders; 3rd Batt. Rifle Brigade; "D" Battery 5th Brigade Royal Horse Artillery; 7th Hussars and Native Corps.

Umbeyla.—At the end of 1863 a somewhat serious trouble again broke out on the borders of the Aklumd of Swat by the fanatical outbreak of Hindustanis, and a very arduous campaign had to be undertaken between October and December before the district settled down. This little war in Umbeyla (Ambela) was responsible for the death of a great number of officers; first the Buner people refused to allow Sir Neville Chamberlain's army to pass, and the commanding points on the mountains were repeatedly taken and retaken, despite the bravery of the Punjabis and the general gallantry of the troops. Around the Eagle's Nest and the Craig Piquet the most desperate fighting took place, and at the latter post Lieutenants Fosbery and Pitcher won the V.C. for their daring, but to little avail until the force was increased to 9,000 men, and then through strenuous opposition it pushed on, and the Hussanzaie tribes were dispersed. In this expedition 36 British and 31 native officers were killed, and 150 British and 689 native soldiers killed or wounded. A bar for this expedition was not authorised until July 1st, 1869.

Bhootan, 1864–6.—The Umbeyla campaign was but the precursor of another, for that restless spirit which seems to thrive so well on the frontier engendered another outbreak, and disturbances among the Bhootanese in 1864 necessitated the sending of an expedition of four columns to invade Bhootan, a district to the north-east of India. The left column met with little opposition at Dhalimcote, where the fort was bombarded and stormed, or at Bhumsong, where the stockaded positions were evacuated without a shot being fired by the enemy, although they made something of a stand at Chamoorchee. Buxa was taken by one column, and while another pressing on to the Barungah Pass found small opposition, the other reached Bisheusing unmolested. The annexation of Bhootan was then formally declared. In 1865, however, the inhabitants endeavoured to free themselves from British dominion, and rather determined fighting ensued; at the various places already mentioned attacks were made upon the garrisons, and it became evident that stronger force was necessary to hold the country. Reinforcements were hurried forward under Brigadier-General Tombs, C.B., V.C., but after the defeat of the Bhootanese at Dewangari, a hill post at the summit of the Dungarah Pass, hostilities came to an end, and with a stronger permanent military establishment peace was ensured, but not effectually until in 1866 a force of 7,000 men was sent into the country. This determined attitude

Naval General Service. New Zealand. Ashantee. Afghanistan. I.G.S., 1895. Sudan, 1899. South Africa, 1900.

SEVEN PHASES OF QUEEN VICTORIA'S LIFE AS REPRESENTED ON THE OBVERSE OF HER WAR MEDALS.

on the part of the Indian Government resulted in the Deb Rajah of Bhootan signing a treaty of peace.

The following regiments were engaged in the campaign, for which a bar was granted on April 28th, 1870: 55th and 80th Foot; 2 batteries of Royal Artillery; 2nd Goorkas; 11th, 12th, 18th, 43rd, and 44th Bengal Infantry; 5th and 14th Bengal Cavalry; 19th, 29th, and 30th Punjab Infantry.

Looshai.—The introduction of tea-planting into Assam, and the consequent immigration of Europeans, encouraged the predatory hill tribes of the vicinity to extend their depredations, and the Looshai expedition had to be organised to punish the tribes which had been making incursions into the fertile districts at the foot of the mountains. Two columns were sent into the district. One under General Bouchier, composed of native troops with artillery sappers, advanced from Cachar, and so harassed the Looshais that they sued for peace, but it was hardly agreed to before they attacked the British column and wounded the commander. The right column, which had set out from Chittagong, joining hands with the left, assailed the stronghold of Lungvel, and taking it by storm burnt it. The payment of an indemnity was then agreed to, likewise the residence of Government agents in the Looshai villages.

The troops engaged in the operations, which extended from December 1871 to February 1872, were native troops only: 2nd and 4th Goorkas; 22nd and 27th Punjab Infantry; 42nd and 44th Assam Infantry; a battery of Peshawar Mountain Artillery; two companies of Sappers and Miners, and 100 Native Police. The troops were under the command of British officers.

ABYSSINIAN WAR

In January 1864 Theodore "The Negus," or Emperor of Abyssinia, fancying himself insulted by the British Government, imprisoned the British Consul,

Captain Cameron, together with the Europeans resident in his territory. Mr. Hormuzd Rassam, with Lieutenant Prideaux and Dr. Blane, was sent by the British Government to obtain the release of the prisoners, but they also were incarcerated, and it ultimately became necessary to send an expedition to secure the release of the prisoners. The Resident at Aden, Colonel Merewether, commanding a reconnoitring party, landed at Mulkutto in Annesley Bay, where he established a base, and, entering into friendly relationship with the local chiefs and tribesmen, prepared the way for the army which under Sir Robert Napier was to march on Magdala and effect the release of the captives.

The force at the command of Sir Robert was about 4,000 British and 8,000 Indian troops, with a transport service and camp followers numbering about 14,000. How splendidly this mass of men was organised, and led with remarkable rapidity across a most difficult country, was a military feat reflecting the greatest credit upon the Indian veteran, who was rewarded with the peerage of Magdala. With this intrepid warrior they had, on scant rations, and each soldier carrying accoutrements, etc., weighing 55 lb., scaled "mountains and descended precipices . . . traversed along the face of deep ravines, where a false step was death!" scorched by the intense sun by day, and chilled to the bone by the cold night air. But the valley of Bachelo was reached; the three brigades formed into one, and then the fighting commenced, when, after descending 3,800 feet, the army attempted to ascend the other side of the ravine, but the 4th with their Sniders—in use for the first time in warfare—and the Rocket Battery of the Naval Brigade, did excellent service; and one body of the enemy being disposed of, the other, amounting to about 5,000 men who made a rush at those in charge of the convoy, was practically decimated by the fire from the battery under Colonel Penn and the Sniders of the infantry. It is said that in one hour each man had discharged ninety rounds of ammunition! Then the guns and rockets were directed upon Magdala, the capital standing 9,000 feet above sea-level. Theodore made overtures for peace, released the captives, and sent a present of sheep and cattle in token of submission; but the fortress was stormed on Easter Monday, April 13th, 1867, the 33rd with the Engineers leading the way, with the 45th in support. The fortress taken, it was discovered that Theodore—deserted by his men—had committed suicide. The fortifications having been destroyed and the huts burned, the troops returned to Mulkutto, and before the end of June 1867 the British force had left Africa.

The following troops were engaged: 4th, 26th, 33rd, 45th; Royal Artillery and Engineers; 3rd Dragoon Guards, and the following Indian regiments: 1st, 2nd, 3rd, 10th, 18th, 21st, 25th, and 27th Bombay Infantry; Bombay and Madras Sappers and Miners; 3rd Bombay Cavalry; 10th and 12th Bengal Cavalry; Scinde Horse and a Naval Brigade (see Naval Section).

Abyssinia Medal.—By a general order dated March 1st, 1869, a medal was granted to all Her Majesty's British and Indian forces, naval and military, who took part in the operations in Abyssinia under Sir Robert Napier between October 4th, 1867, and April 19th, 1868. This is a very distinctive silver medal, being considerably smaller than the others issued. On the obverse, within a beaded circle, is the bust of Queen Victoria facing left, crowned and veiled, encircled by a star of nine points with the letters of the word ABYSSINIA between the angles. The portrait of the Queen is similar to that on the New Zealand medal. On the reverse, within a beaded circle surrounded by a wreath of laurel tied at the base, the recipient's name, rank, and regiment or ship are stamped in relief in the centre. The Indian troops had their names and regiments generally engraved, but I have several in my possession with the names stamped in relief, and also in intaglio. The medal is suspended from a silver swivel ring attached to a crown soldered on to the top of the medal, which is 1¼ in. in diameter, by means of 1½ in. wide ribbon of red, with broad white borders.

ASHANTEE, 1873–4

There had been troubles on the Hinterland and in the West Coast settlements as far back as 1824 and 1826, and again in 1863; but the more serious outbreak of hostilities occurred in 1873, when the Ashantees crossed the River Prah, and, attacking the Assims and Fantees, opened a road for themselves to Elmina and Cape Coast Castle, and in company with the Elminas attempted to capture Fort Elmina, but were beaten off. A force of 20,000 Ashantees, however, sat down before Mampon, 10 miles from the British forts, and there they remained until the advent of Sir Garnet (afterwards Viscount) Wolseley in October 1874, when they abandoned their camp, and fell back behind the Prah. A small force was sent to Elmina to disperse the natives, which it did, while the main force, composed of the 23rd Royal Welsh Fusiliers, the 42nd Black Watch, and 2nd 60th Rifles, pressed forward over the Prah—which the Ashantees asserted no white man could cross—over the Adansi Hills, and across the Bahrein River on to Amoaful, where 20,000 Ashantees under General Amanquatia stubbornly fought for five hours before giving way, and then desultory fighting continued for at least six hours more before the enemy actually retreated, having lost their General and considerably over 2,000 men. In this stubborn fight the 42nd, which had led the attack, and repeatedly charged the enemy, lost Major Baird, who was mortally wounded, and had 7 officers and 104 men placed *hors de combat* by wounds.

Hewing their way through the dense bush, constantly assailed on all sides by the natives, the army pushed on in three columns, and after passing the village of Egginassie encountered considerable opposition, and

BRONZE STAR FOR ASHANTI, 1896.

INDIA 1895 MEDAL FOR THREE CAMPAIGNS.

MAYOR'S STAR FOR DEFENCE OF KIMBERLEY.

later stormed the village of Bequah, where Lord Gifford earned his Victoria Cross. In effecting the bridging of the Ordah, severe skirmishing took place, and at Ordah-su a lengthy battle was fought, the village being ultimately carried and held by the Rifles. In this battle the ill-fated Wauchope, who later met his death at Magersfontein, was severely wounded when serving with the 42nd. Forcing its way onward, with the pipers of the Black Watch playing at its head, the army entered Commassie practically unopposed, and the reign of the barbarous King Koffi Calcali came to an end, a reign in which "murder, pure and simple, monotonous massacre of bound men," was the "one employment of the King, and the one spectacle of the populace."

In the march on Commassie 80 European officers and men were killed, and 261 officers and men wounded, but the number of deaths from disease was very great. The grim spectre of disease is always more to be feared than the bullets of an enemy. In this war Captain (afterwards Lieutenant-Colonel) Reginald Sartorius, who had accompanied the column from Accra in the advance on Commassie, but did not reach there until it was abandoned, earned the V.C. for a daring ride of 55 miles across country in order to communicate with the General Commanding, and at Abogov, under a heavy fire, carrying a wounded Houssa soldier into cover. For his services Sir Garnet Wolseley was offered a baronetcy, which he declined, but accepted a grant of £25,000.

The troops engaged were the 23rd Royal Welsh Fusiliers; 42nd Black Watch; 2nd Batt. 60th Rifles; 150 Royal Engineers; 2nd West India Regiment; Houssa Artillery and a Naval Brigade.

The Ashantee Medal.—By a general order dated June 1st, 1874, Her Majesty granted a medal to her naval and military forces who had been employed on the Gold Coast under Sir Garnet Wolseley during the operations against the King of Ashantee from June 9th, 1873, to February 4th, 1874, with a bar inscribed

COMMASSIE to those who were present at Amoaful, and the actions between that place and Commassie, including its capture, and to those who during the five days of those actions were engaged on the north of the Prah in maintaining and protecting the communications of the main army. This medal has since been given for a number of expeditions in Central Africa, and on the East and West Coast. The recipient's name and regiment, or ship, and the date 1873–4, are engraved in square Roman capitals on the edge of the medal, and blacked in for the campaign 1873–4, but after that date the names are found impressed as well as engraved. On the obverse is the diademed head of Queen Victoria facing left, with plain veil falling over the back of the crown behind, surrounded by the legend VICTORIA REGINA. On the truncation L. C. WYON. The reverse, by Sir Edward Poynter, P.R.A., represents British troops fighting natives in the bush, and while it may be alleged to pictorially represent something of the difficulties met with by the troops—which I do not consider it does—it cannot be said to fulfil the requirements of a decoration for personal wear. The medal, which is $1\frac{2}{5}$ in. in diameter, with a straight clasp and claw clip arrangement, fastened by a rivet through the medal, is suspended from a ribbon $1\frac{1}{4}$ in. wide, yellow with black borders, and two narrow black stripes. The bars are plain, with a raised straight border.

Perak.—Perak, on the western side of the Malay Peninsula, had been independent of Siam since 1882, but the sovereignty of the State had been in dispute. In order to put an end to the constant feuds, the offer of Ismail, one of the claimants, to surrender the sovereignty to Britain was accepted, and Mr. Birch was appointed British Resident in Perak city. Ismail a few months later, desirous of securing the throne, attacked the Residency and murdered the Resident. The neighbouring Rajahs joined with him, and it became necessary to dispatch a punitive expedition. A party of troops and a Naval Brigade therefore proceeded up the Perak River, destroyed the village of Passir Sala, and captured 6 guns, while General Colbourne's column marched through the jungle, so "dense and dark that during all the time not a vestige of sun or sky was visible overhead," and during the advance (on Kintra) the troops—1,300 men of the 80th and a naval brigade from the "Modeste" and "Ringdove" —"were without cover of any kind, and slept in the damp dewy open." On the capture of Kintra, the Maharajah Lela and Ismail fled into the jungle; meantime Brigadier-General Ross, with a detachment of the Buffs and some Goorkas, stormed Kotah Lama and burned it. On January 19th Ismail was decisively worsted, and on March 22nd he was captured and sent to Singapore, and the districts which had been the scene of constant conflict became, under British rule, prosperous and contented.

The troops engaged were: 1st Batt. 3rd; detachments of the 10th and 80th; Royal Artillery and Engineers; a Naval Brigade, and detachments of Indian troops.

An issue of the India General Service Medal 1854 was authorised with a bar for PERAK on September 1st, 1879, for the campaign from November 2nd to March 20th, 1875-6, and by a general order issued in 1881 to those who had been engaged between November 27th and December 10th, 1874.

Jowaki.—Living on the hills of the Afghan frontier, the Jowakis, a branch of the Afridis, long had a bad reputation as marauders, disturbing, robbing, and slaughtering the inhabitants of the Peshawar Valley; finally they attacked a British outpost, and it became necessary to teach them a serious lesson. A force composed of the 9th and 51st Regiments, and two companies of the 4th Rifle Brigade, with Royal Engineers, Artillery, and Indian regiments, was therefore sent against the hillmen under the command of Generals Keyes and Ross. The Jowaki villages, fortified Sangars and crops, were destroyed, and on December 1st, 1877, the stronghold of Jummoo was taken with little opposition. By the end of January 1878 the Jowakis sued for peace; unwilling, however, to accept the terms imposed, the hostilities were resumed, but after a crushing charge by the Bengal Cavalry, on February 15th, their spirit was damped, and they unconditionally accepted the terms laid down for peace.

By another order dated March 1st, 1879, a medal with bar for JOWAKI was awarded to the troops who took part in the operations between November 9th, 1877, and January 19th, 1878.

The troops engaged were the 9th and 51st; two companies 4th Batt. 60th; two companies Royal Engineers; a battery of Field and a battery of Royal Horse Artillery; 20th, 22nd, and 27th Punjab Infantry; 14th Sikhs; 17th Bengal Cavalry.

Campaign against Galekas and Gaikas.—On September 25th, 1877, the Galekas attacked the old allies of the British, the Fingoes, and a party of mounted police at Guadana, but were repulsed. Troops were immediately sent to the scene of hostilities, Kreli's country was invaded and his kraal burnt, and a stiff fight was made at Umzintzani before crossing the Krei. The Connaught Rangers then proceeded to the front, with 50 mounted men of the 24th, and the bandsmen of that regiment as gunners with a 7-pounder gun and Naval Brigades from the "Active" and "Florence." The Galekas, who had joined forces with the Gaikas under Sandilli, were badly beaten at Nynnuxa, but rallied in the Chickaba Valley, where the British column attacked them on January 14th, 1878, and drove them off with considerable loss in men and cattle. On February 25th General Thesiger (later Lord Chelmsford) took over the command from Sir Arthur Cunningham, and Sandilli's force was defeated in the Amotala Mountains by the column under Colonel (now Field-Marshal Sir Evelyn) Wood. In succeeding skirmishes Sandilli was killed; then the Galeka chief, Kreli, surrendered, and an amnesty was proclaimed on

June 28th, 1878, for all who had taken part in the war except the sons of Sandilli.

The Griqua Campaign.—Meanwhile an outbreak in Griqualand West had to be suppressed, and this was done by the colonial troops, without the assistance of the regulars, but not until several sharp engagements between the Griquas and the Colonials had been fought between June 11th and July 20th, 1878, when an amnesty was declared and the war was at an end.

The Basuto War.—Following closely on the suppression of the rising in Griqualand came the Basuto War, occasioned by the refusal of Morosi and his son Dodo, chief of the Baphutis, to pay the hut tax. Morosi had fortified an almost inaccessible mountain near the Orange River, and thither, after a skirmish with the Cape Mounted Rifles, he gathered his people, and for nine months defied the forces and artillery ranged against him in his mountain fastness, which with only one accessible side he had, with considerable skill, fortified and loopholed for musketry fire. After many futile attempts to storm the place Morosi was killed, and what was considered to be an impregnable mountain fortress was taken, and the fortifications blown up. Then it was decided to disarm the natives, and some hard fighting took place against the rebel chief Lethe-rodi, and the rising was not stamped out until the Imperial and Colonial forces in the field totalled 15,000 men. In April 1878 an amnesty was proclaimed, and although the Basutos had been badly defeated they gained practically all they asked for prior to the rising.

THE ZULU WAR

Then a war was entered upon which was to alternately bring disaster and as frequently demonstrate the stern inflexibility of the white man. Zululand, to the north-east of Natal, produced the finest and bravest black warriors against whom Britain has fought. Chaka on becoming chief in 1810 organised the tribe on a martial basis, and to his determined and ruthless efforts to establish a military despotism the splendid fighting qualities of the Zulu were due. Our story commences with the recognition of Cetewayo as King of the Zulus by the British Government in 1872. His father Panda had reverted to the peaceful pastoral pursuits of the tribe, but Cetewayo determined to again organise his countrymen as a military nation, and with such success did he do this that he felt himself in a position to defy white authority. The seizure of two women on British territory, claimed as the runaway wives of a chief, and the refusal of Cetewayo to make reparation for the incursion into British territory, led to the invasion of Zululand on January 11th, 1879. It was decided to invade the country by three columns. The first, or southern column, under Colonel Pearson of the Buffs, was composed of the 2nd Batt. Buffs; six companies of the 99th; 100 Mounted Infantry; 19th Hussars; Royal Artillery with 4 guns; 170 Seamen and Marines from H.M.S. "Active," with a Gatling gun and Rocket Battery; Mounted Colonial Volunteers, and about 1,000 natives. The second, or central column, under Colonel Glyn of the 24th, comprised seven companies of the 1st Batt. and the 2nd Batt. 24th Regiment; Royal Artillery with 2 7-pounders; Natal Mounted Police; Mounted Colonial Volunteers, and 1,000 natives. The third, or northern column, under Colonel Evelyn Wood, consisted of the 1st Batt. 13th; the 90th Regiment; Royal Artillery with 4 7-pounders; 100 Mounted Infantry; 200 Frontier Light Horse; 100 local Volunteers; a body of Boers under Commandant Piet Uys, and 5,000 Swazies.

On January 22nd a body of about 6,000 Zulus were defeated at Inyezane by Colonel Pearson's force, which then marched on to Ekowe (Etschowe), where a depot was formed without opposition, but on the same day the central column, which had crossed the river at Rorke's Drift, where a detachment of the 24th was left, met with overwhelming disaster at the isolated

MEDAL FOR ASHANTI, 1900.

MEDAL FOR TIBET, 1903-4.

(Obverse.)

(Reverse.)

MEDAL AWARDED FOR NATAL NATIVE REBELLION.

hill of Isandhlwana. Lord Chelmsford, with Colonel Glyn, had marched out leaving one company of the 2nd 24th, and five companies of the 1st 24th, with 70 artillerymen and 2 guns, mounted police, volunteers, and native friendlies, to defend the unfortified camp, and while he was reconnoitring in the south-west the Zulu impi of 14,000 attacked the camp at Isandhlwana, where the men, standing back to back, faced the fearful odds until almost every white man had fallen where he stood. A Zulu chief afterwards said, "They turned back to back and fought till they died. Not one tried to escape."

In this desperate struggle Lieutenants Coghill and Melville made a gallant attempt to save the colours of the 24th, and with their precious charge essayed to swim the Buffalo, but Melville's horse being killed Coghill returned to assist him, and his horse being also shot the colours fell into the river, from which they were rescued ten days later. The brave officers were found dead side by side. For this gallant effort the Victoria Cross was posthumously conferred upon the two officers. Around the fatal hill of Isandhlwana 50 British officers, and over 800 non-commissioned officers and men, met their death; the five entire companies of the 1st 24th were annihilated, and 90 men of the 2nd Battalion killed.

Rorke's Drift.—Meanwhile on that fatal day the little force which had been left under Lieutenant Bromhead at Rorke's Drift to guard 35 sick men and the stores, had performed prodigies of valour. About 80 men kept at bay 3,000 Zulus from 5 o'clock on the afternoon of the 22nd until 4 a.m. on the 23rd. Again and again the Zulus tried to rush the hastily improvised fort; they set fire to a building which had been converted into a hospital, and forced the devoted band into the inner defences. Half a dozen times they rushed the meagre entrenchments, but were forced back, until at last on the morning of the 23rd they retired, leaving 370 dead around that well-kept post.

The gallant men who defied that army lost 17 killed and 2 wounded, 5 of the sick men being burnt to death despite the strenuous efforts of their comrades to rescue them.

Eight Victoria Crosses.—In this glorious defence, in which every man was a hero, eight Victoria Crosses were earned: Lieutenant G. Bromhead of the 24th, Lieutenant J. M. R. Chard, R.E., for their brilliant defence; Corporal W. Allen, Privates F. Hitch, Hy. Hook, John Williams, and W. and R. Jones, for their gallant defence of the hospital, and holding it against the enemy while the patients were removed. Corporal Allen and Private Hitch were wounded, but when no longer able to fight served out ammunition to their comrades through that terrible night.

Towards the end of March Lord Chelmsford proceeded to the relief of Ekowe, at the same time instructing Colonel Evelyn Wood to advance on the Zlobani, or Inholbani, mountain, a precipitous and deeply wooded stronghold which he determined to attack. Dividing his force into two columns under Colonel Russell and Colonel (later General Sir Redvers) Buller, he attacked the place on the early morning of March 28th. Buller gained the summit without much opposition, but Russell discovered that an immense impi of Zulus was attempting to cut off the former's retreat, which unfortunately became a rout, as the struggling mass of men were compelled to face the assegais of the Zulus with little opportunity of retaliating. Buller, who was one of the last to leave the "Devil's Pass," won the V.C. for saving the lives of two officers and a trooper during the retreat, which would have had more disastrous results but for Russell's perception and assistance.

Next day the entrenched camp at Kambula was attacked by a force of 25,000 Zulus, but after four hours' fighting, in which assegai frequently clashed with British bayonet, they made off with Colonels Buller and Russell in hot pursuit, and the humiliation of the previous day was avenged, over 3,000 Zulus being killed. Meanwhile, Lord Chelmsford, with 3,000 soldiers and sailors, and 2,500 friendlies, was attacked on his way to Ekowe by a Zulu impi under Dabulamanzi, the half-brother of Cetewayo, at Ginghilova; but after one and a half hour's fighting, during which the Zulus rushed right up to the muzzles of the British rifles, they were driven off with the loss of about 1,000 men. Two days later Ekowe was relieved, and the post abandoned. By this time considerable reinforcements had arrived from England, among the officers being the Prince Imperial of France, who on June 1st, while reconnoitring with Lieutenant Carey of the 98th Regiment and 6 troopers, was killed by a party of Zulus. On July 3rd, Buller, with 500 cavalrymen, was attacked after crossing the White Umvolosi River by nearly 5,000 Zulus, and in this affair Lord William Beresford of the 9th Lancers gained the V.C. for the rescue of a dismounted trooper, and Lieutenant Lysons and Private E. Fowler of the 2nd Cameronians also gained the coveted distinction for their bravery.

Ulundi.—Next day the battle of Ulundi was fought, and despite the repeated charges of the Zulu army of over 15,000 men, the British square remained unbroken, until the enemy gave way after forty minutes' desperate endeavour, and then the 17th Lancers under Colonel Drury Lowe cut them up as they fled. The capture of the King's kraal at Ulundi practically ended the war, although for over a fortnight he eluded the vigilance of his pursuers. On August 28th he was captured, and on September 1st, 1879, the Zulu chiefs accepted the conditions laid down by the British Government. During this campaign 76 officers and 1,007 men were killed; 37 officers and 206 men wounded; 17 officers and 330 men died of disease, while 1,286 were invalided home.

Basuto War, 1878–9.—Sekukuni, who had allied himself with Cetewayo and defied the Colonial Government, stating that the country was his, and he was ready for war, caused trouble by raiding a friendly chief; an expedition was therefore organised to proceed against him, but owing to the need for the concentration of British troops in Zululand, operations were suspended. When, however, Cetewayo had been captured, Sir Garnet Wolseley, who had succeeded Lord Chelmsford in the command, proceeded to attack the strongholds of Sekukuni on November 28th, 1879, and the "Fighting Koppie" was taken after a desperate struggle, during which the chief escaped, but was captured a few days later and taken prisoner to Pretoria. Thus came to an end a succession of wars during the severe trials of which the finest attributes of the British soldier were repeatedly called forth.

The troops engaged from 1877–9 were, in 1877–8–9: 24th, 88th, 90th, 94th; 1878–9: 13th, 80th, 83rd; 1879: 2nd Batt. 3rd; 4th, 21st, 57th, 58th; 3rd Batt. 60th; 3rd Batt. 91st, 97th, 99th; and Naval Brigades.

The Zulu Medal.—For these campaigns a medal similar to that awarded for the Kaffir wars was given to the participants, the only difference being the substitution of a Zulu shield and assegais for the date in the exergue of the reverse. The ribbon is the same. The bars are similar to those given with the I.G.S. 1854 medal, and do not record specific actions, but periods of service. Those serving for three years received the bar inscribed 1877–8–9; two years, 1877–8 or 1878–9; or for one year, 1879. A single bar only was issued with each medal.

AFGHANISTAN, 1878–80

The British had by military and monetary assistance helped Shere Ali on to the throne of Afghanistan in 1869, and, naturally, took umbrage at the overtures made to Russia—the refusal to accept a British Resident at Cabul, the honoured reception of a Russian envoy, and finally the signing of a treaty which placed the Ameer of Afghanistan under the guardianship of Russia.

A mission was consequently sent to Cabul, but it was stopped at Ali Musjid, at the entrance to the Khyber Pass, by threats of attack if it dared to proceed any farther. Sir Nevill Chamberlain, therefore, thought it prudent to return to Peshawar, and on November 21st, 1878, war was declared. Afghanistan was then invaded, and the Ameer declared a Jehad, or holy war, against the British.

Ali Musjid.—The same day the fort of Ali Musjid, armed with 22 guns, was attacked, and the enemy abandoned it after some smart fighting. The column under General Sir S. Browne then pushed on to Jellalabad, having lost in the attack 2 officers and 35 men killed.

The troops engaged at Ali Musjid, on November 21st, 1878, were the 17th, 51st; 4th Batt. 60th; 81st; four Batteries of Artillery; 10th Hussars, and the following native regiments: 1st Sikhs; 4th Goorkas; 6th, 14th, 20th, 27th, and 45th Bengal Infantry; Bengal Sappers and Miners and 11th Bengal Lancers.

Peiwar Kotal.—Major-General (later Field-Marshal Lord) Frederick Roberts, V.C., with the Kurram Field Force, crossed the Thall the same day as General Browne had captured Ali Musjid, and advancing up the Kurram Valley garrisoned Fort Azim, and then proceeded to attack Peiwar Kotal; but finding it too strong to assault in front, General Roberts decided upon a daring march of 10 miles in order to turn the Afghan position, and this he carried out in a masterly and successful manner while the heavy snowstorms made the intense cold less endurable. The attack was successfully delivered, the Afghan defence turned, and then Brigadier-General Cobbe hammered away at the enemy's front with his artillery while Roberts advanced against the main body; and after a contest which fiercely raged for seven hours, the Afghans made off, leaving their tents standing and the whole of their baggage; 6 field and 11 mountain guns were captured, and so distressed was Shere Ali with the defeat that he fled to Balkh, where he died while awaiting permission to make an asylum in Russia. The British losses were: 2 officers killed, and General Cobbe wounded; 90 European and Indian soldiers killed or wounded. In this battle Major J. Cook, of the Bengal Staff Corps, gained the Victoria Cross by charging the enemy and rescuing Major Galbraith, who was attacked by an Afghan. During the operations at Cabul this gallant officer was killed.

The troops engaged were the 8th and 72nd, a squadron 10th Hussars, and the following native regiments: 2nd and 5th Punjab Infantry; 5th Goorkas; 23rd and 29th Bengal Infantry, and 12th Bengal Cavalry.

Charasia.—On the death of Shere Ali, his son Yakoub Khan, who had been imprisoned by his father, ascended the throne, and, recognising the futility of the contest, agreed by the treaty of Gandamak to vest the control of his country in the Indian Government, to allow a British Resident at Cabul, and to accept a subsidy of six lakhs of rupees while the Khyber Pass was to be controlled by us, and the Kurram, Pishi, and Sibi Passes were to be attached to the Indian Empire. On June 17th, 1879, Sir Louis Cavagnari set out with his secretary, Mr. Jenkins, and Dr. Kelly, and an escort of 25 cavalry and 50 infantry of the Guides under Lieutenant Hamilton. Cabul was reached on July 24th, and on the night of September 2nd the Residency was attacked, and after a gallant defence Sir Louis and his officers and almost all the escort were murdered. This outrage necessitated prompt action, and General Roberts hurried up from Simla. Pushing past Peiwar Kotal, the Shaturgardan Pass was occupied before the enemy could reach it. At Kushi the Ameer rode into camp and said that he had been dethroned by his rebellious troops. No one believed him. Pressing as rapidly as possible on to Cabul, Sir Frederick Roberts found that the Afghan army occupied a very strong position at Charasia. On October 6th he stormed the

Edwardian Variety.

Medal for Abor.

INDIA GENERAL SERVICE MEDALS, 1908.

(Reverse.)

(Obverse.)

SILVER MEDAL FOR SUDAN, 1910.

place, and after a determined defence the Afghans fled towards Cabul in great confusion, leaving behind them 20 guns and 2 standards. Next day the British army marched on Cabul, and entered with the band of the 67th playing at its head; by October 8th General Roberts was occupying the fortified cantonments of Sherpur, in which he found 73 guns. The British losses were 20 killed and 70 wounded.

The regiments engaged were the 67th, 72nd, and 92nd; one battery Royal Horse Artillery; one battery Royal Field Artillery; two Mountain Batteries, and 9th Lancers. Native regiments: 5th Goorkas; 5th Punjab Infantry; 5th Punjab Cavalry; 12th Bengal Cavalry; 14th Bengal Lancers; Bengal Sappers and Miners.

Cabul.—Once inside the city the army was practically shut in, for the whole country was hostile to the British. Colonel Sir Hugh Gough, V.C., relieved the Shaturgardan garrison, and the lines of communication were directed to Gandamak and the Khyber. Throughout December General Roberts was kept busy with continual fighting; during this time, in an engagement with the Kohistanee on December 11th, the 9th Lancers suffered severely, and for rescuing some of the unfortunate troopers who had fallen into a deep ditch the Rev. J. W. Adams was awarded the V.C. Standing up to his waist in water, and under a heavy fire, he stuck to his task until the Afghans were within a stone's throw of him. On this unfortunate day three guns had to be spiked and abandoned. After this affair the Afghans made direct for Cabul, but were checked on their advance on Sherpur by the steady conduct of the 72nd Highlanders. On the 14th the Asmai heights were occupied by the enemy, but after a determined fight they were driven off; almost immediately, however, a force of over 15,000 Afghans, led by Moollahs shouting "Allah Ya Allah!" dashed with reckless impetuosity upon the place, and, despite the utmost gallantry on the part of the British troops, a portion of the position and a couple of guns had to be abandoned. Then the increasing number of the enemy compelled the concentration of the army within the cantonments of Sherpur, which were fortified to resist siege, and by the intuition of Sir Frederick Roberts had been well provisioned to sustain the defenders.

At 6 o'clock on the morning of December 23rd, 1879, just before the day broke, 30,000 Afghans flung themselves at the British defences, and maintained the assault until 1 o'clock, then realising that further efforts were fruitless, and suffering from heavy losses, they began to retire, when the cavalry sallied forth and put them to rout. In this defence 2 officers and 8 men were killed, and 5 officers—including Colonel Gough—and 41 men wounded. It is estimated that the Afghans had 3,000 men placed *hors de combat.* During the night the enemy melted away, and when General Gough arrived with reinforcements not an Afghan was to be seen. Cabul was then reoccupied.

The regiments engaged were the 9th, 67th; 72nd Seaforth Highlanders; 92nd Gordon Highlanders; 9th Lancers. Native regiments: 2nd, 4th, and 5th Goorkas; 5th Punjab Infantry; 23rd and 28th Bengal Infantry; 12th Bengal Cavalry, and 14th Bengal Lancers; 5th Punjab Cavalry; two batteries Punjab Artillery, and Bengal Sappers and Miners.

Ahmed Khel.—General Sir Donald Stewart set out from Kandahar in April 1880 to occupy Ghuznee, and to open up communication with General Roberts at Cabul. On April 19th he met the enemy, 15,000 strong, at Ahmed Khel, 23 miles south of Ghuznee. There a mad charge of 4,000 Ghazis threatened to exterminate Stewart's little force, but standing firm they withstood the shock of the fanatics, who as they rushed at the rallying squares of Britisher, Sikh, Punjabis, and Bengalese, were mowed down by their steady fire. In this conflict, lasting only one hour, 1,000 Ghazis were killed, but of the British force only 17 were killed, and 6 officers with 120 men wounded. Marching on to Ghuznee, they reached the city on the 20th. Sir Donald

Stewart then assumed command at Cabul, and Abdul Rahman was recognised as Ameer.

The regiments engaged at Ahmed Khel were 59th; 2nd 60th Rifles, and a battery of artillery; native regiments; 2nd Sikhs; 15th, 19th, and 25th Bengal Infantry; 19th Bengal Lancers, and 1st Punjab Cavalry.

March to "Kandahar."—When Abdul Rahman, who had been living in Russian Turkestan, returned to Cabul, and his authority appeared to be established, the British army was ordered back to India. They were, however, delayed by the determination of Ayoub Khan, a younger brother of the deposed Ameer, to make a bid for the Ameership, and he set out with the object of seizing Candahar, which had been left in the command of General Primrose with a garrison of about 3,000 men, including the 7th Fusiliers, the 66th, and a number of native troops. To check the advance of the Afghans, about 2,000 men were sent out with 6 guns under General Burrowes as far as the Helmund, but encountering an enormous army at Maiwand he assumed the defensive in a bad position, and the Ghazis drove the British troops from the field with the loss of the colours of the 66th and 2 guns. Thirteen hundred of the British force lay dead on the field. The remnants of that unfortunate army got back to Candahar after suffering great privations, and General Primrose concentrated his troops in the citadel, where they were hemmed in by the enemy, and from August 11th to the 31st waited for relief.

General Sir Frederick Roberts then offered to march on Candahar, and with 10,000 men and 8,000 camp followers set out on August 8th to march over 300 miles through a mountainous and hostile country. His army consisted of the 2nd 60th Rifles; 72nd and 92nd Highlanders; 9th Lancers; three Batteries of Artillery; 2nd, 4th, and 5th Goorkas; 2nd and 3rd Sikh Infantry; 15th Sikhs; 24th and 25th Punjab Infantry; 23rd Pioneers; 3rd Bengal and 3rd Punjab Cavalry, and Central India Horse, formed into three brigades under

Brigadiers Baker, Macgregor, and Macpherson. Ghuznee was reached on August 15th; on the 23rd Khelat-i-Ghilzie was relieved, and after a rest the force marched on to Candahar, and by August 27th was in communication with General Primrose. In face of the relieving force Ayoub Khan withdrew to the hills, and General Roberts entered Candahar on the 31st.

Battle of "Kandahar."—Wasting no time, he made a reconnaissance, and driving the enemy at the point of the bayonet from several of their positions, proceeded next morning to give battle to Ayoub. The village of Gundi was carried by the 92nd Highlanders and the 2nd Goorkas; the 3rd Sikhs and the 5th Goorkas, working round the end of the Pie Paimal, carried several villages, and despite the tigerish fighting of the Ghazis the village of Pie Paimal was taken at noon, and then the final stand was made by the enemy; but they could not withstand the charge of the 92nd Highlanders and Goorkas, and four hours after the battle commenced the Afghans were in full flight, leaving behind them 31 guns including the 2 Royal Horse Artillery 9-pounders taken at Maiwand. In front of one of their tents lay the bleeding body of Lieutenant Maclaine of the Royal Horse Artillery, who, taken prisoner at Maiwand, had been reserved for the indignity of having his throat cut when his foes were beaten and release seemed imminent. The 72nd Highlanders suffered most severely in the battle of "Kandahar," losing among their dead Colonel Brownlow. Two other officers were killed, and 11 wounded, the loss in men being 46 killed and 200 wounded.

The following troops were engaged at "Kandahar" on September 1st, 1880: 7th, 60th, 66th, 72nd, and 92nd; three batteries artillery and 9th Lancers. Native regiments: 1st, 4th, 19th, 28th, and 29th Bombay Infantry; 2nd and 3rd Sikhs; 2nd, 4th, and 5th Goorkas; 15th, 23rd, 24th, and 25th Bengal Infantry; 3rd Bengal, 3rd Bombay, and 3rd Punjab Cavalry; 3rd Scinde Horse; Poona and Central India Horse.

CONSPICUOUS SERVICE CROSS.

DISTINGUISHED SERVICE ORDER.

DISTINGUISHED SERVICE CROSS.

A medal without bar was given to the following regiments: 12th, 14th, 15th, 17th, 25th, 31st, 53rd, 63rd; 6th Dragoon Guards; 8th and 15th Hussars. A number of men of the 65th were engaged as signallers, and took part in several of the actions.

The Afghan Medal.—With the Afghan medal, which was granted on March 19th, 1881, were issued six bars for ALI MUSJID, PEIWAR KOTAL, CHARASIA, AHMED KHEL, KABUL, and KANDAHAR. The medal bears on the obverse a crowned head of Queen Victoria, which is not truncated like her effigies on the other medals, but the shoulders are couped by the bevel of the medal, which is $1\frac{2}{5}$ in. in diameter. The bust is surrounded by the legend VICTORIA REGINA ET IMPERATRIX. This obverse is the fourth of the series facing page 188. On the reverse a column of Anglo-Indian soldiery is depicted on the march; an officer leads the way, accompanied by an infantrymen; an elephant bearing a mounted gun on its back makes the central feature, with native cavalry to fill up the foreground of a picture which has a fortress-capped mountain for background. AFGHANISTAN is impressed above in raised letters, and the date 1878–9–80 in the exergue. The names of the British recipients were engraved in rather squat Roman capitals, but the medals given to native troops were generally engraved in slanting writing letters. The suspenders to this medal are straight, like the General Service medals, the bars rather broad with square ends; the ribbon, $1\frac{1}{4}$ in. wide, is green with broad dark-red edges.

Star for Kabul-Kandahar, 1880.—Her Majesty also bestowed a bronze decoration on the troops who accompanied Major-General Sir Frederick Roberts on his 318-mile march from Kabul to Kandahar. It was subsequently granted to the troops which then composed the garrison of Khelat-i-Ghilzie, and who accompanied the force from that place to Kandahar. The decoration is a five-pointed radiated star, with small balls between the inner angles, and was made from the bronze guns taken at the battle of Kandahar on September 1st. In the centre is the Imperial monogram V.R.I., encircled by a band, with raised border inscribed **KABUL TO KANDAHAR**, having **1880** below with a twig of laurel on either side. At the top is the Imperial crown, from which is a ring for suspension. The reverse is plain with a hollow centre, round the edge of which the name and regiment of the British recipient are indented in skeleton block letters. But those given to the native troops were generally engraved in a sloping script, as on the India 1895 medal, and occasionally in block capitals. The ribbon, $1\frac{1}{2}$ in. wide, is of the rainbow pattern, but unwatered.

THE EGYPTIAN CAMPAIGNS

In 1882, owing to the rebellion of Arabi Pasha, British troops and ships were sent to Egypt. The bombardment of Alexandria took place on July 11th, followed by the decisive battle of Tel-el-Kebir, on September 13th. During this campaign the famous midnight charge of the heavy brigade at Kassasin took place. The British army of about 11,000 infantry, seamen, and marines, 2,000 cavalry, with 60 guns, marched on Tel-el-Kebir, under Sir Garnet Wolseley, who finding the rebel entrenchments reaching at least 4 miles from the canal out into the desert, and that his force would have to attack without a scrap of cover, decided to make a night march from Kassasin to get within charging distance of the enemy's position. The march was directed by Lieutenant Rawson, R.N., who steered by the stars, and the " monotonous tramp, the sombre lines, the dimly discerned sea of desert faintly lighted by the stars, were at once ghostly and impressive." When about 300 yards from the enemy's trenches the Highland Brigade, just before daybreak, rushed forward, and with a cheer went over the trenches, where after a stubborn resistance of about half an hour the rebels fled in confusion towards Zagazig, and the capture of Arabi at Cairo concluded the war. Nine British officers and 48 men were killed, 27 officers and 385 men wounded or missing. For his brilliant conduct of the war Sir Garnet Wolseley was raised to the peerage.

The following troops were engaged at Tel-el-Kebir: One batt. Grenadiers, Coldstreams, and Scots Fusilier Guards; 18th, 42nd, 46th, 60th, 72nd, 74th, 75th, 79th, 84th, and 87th; detachments of the 1st and 2nd Life Guards and Royal Horse Guards; 4th and 7th Dragoon Guards; 19th Hussars; a batt. of Marines; Marine Artillery, and a brigade of Seamen. The following Indian troops were also engaged: 7th Bengal and 20th Punjab Infantry; 29th Beloochees; 2nd, 6th, and 13th Bengal Cavalry.

Egyptian Medal, 1882.—The medal is a simple one, $1\frac{2}{5}$ in. in diameter, depending from a suspender like that on the Ashantee medal; the ribbon is $1\frac{1}{4}$ in. wide alternately blue and white, three blue stripes and two white. On the obverse is the same Queen's head as on the Ashantee medal, but the reverse has a well-modelled Sphinx, with EGYPT in block letters above and 1882 in the exergue, while bars were given for the two engagements, ALEXANDRIA 11TH JULY and TEL-EL-KEBIR. They are wide and square, and set rather far apart, as in the Afghan medal of 1878–80. The names are engraved on the edge in slanting Roman capitals; in a few instances the naming is done in upright capitals.

The Khedive's Star.—Prince Twefik gave to all who took part in the campaign a bronze star of five points bearing on the obverse in the centre a front view of the Sphinx, behind which are three pyramids in the distance encircled by a band bearing the inscription EGYPT 1882, and below in Arabic characters "The Khedive of Egypt" and the year of the Hegira "1299." On the reverse within a raised border is the Khedive's monogram T. M. (Twefik Muhammad), surmounted by a crown with crescent and star above. The star is suspended from a laureated bar bearing a crescent and star by means of a dark-blue ribbon 1½ in. wide.

Against the Dervishes.—Quite a different type of enemy were the followers of the Mahdi, who after practically exterminating the Egyptian army of 7,000 men led by Hicks Pasha, and under the generalship of the slave-trader, Osman Digna, had made short work of Baker Pasha's 5,000 Egyptians at El-Teb on February 1st, 1884, met at the same place on the 29th, and fought the force of 4,000 men which Major-General Graham had got together. The battle, which was of a determined character (see also page 321), the brunt falling on the "Black Watch," the 65th, and Naval Brigade, resulted in the defeat of the Arabs with a loss of 2,000 killed; 22 British officers and 167 men were killed or wounded. For this battle those who already possessed the Egyptian medal received the bar inscribed EL-TEB.

The following troops were engaged at El-Teb: "Black Watch"; Gordon Highlanders; Irish Rifles; King's Royal Rifle Corps; 1st Batt. York and Lancaster Regiment; 100 Royal Engineers; 126 artillerymen; 328 men of 10th Hussars; 410 of 19th Hussars; 125 Mounted Infantry; Marine Light Infantry and Marine Artillery, and a Naval Brigade of 162 men.

Tamaai.—Fourteen days later the British force again met the Mahdi's hosts, and at Tamaai routed the Arabs with a loss of over 3,000. It was at this battle that a few Arabs first got into the British square. The "Black Watch," whose quick movement on being ordered to charge, left a gap between themselves and the York and Lancaster Regiment, losing Major Aitken, 8 sergeants, and 50 men killed, 3 officers and 26 men wounded. The British losses were 5 officers and 104 men killed, and 8 officers and 120 men wounded or missing. (See also Naval Section, page 322.)

The following troops were engaged at Tamaai: "Black Watch"; Gordon Highlanders; Royal Irish Fusiliers; York and Lancaster Regiment; King's Royal Rifle Corps; Artillery; 26th Company Royal Engineers; a squadron of the 10th and 19th Hussars; Royal Marines and a Naval Brigade.

A bar for TAMAAI was given for this battle, and to those who took part in that of El-Teb also a bar bearing the words EL-TEB—TAMAAI. The battle of Suakin on March 27th brought the first Soudan War to a close. Those who landed at Suakin or Trinkitat between February 19th and March 26th, and had taken part in the 1882 campaign, were granted the bar for SUAKIN 1884. It was further approved that a bar be issued to all those who were actually present at either or both of the actions on February 29th and March 13th; this bar to be inscribed EL-TEB—TAMAAI for those who were in both actions, and EL-TEB OR TAMAAI for those who were in one or the other, but not both.

To Relieve Gordon.—The Mahdi, however, was not conquered; he had merely run away to fight another day. The defence of Khartoum for ten months by General Gordon, and the belated and unsuccessful attempts to relieve him, are matters of history which will not be easily forgotten; but for our purpose they recall the gallant Britisher who refused to yield, and the splendid efforts which were made by the officers and men who went up the Nile in 1884–5. They started from Cairo in September under Lord Wolseley, and reached Dongola early in November, when the gallant Sir Herbert Stewart started off with his little army of 1,600 men to reach Khartoum, by first marching 170 miles and then travelling 100 miles by steamer.

Abu Klea.—It was during this march that the battle of Abu Klea—"where the rifles jammed and the shoddy bayonets twisted like tin"—was fought on January 17th, 1885, and the survivors of the little band of about 1,200 who took part received the bar ABU KLEA. Here the genial giant Colonel Fred Burnaby was thrown from his horse in riding outside the square to assist some skirmishers, and killed as he lay on the ground, and here Lord Charles Beresford had many hair-breadth escapes from death (see page 325), while the Arabs achieved the remarkable feat of breaking a British square and getting inside, killing the wounded and a number of camels ere they were themselves dispatched. This, one of the most strenuous of the battles in the Sudan, resulted in a loss to the enemy of over 2,000, and to the British of 9 officers killed and 9 wounded (2 dying of their wounds), 65 non-coms. and men killed, and 85 wounded; but the enemy were still active, and as Sir Herbert Stewart marched on he found his way to the Nile barred by them, so a zareba was formed at Abu Kru, and here he was mortally wounded. The enemy were driven off; the harassed troops pressed on, reached the Nile, and heard that Khartoum had fallen! Their pluck, perseverance, and privations were in vain. In the fight between Abu Klea and the river 1 officer and 22 men were killed, and 8 officers and 90 men wounded; this was called the battle of Gubat, but no bar was given for it.

The following troops were engaged at Abu-Klea: Sussex Regiment; Mounted Infantry; Royal Artillery and Royal Engineers; Heavy Camel Corps; 19th Hussars, and a Naval Brigade.

Kirbekan.—During the campaign the battle of Kirbekan was fought on February 10th, and Major-General Earle, the General in Command, killed in a manner which calls to mind Forbes Mitchell's description of how Hodson, of Hodson's Horse, met his death. The battle had been won, but odd parties of survivors were being routed out; the General, despite warnings, went up to a stone hut full of Arabs, and essayed to enter it, but was shot dead. Needless to say his death was avenged—indeed, of the force of 2,000 Lord Wolseley said "scarcely any can have escaped." Two colonels were also killed at this battle, which raged for five hours. The "Black Watch" lost Lieutenant-Colonel Coveney killed, and Lieutenant-Colonel Wauchope wounded. The bar inscribed KIRBEKAN was well earned. It should be noted that no medal for this campaign is genuine which has a single bar inscribed

Kirbekan; it was only given in combination with the bar THE NILE 1884–85. There are, of course, other combinations, but these were due to the fact that the recipient had taken part in the earlier campaigns. Four-bar medals are rather rare, and five-bar medals very uncommon.

The following troops were engaged at Kirbekan: South Staffordshire Regiment; 1st Batt. Black Watch, squadron 19th Hussars; 2 gun crews of artillery; small naval detachment with maxim gun—about 1,200 men.

The following troops received the bar for the Nile: "Black Watch"; Gordon and Cameron Highlanders; 18th Royal Irish; Duke of Cornwall's Light Infantry; Royal West Kent Regiment; Essex Regiment; South Staffordshire Regiment; 19th Hussars; Heavy Camel Corps, and a Naval Brigade.

To Recover the Sudan.—The fall of Khartoum led to the organisation of another campaign in order to recover the Sudan from the power of Osman Digna, and for this arduous campaign those who did not receive the new Egyptian medal had a bar added to the one they possessed for SUAKIN 1885. During this campaign a New South Wales contingent of 800 took part. A battle which came as a surprise, and resulted in heavy losses, was fought on March 22nd, 1885, and for this a bar inscribed TOFREK was given. The only entire regiment present was the 1st Batt. Royal Berkshire. The losses of the Hadenowas were over 1,500; the British lost 5 officers and 94 men killed, 6 officers and 136 men wounded, 1 officer and 70 men missing, and of the Indian contingent 120 were killed or wounded.

The following troops were present at Tofrek: the Berkshire Regiment; a detachment of Royal Engineers; a squadron of the 5th Lancers and 20th Hussars; a battalion of Marines; a Naval Brigade, and an Indian Brigade.

By a General Order dated September 1st, 1885, a medal of the 1882 pattern was granted to all Her Majesty's forces employed in the operations in the Sudan, in commemoration of their arduous labours in the ascent of the Nile, and their gallantry in the operations which ensued, and also for the operations in the Eastern Sudan in the vicinity of Suakin; and Her Majesty further approved of the issue of the following bars: THE NILE 1884–85, ABU KLEA, KIRBEKAN, SUAKIN 1885, and TOFREK. A General Order dated June 1st, 1886, signified that the Egyptian medal be granted to the troops engaged in the operations of the Upper Nile who had not previously received it.

Gemaizah, 1888.—The neighbourhood of Suakin was again the scene of battle when, on December 20th, 1888, the Dervishes advanced on the place with the determination to invest it. General Grenfell, however, forced the issue, and on December 20th, 1888, after one and a half hour's fighting, totally routed them; for this battle the bar for GEMAIZAH 1888 was awarded by an Army Order in January 1890, and the medal to all who were employed at, or south of, Korosko on August 3rd, 1889.

The following troops were engaged at Gemaizah: King's Own Scottish Borderers; the Welsh Regiment; Royal Irish Rifles; 137 officers and men 20th Hussars; Mounted Infantry; 17 officers and men of the 24th Company Royal Engineers; Naval Brigade of men from H.M.S. "Starling" and "Racer," and Egyptian troops.

Toski.—After this battle the expedition was withdrawn, and the Sudan was left to the mercy of the Dervishes, Suakin alone being held by the British and Egyptians. Ever on the war-path, the followers of the Mahdi resolved to invade Egypt, and were actually permitted to penetrate northwards for about 50 miles. At Toski, however, they were seriously encountered by Sir F. Grenfell, the Sirdar of the Egyptian Army, and on August 3rd a terrific but comparatively bloodless battle for the British and Egyptians was fought. The Dervishes, however, were routed, and almost every leader was killed. For this battle a bar inscribed TOSKI 1889 was awarded to all troops who were present at the action. Men who had received bars for former Egyptian campaigns, on receiving new ones, returned the old ones to the Commissary-General of Ordnance, Woolwich Arsenal; the rivets of all genuine medal bars are, therefore, always the same. The names and regiments are usually engraved in slanting Roman letters, but those given to the navy are generally stamped in rather large skeleton upright Roman letters.

The following were engaged at Toski: 98 officers, non-commissioned officers, and men of the 20th Hussars, and Egyptian troops.

Tokar.—The battle of Tokar was fought in 1890, but the British silver medal was not awarded for this engagement. The Egyptian Government, however, issued a bar inscribed in Arabic characters "Tokar" and "1308 H" (**TOKAR 1890**), to be worn with the Egyptian Star, which, except for variations in the date, had been awarded to all who took part in the campaigns. Officers and men from the "Dolphin" and "Sandfly" received this star and British officers who were serving with the Egyptian Army.

The Khedive's Star.—For the first campaign the Star bears in Arabic and English **EGYPT 1882**; for the campaigns of 1884 and 1885 **EGYPT 1884-6**; but for the campaigns 1888–9, which include the fighting round Suakin and up the Nile, the Star was issued without a date.

Riel's Rebellion.—Whilst we were busy in Egypt, "Riel's Rebellion" was engineered in 1885 among the half-breeds and Indians on the North-West Territories. It was a nine-days' wonder, but nevertheless a troublesome one. Riel was attacked at Batoche on May 11th, defeated by General Middleton, and after trial executed.

By Militia General Orders dated September 18th, 1885, the Canadian troops which took part in suppressing the rebellion were awarded a medal inscribed on the reverse in fancy letters NORTH-WEST CANADA 1885, the date being placed in the centre, with North-West above and Canada below, and the whole surrounded by a wreath of maple leaves, the terminals of which are at the bottom of the medal. The obverse is the same as the Egyptian 1882 medal, and the bar similar to that used with that medal. One bar for SASKATCHEWAN was issued with the medal to those who took part in that engagement. The medals were issued unnamed, but they are found both engraved and impressed; the ribbon is blue-grey, with two red stripes.

The Canada Medal.—In January 1899 the Home Government approved of the issue of a medal by the Canadian Government to the Imperial Forces and Canadian Militia which had taken part in the campaigns occasioned by the Fenian Raids, and in the Red River expedition, 1870, so ably organised by Viscount Wolseley. This medal, which I illustrate facing page 148, had the recipient's name and regimental number, as well as the regiment, generally impressed in square block letters. The obverse is as on the Egyptian 1882 medal. The ribbon is of scarlet, with a broad white stripe down the centre. When the medals were struck in 1899, 16,120 were issued; of these only 365 had the single bar for RED RIVER; 150 were issued with the two bars FENIAN RAID 1866 and RED RIVER 1870; 30 only were issued with the combination of RED RIVER 1870 and FENIAN RAID 1870, and twenty with three bars for FENIAN RAID 1866, FENIAN RAID 1870, and RED RIVER 1870.

Annexation of Burma.—Britain also became involved in another Burmese War while the Egyptian campaigns were proceeding. King Theebaw, who commenced his reign in a murderous manner, impelled by the financial straits into which his extravagance had led him, began to extort money from British subjects, and moreover laughed at the demand that a British agent should be installed at Mandalay, expressing his contempt for the "barbarians" in a very decided and provocative manner. As a result an expedition under Major-General (afterwards Sir H.

N. D.) Prendergast, V.C., was sent into the country; the dethronement of the King was declared by the British, Minhla was stormed and carried, and the victorious troops marched to Mandalay, where Theebaw and his army surrendered. On January 1st, 1886, Lord Dufferin proclaimed the annexation of Burma. The British troops had plenty of work, however, after this, for the dacoits gave a deal of trouble; several pitched battles were fought, and the country remained in an unsettled state for some time. Medals and bars were given for two periods during this expedition and long-drawn guerilla warfare. For the first period the Indian General Service Medal 1854 was awarded, with a bar inscribed BURMA 1885–87, and it is noteworthy that for the first time a medal of bronze was given to all authorised camp followers. To those who took part in the second period of fighting a bar for BURMA 1887–89 was awarded; the bars only, in both instances, being given where the soldier already possessed the medal.

The following regiments took part in the Burma campaign, 1885–7: Royal West Surrey Regiment; Liverpool Regiment; Somerset Light Infantry; Royal Welsh Fusiliers; Hampshire Regiment; Yorkshire Light Infantry; Royal Munster Fusiliers; 1st Battery 1st Brigade East Division; 5th Battery 1st Brigade South Division; 7th Battery 1st Brigade North Division; 8th Battery 1st Brigade London Division; 9th Battery 1st Brigade Cinque Ports Division Artillery. The following native regiments were also engaged: 1st, 2nd, 5th, 11th, 12th, 13th, 15th, 16th, 18th, 26th, 27th, 44th Bengal Infantry; 3rd, 12th, 13th, 15th, 16th, 17th, 21st, 23rd, 25th, 27th Madras Infantry; 1st Madras Pioneers; 1st, 5th, 7th, 23rd, 25th, 27th Bombay Infantry; 3rd Goorkas; 2nd and 3rd Infantry Hyderabad Contingent; 1st, 3rd, 4th, 6th Companies Madras Sappers and Miners; 2nd, 4th, and 5th Companies Bengal Sappers and Miners; 2nd Company Bombay Sappers; 1st Madras and 1st Bombay

Lancers; 2nd Squadron 2nd Madras Cavalry; 3rd Cavalry Hyderabad Contingent; 7th Bengal Cavalry; No. 1 Bombay Mountain Battery; No. 4 Punjab Mountain Battery.

Naga, 1879–80.—To those engaged in the operations against the Nagas, an uncivilised race of hillmen in the north-east of Assam, during March 1875, December 1879, and January 1880, a bar inscribed NAGA 1879–80 was awarded. Only British officers and native troops received this medal or bar, and it has so far been seldom met with in England. The medals realise in the sale-room, according to condition, anything between 10*s.* and 30*s.* apiece.

The Naga expeditionary force comprised the 42nd, 43rd, and 44th Goorkas; 18th Bengal Native Infantry, with the crews of two mountain guns.

Sikkim, 1888.—Another expedition into Sikkim became necessary in March 1888, owing to an attempt on the part of the Thibetans to annex the district which lies to the north of Darjeeling. The British field force under Colonel (afterwards General) T. Graham had to march over rough mountain tracks, and through dense jungle, where pestilent leeches, attaching themselves to man and beast, added to the hardships of the campaign. It was not until September 24th that any really big action was fought, and that was decisive; the Thibetan army of 10,000 men being attacked on the Tukola ridge, and within a short time totally routed by the British force of 2,000 with comparatively little loss to the victors.

The force engaged in this campaign was mainly composed of Goorkas, the 32nd Bengal Pioneers, and two companies of the 2nd Batt. Derbyshire Regiment. The bar for SIKKIM 1888 was added to the medal already won by those who had taken part in the Indian wars, and the medal with bar was granted to those who did not possess one. Bronze medals with the bar were awarded to authorised followers.

The Black Mountain Expedition.—In 1888 a punitive

expedition under Brigadier-General M'Queen, C.B., was sent into the Hazara or Black Mountain district on the left bank of the Indus, beyond the north-west frontier of India, to punish Khan Khel and to chastise the Azakis and Hassanzais for the murder on June 18th of Major Battye, Captain Thurston, and a number of the native surveying party which they led. The district lies in a perfect maze of mountains, on the crags and precipices of which watch-towers and villages were perched, frequently 9,000 feet above sea-level; indeed, the Northumberland Fusiliers and the Kybeerees actually carried the Gorapher Peak 9,500 feet above sea-level, thus establishing a record in altitude for an assault by British troops. The expedition entered the district from two directions, and the fourth column concentrating on October 1st prepared to act on the offensive. On the 4th they proceeded against the enemy, who was found in great force—official figures say 4,000—at the village of Kotkai, which, built upon a great mass of broken rocks on a steep spur, was defended by well-built sangars, but this was taken after some hard fighting, in which the 18th Royal Irish met a charge of Ghazis, who rushed from cover and surprised the Irishmen, but they "went at them with a will, bayoneting or shooting every Ghazi within reach." In this rush 88 fanatics paid the penalty of their madness.

Leaving the 29th Punjab Infantry in possession of Kotkai as an advance post, the expedition pushed on and blew up the hill fortress of Maidan; meanwhile the other column had been pressing forward from the Agror Valley, climbing mountains and destroying watch-towers and hostile villages, so that the pressure of the two forces, together with the losses they had sustained, compelled the hillmen to sue for peace, promising better behaviour in the future, and paying the heavy indemnities imposed. In this campaign, which concluded on November 9th, the Hazara field force lost 2 officers mortally wounded, and about 100 men killed and wounded.

The following regiments were engaged in the Hazara or Black Mountain expedition: 1st Batt. Northumberland Fusiliers; 1st Batt. Suffolk Regiment; 2nd Batt. Royal Irish Regiment; 2nd Batt. Sussex Regiment; 2nd Batt. Seaforth Highlanders; 240 men of the Scottish Division Royal Artillery; 4th Goorkas; 3rd, 14th, and 45th Sikhs; 4th, 24th, and 29th Punjab Infantry; 4th Bengal Native Infantry; 34th Pioneers; 15th Bengal Cavalry, and a Native Mountain Battery.

Hazara, 1888.—To those not already in possession of the India medal it was awarded with the bar HAZARA 1888, while the bar was added to the medals of those who already possessed the I.G.S. 1854 medal. Bronze medals were awarded to the followers.

Chin-Lushai, 1889–90.—In February 1889 an attack was made upon another surveying party under Lieutenant Stewart, in which he and his men were killed. A punitive expedition was consequently formed to proceed against the tribes inhabiting the Chin Hills and Lushai. The expedition advanced on November 15th in two columns, Brigadier-General Symons proceeding against the Chin tribes, and Colonel Tregear against the Lushais. The little forces had to make their way through the roadless and pestilent jungle, which caused the troops much suffering from disease, until, having destroyed the enemy's villages and crops, and captured a few stockades, the chiefs of the tribes thought it expedient to submit, which they did, and on April 30th, 1890, the expedition was disbanded.

The following troops took part in the expedition: those under General Symons were the King's Own Scottish Borderers, detachments of the 1st Cheshire Regiment and the Norfolk Regiment, who with the 2nd Batt. 4th and 24th Goorkas, and two companies of the Queen's Own Madras Sappers and Miners, constituted the Burma Field Force. The native troops under Brigadier Tregear consisted of the 2nd Goorkas, 3rd and 9th Bengal Infantry, 28th Bombay Pioneers, and one company of Bengal Sappers and Miners. The

India Medal 1854 with bar for CHIN-LUSHAI 1889–90 was given to those who did not already possess the medal, while to those already issued the bar was added. A bronze medal and bar was given to all authorised camp followers.

Samana, 1891.—The Meeranzies, a Pathan tribe, forgetful of the lesson given them by Sir Neville Chamberlain in 1855, again caused considerable trouble in 1891, and Sir William Lockhart marched from Kohat on April 5th to chastise them. Being defeated with considerable losses at Mastaon, they were glad to submit, and by May 25th the war was concluded, and native troops were left to garrison Samana. The British losses in this expedition were over 100 officers and men killed and wounded.

The following regiments were engaged: 1st Batt. King's Royal Rifles, two companies of the 2nd Batt. Manchester Regiment and No. 3 Mountain Battery of the Royal Artillery; 1st, 2nd, 4th, 5th, and 6th Punjab Infantry; 3rd Sikh Infantry; 1st Batt. 5th Goorkas; 15th Sikhs; 19th, 22nd, 27th, and 29th Bengal Infantry; 23rd Punjab Pioneers; 19th Bengal Lancers; Bengal Sappers and Miners, and No. 3 Peshawar Mountain Battery.

The India Medal 1854, with bar for SAMANA 1891, was awarded for this campaign; bronze medals with bar being given to authorised followers.

Hazara, 1891.—In 1891 trouble again broke out in the Black Mountain district, and another punitive expedition under Major-General Ellis was necessary in order to compel that respect for British authority which it was believed the campaign that had only concluded about sixteen months previously had succeeded in achieving. The expedition set out on March 12th, and took a couple of months to bring the tribes to their senses. On May 16th they surrendered unconditionally, hostilities were concluded, and the troops returned.

The Hazara Field Force of 1891 comprised the

1st Royal Welsh Fusiliers; 2nd Seaforth Highlanders; 1st King's Royal Rifles; 6 men of the 2nd Manchesters; Nos. 1 and 9 Mountain Batteries of the Royal Artillery; 4th Sikh Infantry; 5th Goorkas; 11th, 19th, 27th, 28th, and 37th Bengal Infantry; Bengal Sappers and Miners; 32nd Punjab Pioneers; Infantry of the Guides Corps, and the 2nd Derajat Mountain Battery. The I.G.S. 1854 Medal was given in silver to the troops, and in bronze to authorised camp followers, with bar inscribed HAZARA 1891. Those already in possession of the medal had the bar attached thereto.

Burma, 1889–92.—During the years 1889–90–1–2 a series of expeditions was organised to put down disturbances in Burma and Lushai; for the services rendered by the various forces the India Medal 1854, with bars for BURMA 1889–92 and LUSHAI 1889–92, were issued to the troops who took part in the following expeditions: The Pokhau expedition from April 16th to May 16th, 1889; Touhon expedition from September 17th, 1889, to April 1890; Chinbok Column, January 1st to 20th, 1890; Thetta Column, January 1st to 4th, 1891; Chinbok Column, January 8th to February 14th, 1891; Morneik Column, January 27th to March 28th, 1891; Wuntho Field Force, February 18th to May 7th, 1891; Tlang-Tlang Column, March 29th to April 3rd, 1891; Baungshe Column, December 25th, 1891, to February 29th, 1892; Irrawaddy Column, December 15th, 1891, to April 18th, 1892; North-Eastern Column, December 15th, 1891, to April 7th, 1892.

Lushai, 1889–92.—The medal with bar LUSHAI 1889–92 was awarded to the officers and men comprising the Lushai Expeditionary Force from January 11th to May 5th, 1889; the relieving force under Lieutenants Swinton, Cole, and Watson, which went to the assistance of the forts Ayal and Changsil when attacked by the Lushais, September 9th to December 9th, 1890; the expedition under Captain Hutchinson against the

ROYAL NAVAL RESERVE.

H.E.I. CO.'S MERITORIOUS SERVICE MEDAL.

EDWARDIAN ARMY.

LONG SERVICE AND GOOD CONDUCT MEDALS.

Jacopa Village to avenge the attack on Mr. Murray, February 20th to March 3rd, 1891; the expedition under Captain Lock from March 1st to June 1892, necessitated by the general rising of the Eastern Lushais, and the attack at Lalbura upon a force under Mr. M. Cabe; also the force under Captain Shakespear, which advanced into the South Lushai hills between March 16th and May 13th, 1892. All dates are inclusive.

Two divisions took part in this campaign, and European troops participated only with the Burma section. 1st King's Own Scottish Borderers, and detachments of 1st Batt. Cheshire and Norfolk Regiments. The native troops were: 2nd Batt. 4th and the 42nd Goorkas; two companies of the Queen's Own Madras Sappers and Miners. The troops engaged with the Chittagong column were: 2nd Goorkas, 3rd and 9th Bengal Infantry; 28th Bombay Pioneers, and a company of Bengal Sappers and Miners.

North-East Frontier, 1891.—Early in 1891 the British Resident at Manipur was attacked, and columns from Tamu on March 28th, Silchar on April 15th, and Kohmia on April 20th were sent forward to secure the submission of the Rajah and the punishment of those responsible for the attack on the Resident. The Rajah submitted to the terms laid down, and by May 7th, 1891, the campaign was brought to a conclusion.

The Manipur Field Force consisted of about 300 of the 4th King's Royal Rifles; 1st Batt. 2nd Goorkas; 2nd Batt. 4th Goorkas; detachments of the 42nd, 43rd, and 44th Goorkas; 18th Bengal Infantry, and the 8th Bengal Mountain Battery. For this expedition the bar for the N.E. FRONTIER 1891 was added to the India General Service Medal 1854.

Hunza, 1891.—Lieutenant-Colonel Algernon Durand had foreseen that trouble was brewing, and made very praiseworthy efforts not only to circumvent it, but to ensure that should hostilities occur he would be in a position to defend himself against the robbers and slave-dealers who surrounded him. An attack had been made on the Chalt garrison, and further trouble was anticipated when Durand informed the chiefs of Hunza and Nagar, without referring to their misconduct, that the British intended having free access to their territory in order that "all requisite arrangements might be made for holding the frontier, and that unless they complied with our demands troops would enter their country, and make roads in spite of them." The tribesmen armed, and "confident in the strength of their defiles . . . and to defeat Kashmir troops as of yore," they collected, and Colonel Durand concentrated at Chalt, from whence on November 29th an ultimatum was sent to the chiefs; but the envoy was insulted and robbed of his horse, so the British force crossed the Hunza River on December 1st by a temporary bridge thrown by Captain Aylmer, V.C., and the war which Colonel Durand had striven to prevent commenced, and with a little force of 1,000 men and 2 guns he opposed himself to about 5,000 or 6,000 men.

The force was made up of 180 of the 5th Goorkas, 400 Goorkas and Dogras of the Kashmir Body-guard, 250 Dogras of the Kashmir Raga Pertab Regiment, 150 Punyali levies, and the commander's personal escort—20 men of the 20th Punjab Infantry. The only Europeans present were officers. The forts of Nilt were assaulted and captured after very desperate fighting, in which Captain Aylmer and Colonel Durand were severely wounded, and in the moment of victory the command passed out of the latter's hand. In two hours the place was captured, but owing to the failure to advance 4 British officers had been lost, and nearly 40 men killed and wounded after the successful assault. The cliff opposite Nilt, a very difficult position, was next scaled after "a little Dogra sepoy named Nagdu one night succeeded in climbing 1,200 feet up the cliff, and found a practicable path." Then the place was stormed by the men under the command of Captain Colin Mackenzie of the Seaforths; Lieutenant Manners-Smith led a little body of Kashmir troops up a cliff 1,000 feet high, and within two days Nagar was occupied, and the Hunza chief with Uzr Khan fled across the frontier. It is noteworthy that from the men of the district against whom we had fought, Colonel Durand raised the Hunza-Nagar levies who later fought so well for us.

In this little hard-fought campaign three Victoria Crosses were won, by Captain (now Major-General C. B. Aylmer, Lieutenant (now Major) Boisragon, and Lieutenant (now Lieutenant-Colonel C. I. E.) Manners-Smith, while several of the native troops earned the Order of Merit for their gallantry and devotion.

For this arduous expedition the I.G.S. Medal 1854, with bar for HUNZA 1891, was issued to those participants who had not already received the medal for other services. Those who possessed the medal received the bar only.

East and West Africa.—A great deal of fighting had been forced upon us by the African tribes, against whom expeditions or columns had been sent from 1887 to 1892. First against the Yonnie Tribe, when the coloured men of the 1st West India Regiment fought side by side with men from the crews of H.M.S. "Acorn," "Icarus," and "Rifleman," from November 13th, 1887, to January 2nd, 1888. Next an expedition had to be sent up the Gambia River, the men of H.M.S. "Swallow" composing the force which operated between December 29th, 1891, and February 5th, 1892. Then the 1st West India Regiment again took the field on March 8th, 1892, against Tambi, men of H.M.S. "Alecto," "Race," "Sparrow," "Thrust," and "Widgeon" co-operating, and in the same year the native troops took part in the expedition against Toniataba, and in the expedition against the Jebus. For these services a medal similar to that issued for the Ashantee War was granted on November 1st, 1892, but, as in the case of the Zulu War medal, dates upon the bars are the only indication of the particular service, or services, rendered. Three bars were sanctioned, 1887–8, 1891–2, and 1892, so that the last bar was issued for three different expeditions. This is known as the East and

West Africa medal, but it was not a distinct issue, for those who possessed the Ashantee medal were only given the bar or bars. The ribbon is the same as used for suspension with the Ashantee medal, and the names were either engraved or impressed on the edge of the medal.

Central Africa Medal.—In 1895 a series of operations were found necessary in Central Africa, and to decorate the native troops engaged the same medal was awarded without bars. It was, however, suspended from a brown, white, and black ribbon, 1¾ in. wide, by means of a swivel ring, so that the ribbon was run through the ring as in the case of the Abyssinian medal. This slight distinction, and the issue of a new ribbon, constitutes the medal a distinct decoration, and those who had already received the West African medal were entitled to it. It, moreover, distinguishes the Central Africa medal from the East and West Africa medal, which again had name bars added for WITU 1890 to the crews of ten of H.M. ships; LIWONDI 1893 to 3 officers and 34 men of H.M.S. "Herald" and "Mosquito"; WITU AUGUST 1893 to 200 seamen and 36 marines; JUBA RIVER 1893 to 1 officer and 40 men of the "Blanche"; LAKE NYASSA 1893 to 100 Sikhs and parties of men from H.M.S. "Adventurer" and "Pioneer"; a bar inscribed 1893–94 was issued to those who took part in the operations against the Sofas from November 1893 to January 1894; GAMBIA 1894 was inscribed on the bar to 50 of the 1st W.I.R. and parties of seamen and marines from four of H.M. ships; BRASS RIVER 1895 to men of four of H.M. ships, and NIGER 1897 to the troops taking part in the expedition to Egbon, Bida, and Ilorin or garrisoning Fort Goldie and Lokoja from January 6th to February 25th, 1897. The bars are square and set somewhat far apart; the name and rating, etc., of the recipient impressed on the edge.

Between January 1894 and April 1898 a series of operations and expeditions was found necessary against various chiefs in British Central Africa, and for these a bar inscribed CENTRAL AFRICA 1894–1898 was added to the Central Africa medal, which when issued in 1895 had a swivel ring instead of a clasp. When the bars were added in 1899 the straight-bar clasp with claw clutch—as used with the first issue of this medal, the Ashantee—replaced the swivel ring.

Ashanti, 1896.—In 1895 trouble in Ashantee necessitated the sending of a small force, comprising the 2nd Batt. of the West Yorkshire Regiment, and a composite battalion made up of 20 men of each of the three Guards Regiments, and the following eight regiments: 1st Yorkshire Light Infantry, 1st Leinster Regiment, 1st Northumberland Fusiliers, 2nd Royal Irish Fusiliers, 2nd Devons, 2nd Shropshire Light Infantry, 3rd Batt. Royal Rifles, 2nd Batt. Rifle Brigade, together with the necessary detachments of Artillery, Engineers, Army Service, Ordnance, and Medical Corps. For this campaign against King Prempeh, a very distinctive decoration was given: a St. Andrew's Cross bisects the corners of a four-pointed star, over which is a circular centre containing the Imperial crown surrounded by a plain band inscribed ASHANTI above, and 1896 below. On the back, in raised letters, is the inscription FROM THE QUEEN. The star is suspended from a yellow ribbon, with rather broad black stripes near the edges, by means of a ring; it was issued unnamed, but the Colonel of the Yorkshires had the name of each of his men and that of the regiment engraved upon their stars.

West African Medal.—For their services against Mwele a number of men from four of H.M. ships were awarded the West African medal, but, instead of a bar, MWELE 1895–6 was impressed on the edge, on either side of the clasp claw, together with the name and ship of the recipient. Those already in possession of the medal had Mwele and the date engraved on the edge. The 24th and 26th Bombay Infantry were also engaged, and they likewise received the medal: the latter have the names, etc., engraved inslanting script like India 1895 medals. A bar for NIGER 1897 was issued later, and in 1898 bars for BENIN 1897 were issued to men from nine of H.M. ships, and a bar for DAWKITA 1897 was also added for the members of the Gold Coast Constabulary who took part in the defence of Dawkita. In 1899 authority was given to add the bar for SIERRA LEONE 1898–99 to the East and West Africa medal for those engaged in the country between February 1898 and March 1899. Men of the West African Rifles and 1st West India Regiment were engaged, also some Royal Garrison Artillerymen. The name and date are arranged in two lines, and the bar is consequently rather wider than the others. The named edges are engraved in capital Roman letters, but some issued to Europeans are impressed. In 1900 it was decided to issue four more bars, and to revert to the use of dates instead of names; these bars cover 1896–8, 1897–8, 1898, and 1899. In the following year the bar inscribed 1896–99 was issued to those who had been employed in the northern parts of the Gold Coast, or in the hinterland of Lagos; later those who had been engaged in the Kadima and Munchi expeditions were awarded the medal and bar for 1900.

The Relief of Chitral.—Early in 1895 the war cloud was gathering in Chitral; the Nizam-ul-Mulk, the nominee of the British, had been shot at the instigation of his supposedly idiot half-brother, Amir-ul-Mulk, and the Indian Government was in no great haste to recognise the succession attained in such a way, especially as the Mehtar indulged in the most savagely murderous deeds. Later came the news to Surgeon-Major (afterwards Sir George) Robertson that on the night of January 22nd Umra Khan had crossed the Lowari Pass into Chitral territory with 3,000 or 4,000 followers, demanding that Amir-ul-Mulk should join him. Then Robertson made up his mind to reach Chitral, and occupy the fort in the valley, as speedily as possible. The little force pressed on, many an unfortunate man getting frost-bitten during the terribly cold nights. An ultimatum was then sent to Umra Khan, indicating that he must at once leave Chitralis territory. Of course he did not; one or two unfortunate affairs happened—the disaster at Mastuj and at the Koragh defile, where a number of the surviving Sikhs, after existing without food or water for seven or eight days, were cajoled, by the promise of life, from the little sangars in which they had taken refuge, and then cruelly murdered. The gallant defence of Rashun under Edwardes and Fowler, until they were captured by the cunning of Muhammad Isa, must be mentioned *en passant.*

On March 27th the famous march to Chitral began, and to join up with the force marching from India Colonel J. G. Kelly set out with his little force of 400 of the 32nd Pioneers and 40 Kashmir Sappers, with two mountain guns, from Gupis. Laboriously climbing the hill tracks, a foot deep in snow, until almost snow-blind, his men reached Ghizr, 10,000 feet above sea-level, where a body of Hunza-Nagar levies were encamped. These he added to his force; then Kelly endeavoured to push through the great carpet of snow 3 feet thick. It was of no avail, for the pitiless snow blinded his men, and he had to return to Ghizr until it stopped, which it did on April 3rd. Of the heroic march of these men, of the difficulties they faced and undauntedly overcame, one is tempted to write at great length, but Sir George Robertson does fitting justice to this party of his rescuers in "Chitral, the Story of a Minor Siege."

The main force from India under Major-General Sir Robert Low advanced across the Malakund Pass after a smart fight on April 3rd, in which the Gordons and Scottish Borderers, "racing for the honour of their regiments . . . swarmed over the Dechoities at a wonderful speed, and breathless but triumphant had crowned the pass by 2 o'clock, the 60th Rifles being close up." Here the British force lost 11 killed, and 47 wounded. The Jandol Valley was attacked on April 17th, and occupied next day. On the 19th General Gatacre started with a couple of battalions and some guns for

Dir and Chitral; pushing across the rugged Janbatai ridge, he learned that the garrison of Chitral was almost at its last gasp—and five days' march lay between him and his goal! He therefore asked permission from Sir Robert Low to make a dash for Chitral with 500 of the Buffs, a few native sappers, and a couple of mountain guns; but when sanction was obtained it was learned that Kelly had reached Chitral, and that there was no need to hurry.

The Defence of Chitral.—Meanwhile Surgeon-Major Robertson had "directed" by a gesture Amir-ul-Mulk to leave the Mehtarship, and had raised in his stead his little brother Shuja-ul-Mulk. Then on March 3rd an unfortunate and desperate fight took place, in which Surgeon Whitchurch earned the V.C. by one of the most unselfish actions on record, and some devoted Goorkas gained the Order of Merit for the rescue of a mortally wounded officer. On March 4th the siege began in earnest, and preparations were made for the defence. Within the walls were 550 persons to be fed; to defend the place about "340 riflemen, but, excluding those in hospital, only 83 Sikhs—good shots, and trustworthy soldiers," and 52 Chitralis. How well they held the fort until the enemy melted away on April 18th, 1895, is one of the glorious records of British pluck and Indian devotion and gallantry. During the siege, and the reconnaissance of March 3rd, 41 were killed, and 62 wounded, the latter including the commander.

The following took part in the defence: Sir George Robertson, K.C., S.I., and 6 officers; Surgeon-Captain H. F. Whitchurch, 90 of the 14th Sikhs, 300 of the 16th Punjabis, and 4th Kashmir Rifles, assisted by about 40 servants and followers, who likewise received the India Medal 1895, with bar for the DEFENCE OF CHITRAL 1895. It is noteworthy that the new medal is thicker than the one it replaced, and weighs 1¼ oz. instead of 1 oz.

The India Medal, 1895.—For the Defence and Relief of Chitral it was decided to issue a new India medal, and as a result that illustrated facing page 192 was designed "to commemorate the military operations in, and on the frontier of, India, and to be in future known as 'The India Medal 1895.'" The bars are the same in design as those used on the old India medal.

In 1898 another bar was awarded to those who had taken part in the operations on the North-West Frontier between June 10th, 1897, and April 6th, 1898, and to those who formed the Tirah expeditionary force which proceeded beyond Kohat or Peshawar between October 1897 and April 6th, 1898. This bar was inscribed PUNJAB FRONTIER 1897–8. In 1897, as a result of the efforts of fanatical priests, the hillmen in the Tochi Valley made an attack upon a British officer and his escort, and then, following an outbreak in the Swat Valley, Malakand was attacked by thousands of the natives, but through the gallant efforts of the little garrison the place was held until the arrival of the relieving force. The troops who took part in the defence and relief of Malakand and Chakdara received the bar inscribed MALAKAND 1897. During the same year it was found necessary to again take up arms against the Pathan Meeranzies, and the troops engaged received the bar for SAMANA 1897. In August 1897 the Afridis, Mohmands, and Orakzais combined against the British, and troops were sent to the frontier. That portion of the Tirah expeditionary force which was also engaged beyond Kohat and Peshawar, likewise the Kurram Movable Column and the Peshawar Column, between October 1897 and the early part of April 1898, received the bar for the TIRAH 1897–98. This campaign is made famous by the storming of the heights of Dargai, and the coolness of Piper Findlater of the Gordon Highlanders, who continued to play his pipes when a bullet had crippled him. He was awarded the V.C.

The following troops were engaged in the various forces employed: the Mohmand Field Force comprised, in the 1st Brigade, 2nd Batt. Argyll and Sutherland Highlanders; 1st Punjab Infantry; 1st Sikhs; 33rd

Bengal Infantry; 2nd Company of the Bengal Sappers and Miners; 6 guns of the Peshawar Battery, and a detachment of the Hospital Staff. The 2nd Brigade included the Somerset Light Infantry; 3rd Batt. Rifle Brigade; 6th Bengal Infantry; 14th Sikhs; 25th Punjab Infantry; a squadron of the 1st Punjab Cavalry; 13th Bengal Lancers; 4 guns of the 6th Bombay Mountain Battery, and 51st Field Battery.

The Malakand Field Force comprised, in the 1st Brigade, 1st Somerset Light Infantry; 2nd Batt. 1st Goorkas, and the 21st Bengal Infantry. The 2nd Brigade included 2nd Oxfordshire Light Infantry; a regiment of the Imperial Service Troops; 9th Goorkas; 37th Bengal Infantry; 28th Bombay Infantry; two squadrons 11th Bengal Lancers, and the 13th Bengal Lancers; 3rd Mountain Battery of the R.A.; No. 5 Bombay Mountain Battery; No. 8 Bengal Mountain Battery; No. 4 and No. 5 Companies Bengal Sappers and Miners. The 3rd Brigade consisted of 1st Royal West Surrey Regiment; two squadrons of the 4th Dragoon Guards and 11th Hussars; K Battery Royal Horse Artillery; No. 1 Mountain Battery R.A.; 22nd and 39th Bengal Infantry; No. 3 Company Bengal Sappers and Miners; two squadrons of the 11th Bengal Lancers.

The Samana Force comprised, in the 1st Brigade, 1st Royal West Kent Regiment; 24th and 31st Punjab Infantry; 45th Sikhs. Included in the 2nd Brigade were 1st Buffs; 35th Sikhs; 38th Dogras; Guides Infantry; two squadrons 11th Bengal Lancers; 5th Bengal Mounted Battery; 5th Company of Bengal Sappers and Miners. The 3rd Brigade was made up of 1st Batt. Queen's; No. 1 Battery R.A.; 22nd Punjab Infantry; 38th Bengal Infantry; 39th Gahrwalis; 3rd Company Bengal Sappers and Miners; half company of Madras Sappers; a squadron of the 10th Bengal Lancers, and 2 of the 11th.

The Tirah Field Force was made up of two divisions of two brigades each. The 1st Division comprised:

2nd Devons; 2nd Yorkshire Regiment; 1st Royal West Surrey; 2nd Royal Irish; No. 1 Mountain Battery R.A.; 2nd Batt. 1st Goorkas; 2nd Batt. 4th Goorkas; 3rd Sikh Infantry; 28th Bombay Infantry; 30th Bengal Infantry; Kapurthala Infantry; two squadrons of 18th Bengal Cavalry; No. 1 Kohat Mountain Battery; No. 2 Derajat Mountain Battery; 3rd and 4th Companies Bombay Sappers and Miners; Maler Kotla Sappers. The 2nd Division included: 1st Gordons; 1st Dorsets; 1st Northamptons; 2nd King's Own Borderers; Nos. 8 and 9 Batteries R.A.; 1st Batt. 2nd and 3rd Goorkas; 36th Bengal Infantry; Jhind Infantry; 21st Madras Infantry; two squadrons 18th Bengal Lancers; No. 4 Company Madras Sappers and Miners; Sirmoor Sappers.

The lines of communication were kept by the 22nd and 39th Bengal Infantry; 2nd Batt. 2nd Goorkas; 2nd Punjab Infantry; 3rd Bengal Cavalry; Jeypore and Gwalior Transport Corps.

The Kurram Column consisted of: 4 guns 3rd Field Battery R.A.; 12th Bengal Infantry; Nabha Infantry; Central India Horse; 6th Bengal Cavalry.

The Peshawar Column was composed of: 2nd Royal Inniskillen Fusiliers; 2nd Oxford Light Infantry; 3rd Mountain Battery R.A.; 57th Field Battery R.A.; 9th and 45th Bengal Infantry; 9th Bengal Cavalry; No. 5 Company Bengal Sappers and Miners.

In reserve was the Rawul Pindi Brigade, composed of: 2nd Yorkshire Light Infantry; 1st Duke of Cornwall's Light Infantry; 2nd Hyderabad Infantry; 27th Bombay Infantry, and the Jodhphur Lancers.

Matabeleland, 1893.—The first Matabele War, which broke out in 1893, will always be remembered for the brave but unavailing stand made by Major Wilson's little band at the Shangani River. Then Lobengula was defeated, and things settled down—as they generally did in South Africa—for a time, for within three years a second Matabele War had to be contended with until by December 31st, 1896, the natives in Rhodesia had been quietened again. In this latter war Prince Alexander of Teck took part with his regiment, the 7th Hussars.

Bechuanaland.—At Christmastide a rebellion broke out at Pokwani in Bechuanaland, about 40 miles north of Kimberley. The Cape Mounted Police and the Diamond Fields Police set out on Christmas Eve to deal with the recalcitrant natives, and after manœuvring in the pouring rain all Christmas Day, the chief's stronghold was attacked and carried on Boxing Day, and the rising was entirely crushed by December 27th, 1896. For this short campaign the Cape Government, with the approval of the Home Government, issued a medal in 1900 to those who had participated in this and other campaigns against the natives, and bars were issued for service in Basutoland in 1880–1, and Transkei, 1880–1, including Griqualand East and Tembuland.

The Cape of Good Hope General Service Medal.—The medal awarded is entitled the Cape of Good Hope General Service Medal; it is $1\frac{7}{16}$ in. in diameter, and bears on the obverse the bust of Queen Victoria facing left, with the inscription VICTORIA REGINA ET IMPERATRIX. On the reverse are the arms of Cape Colony, and the motto SPES CONA. It is suspended from a $1\frac{1}{4}$ in. blue ribbon, with a broad yellow stripe down the centre. The bars issued with the medal are, reading upwards from the medal, BASUTOLAND, TRANSKEI, BECHUANALAND. The only soldiers connected with the Imperial forces who received this medal with any of the bars were officers serving with the Colonial troops. A medal with the single bar "Transkei" has realised £3 2*s.* 6*d.* in the sale-room; with two bars, "Transkei" and "Basutoland," £2 15*s.* The medals are generally engraved in rather large square Roman capitals.

The Dongola Expedition.—The Dongola expedition took place in 1896, and for this the Khedive of Egypt ordered a medal to be issued to all the British and native troops who served at, and south of, Sarras between March 30th and September 23rd, 1896, and at Suakin under Brigadier-General Egerton during the same period. Two bars were awarded; one inscribed HAFIR for services south of Fareia on September 29th, and the other FIRKET for those engaged south of Akashen on June 7th, each name being inscribed in English and Arabic on a single bar. The medal bears on the reverse an oval shield, charged with three crescents, and stars above, backed by a trophy of arms and flags, and a panel below bearing an inscription in Arabic reading "The recovery of the Sudan, 1314 H." On the obverse is the cypher of the Khedive Abbas Hilmi II, and 1314 H.—A.D. 1897. It is suspended from an orange-coloured watered ribbon, with a broad blue stripe down the centre, by means of a straight clasp. The medal was issued unnamed; bronze medals without bars were issued to camp followers and non-combatants. This being a foreign decoration it was necessary for Queen Victoria to give her troops permission to wear the medal. The first British regiment to receive it was the 1st Batt. North Staffordshires. Later the following regiments received the medal without bars, which were added afterwards: 1st Seaforth and Cameron Highlanders; half a battalion of the 1st Lincolns; 21st Lancers; a detachment of the 16th Company Eastern Division R.A.; 32nd Field Battery R.A.; 2nd Company Royal Engineers. To those who took part in the operations at and about Kerma, on the Nile, in 1897, the Khedive's medal was given with a bar inscribed SUDAN 1897, and to those who were engaged at and near Assouan and Abu-Hamed the medal with bar for ABU-HAMED was awarded; the bars only being issued to those already in possession of the medal.

Matabeleland, 1893, and Rhodesia, 1896.—In 1896 Queen Victoria authorised the granting of a medal by the British South Africa Company to its own forces and details from British regiments and the local Mounted Rifles and Police. All who took part in hostilities between October 16th and Christmas Eve 1893 were awarded the medal for MATABELELAND, and those only of the Queen's officers who had received permission to take part in the operations were allowed to accept and wear the medal. It is $1\frac{7}{16}$ in. in diameter, and has on the reverse a representation of the British lion boldly charging, though wounded in the chest with an assegai. In the exergue, in tall thin Roman capitals, is BRITISH SOUTH AFRICA COMPANY, and above either MATABELELAND 1893 or RHODESIA 1896. On the obverse is a rather poor effigy of Queen Victoria. The suspender is very ornate, the rose, shamrock, and thistle having been used in a weak sort of manner to decorate it. This clasp or suspender is the most unsuitable and least beautiful of any in the whole series. The ribbon, $1\frac{1}{4}$ in. wide, is composed of three blue and four yellow stripes. It is the first instance of a chartered company being authorised to award a medal since the H.E.I. Co. gave the Mutiny medal. A bar for RHODESIA 1896 was given to those already in possession of the medal. The bar for MASHONALAND 1897 was added the following year, but those who did not possess the medal were awarded one with the name and date on the reverse as explained above. In this campaign the only Imperial troops employed were detachments of the 2nd Hampshires and the 7th Hussars.

Regiments Engaged.—Among those who received the medal for Matabeleland were a company of the 2nd W. Riding Regiment; 1 officer and 18 men of the 1st Batt. "Black Watch"; 3 men of the 2nd York and Lancaster Regiment; a detachment of the 3rd Dragoon Guards; Cape Mounted Rifles and British Bechuanaland Police. Those entitled to wear the medal for services in Rhodesia between March 24th and December 31st, 1896, were details of the 2nd, 3rd, and 4th Royal Rifles; 2nd and 4th Rifle Brigade; 1st Royal Irish; 1st Royal Dublin Fusiliers; 2nd Royal Irish Fusiliers; 1st Derbys; 1st Leicesters; 2nd Norfolks; 2nd Hampshires; 2nd Royal West Kent; 2nd W. Riding Regiment; 2nd York and Lancasters;

24th, 25th, and 26th Western Division Royal Artillery; Army Ordnance and Army Service Corps. The 7th Hussars was the only cavalry regiment employed.

THE SUDAN

The battles of Toski in 1889, or Tokar in 1890, did not settle the unrest in Egypt. It smouldered and again broke into serious flame in 1897. Thanks, however, to the admirable organisation of the Egyptian army by Earl (then Sir Herbert) Kitchener, and the prompt measures which he took to deal with the Dervishes, the British flag and the Egyptian crescent soon flew over what remained of the Residency of Khartoum. Before, however, this was achieved the battles of the Atbara and Omdurman had to be fought, and for these bars bearing the legends THE ATBARA and KHARTOUM were issued with the medal to those who took part in the fights.

The Atbara.—The battle of the Atbara was fought on the morning of Good Friday 1898, the attack of the Anglo-Egyptian army of about 12,000 men being made at sunrise after a night march, during which the troops had plodded across the desert from Umdabieh in brilliant moonlight; they halted just before 5 o'clock for a rest, during which they shiveringly discussed the prospects of the battle, for if the days are scorching in the Sudan, the nights are icily cold. Then the squares formed, and the army marched forward to battle. At first the artillery played upon Mahmud's camp, but the Dervishes seemed to take little heed of it, for it is said it was fully half an hour before they began to reply with their rifles. Then at 7.30 a.m. the great guns stopped; the bugles rang out "Advance!" and as the pipers played "The March of the Cameron Men," the infantry advanced to the attack. Again the enemy took little heed, until the Cameron men reached the top of the ridge overlooking the zareba, and then the rifles of the Dervishes rang out, but they could not stop the Camerons' onslaught; followed by the Lincolns, the Seaforths, and the Warwicks, they went clean through the zareba. General Gatacre and Private Cross of the Camerons distinguished themselves by being the first in. The onslaught of the British and Sudanese was irresistible, and whoever could of the fighting Dervishes rushed riverward. Osman Digna again decamped, but a number of other leaders, including Mahmud, were either killed or taken prisoners; 4,000 Dervishes were taken prisoners, and 3,000 killed. The Camerons lost Captain Finlay and Urquhart, and the Seaforths Lieutenant Gore, the three Scots officers falling as they led their men into the zareba; 18 men were also killed, and 88 wounded. Fourteen native officers were killed, while the native regiments also lost 50 killed, and 319 wounded. The Sirdar at the conclusion of the battle highly complimented the Colonel of the Camerons upon the steadiness of their advance in face of the fire which greeted them at the zareba.

Omdurman.—This battle quieted the Dervishes until August, when preparations were made to give them the *coup de grâce.* This happened on September 2nd, 1898, at Omdurman—made famous by the maiden charge of the 21st Lancers under Captain Martin against enormous odds—when the enemy lost 11,000 killed, 16,000 wounded, and 4,000 made prisoners. The British infantry division was again commanded by General Gatacre, the Brigadiers being Generals Wauchope and Lyttleton. The 1st Brigade consisted of the Camerons, Seaforths, Lincolns, and Warwicks, and a Maxim Battery. The 2nd Brigade comprised 1st Grenadier Guards; 1st Northumberland Fusiliers; 2nd Lancashire Fusiliers, and 2nd Rifle Brigade. Macdonald, Maxwell, and Lewis again commanded three brigades of the Egyptian army, and Collinson Bey a fourth. There were 7,500 Britishers, and 12,000 Egyptians.

On April 2nd the Dervish army advanced against the British, their front extending for about 3 miles. The British formed an obtuse angle, and at 6.30 a.m. opened fire upon the advancing hosts with such precision that hundreds were mown down, but the Dervishes

VICTORIAN VOLUNTEER OFFICER'S DECORATION.

EDWARDIAN COLONIAL AND AUXILIARY FORCES DECORATION.

VOLUNTEER OFFICERS' DECORATIONS FOR LONG SERVICE.

did not falter. As Steevens said, "No white troops could have faced that torrent for five minutes, but the Baggara and the blacks came on." There was no fear; the black-eyed houris were waiting with outstretched arms to receive such doughty warriors. Still they could not stand against their foe, but they retired with such caution that they were able to re-form. The pressure of their well-drilled enemy, however, was too great, and they began to flee. The bravest of them, however, planted their standards as rallying-points, and stood their ground until they were killed; the others appeared to be making for Omdurman, 5 miles from the scene of battle, and to prevent this the 21st Lancers were ordered to charge. They hacked their way through the dense mass of men who feared nothing. Then the Khalifa's stalwarts made a last effort, but Macdonald's Sudanese steadily met the onslaught; repulsed it, sent the Dervishes flying in all directions, and the battle of Omdurman was won. Mahdism was worsted, but Osman Digna had again eluded the victors. At last Khartoum was taken, the captives set free, and the dominion of the rebel slave-trader overthrown.

Khartoum.—To those who took part in this battle, which ended in the taking of Khartoum, a bar for KHARTOUM was added to the Khedive's medal, and a bar for GEDAREF to all engaged in the capture of the place, and the fighting which followed in the district.

The following British regiments were engaged in the campaign: 1st Grenadier Guards; 1st Seaforth and 1st Cameron Highlanders; 1st Warwicks; 1st Northumberland Fusiliers; 2nd Lancashire Fusiliers; 1st Lincolns; 2nd Rifle Brigade; a detachment of 1st Royal Irish Fusiliers, with 4 Maxim guns; 16th Company Eastern Division R.A.; 32nd and 37th Field Batteries R.A.; 21st Lancers; a detachment of Royal Engineers; Army Service and Army Ordnance Corps; Royal Army Medical Corps.

Sudan, 1899, and Gedid.—Early in 1900 a bar bearing the record SUDAN 1899 was awarded to those who had taken part in the second Dongola campaign, and a medal with bar for GEDID was issued to those who were engaged in the actions at that place on November 22nd and 24th, 1899, which resulted in the defeat of the Khalifa Abdulla El Taaishi. Bronze medals were issued to the civilian syces and authorised camp followers. The number of bars issued with the Khedive's Sudan medal varies from one to eight, but it is seldom that European soldiers received medals with more than two bars. European officers in some instances received more, but Major-General Sir Archibald Hunter, I think, has the distinction of possessing one bearing six bars.

Three Later Bars.—In 1905 and 1906 it was decided to add three more bars to the Khedive's Sudan medal, one for BAHR-EL-GHAZEL 1900–2 for services rendered in the reconquest of that province; one for JEROK for operations in the Blue Nile province against Wad-el-Mahmud in 1904; one for NYAM-NYAM to those employed against the Nyam-Nyam tribesmen in the province of Bahr-el-Ghazel in 1905.

Queen's Sudan Medal.—In March 1899 Her Majesty Queen Victoria approved of a new medal being struck to commemorate the military operations in connection with the reconquest of the Sudan. The medal, $1\frac{2}{5}$ in. in diameter, in silver was granted to all officers, warrant officers, non-commissioned officers, and men of the British, Indian, and Egyptian forces, and native allies who were employed in the military operations in Egypt resulting in the capture of Abu-Hamed, the reconquest of the Province of Berber, the defeat of Emir Mahmud's army on the Atbara, and the final operations resulting in the overthrow of the Khalifa's troops at Khartoum. Civilian syces, civilian servants of officers, and authorised followers were entitled to the medal in bronze. No bars were issued with this medal. On the obverse is a half-length effigy of Queen Victoria, facing left, crowned, with a lace veil flowing behind, on which a small Imperial crown is placed, wearing the Ribbon and Star of the Order of the Garter, the Royal Order of

Victoria and Albert, and the Imperial Order of the Crown of India, holding a sceptre in her right hand. Legend, VICTORIA REGINA ET IMPERATRIX. On the reverse is a figure of a winged Victory seated, having a palm-branch in her outstretched right hand, and a laurel wreath in her left. There are two flags behind on either side—the British being on the left and the Khedive's to the right; beneath on the pedestal which supports the figure is the word SUDAN in intaglio, the space beneath the pedestal being filled by the decorative disposition of three lilies of the Nile in leaf. This is, in my opinion, the most symbolic and beautiful medal issued since Wyon took for his model a Greek coin, and based his figure of Victory—which adorns the Waterloo medal—upon it. The name of the recipient, together with his rank and regiment, was generally engraved upon the edge of the medal in neat Roman capitals. I have seen some with the lettering impressed in very small Roman capitals. The medal is suspended by a straight bar from a black and yellow ribbon 1¼ in. wide, divided down the centre by a narrow crimson stripe.

East and Central Africa Medal.—A medal to replace the old one with swivel ring was issued to those who took part in the troubles which occurred in the Uganda Protectorate during 1897 and 1898. The medal, 1⅖ in. in diameter, had on the obverse the same bust of Queen Victoria as the Sudan medal, and on the reverse Britannia with the British lion beside her, holding out in her left hand a palm branch and a scroll toward the rising sun, and in her right a trident. In the exergue, in Roman capitals, is the denomination EAST AND CENTRAL AFRICA. The medal for Somaliland 1902–4 (illustrated facing page 296) is the same but for the inscription in the exergue. Two bars were issued with the medal, LUBWA'S and UGANDA 1897–8. A bar with the date 1898 was issued to those who fought against the Ogaden Somalis from April to August 1898, and a bar for UGANDA 1899 was awarded to those employed in the military operations against Kabarega in the Protectorate between March and May 1899. The ribbon, $1\frac{1}{4}$ in. wide, is half red and yellow, and the names, etc., of recipients are impressed or engraved on the edge in light Roman capitals.

Royal Niger Co.'s Medal.—For participation in military operations in the Niger Company's territories during the period of 1886 to 1897 the Company awarded a silver medal to the members of its executive staff, and bronze medals to the natives. The names of the recipients were impressed upon the silver medals, but the native's number only was indented upon the edge of the bronze medals issued to those who were actually serving on December 31st, 1899. These medals, which were struck by Messrs. Spink & Son, bear on the obverse the diademed bust of Queen Victoria, and on the reverse a shield bearing a Y-shaped elevation, with the words ARS, JUS, PAX; behind the shield is a trophy of arms and flags. This medal, $1\frac{1}{2}$ in. in diameter, is suspended from a ribbon $1\frac{1}{2}$ in. wide, composed of three equal stripes of yellow, black, and white, by means of a straight clasp. The bars, $1\frac{2}{5}$ in. long by $\frac{3}{10}$ in. wide, have the record NIGERIA 1886–1897 attached to the silver medals, but NIGERIA only on the bronze.

Ashanti, 1900.—After the rebellion of certain native tribes had been quelled, and the siege of Kumassi raised, King Edward VII approved of a medal—the first bearing his effigy—being issued to those who had taken part in the operations, and a bar inscribed KUMASSI also to those who had defended or relieved the city. On the obverse is the bemedalled bust of King Edward in Field-Marshal's uniform, which was used later on the South Africa medal 1901–2; on the reverse is modelled a lion looking toward the left of the medal, where the rising sun is depicted behind a hill. The lion is statant gardent in front of an African shield, whereon are two assegais, one being significantly broken; below is a plain panel with the word ASHANTI. The ribbon is alternately black and green—three black and two

KING EDWARD'S MILITIA LONG SERVICE MEDAL.

KING EDWARD'S MILITIA LONG SERVICE AND GOOD CONDUCT MEDAL. (Reverse.)

KING GEORGE'S TERRITORIAL EFFICIENCY MEDAL.

green stripes. The suspender is straight, and the names, etc., of recipients—mostly native soldiery—indented in small block letters on the edge of the medal, a few were issued engraved. The Flying Column was composed of 100 of the West African Regiment, 250 of the 2nd Central African Regiment, 350 West African Frontier Force, and 40 Sierra Leone Police with 2 guns. Some men of the 3rd W.I.R. also received the medal with bar for Kumassi.

CHINA, 1900

The Chinese society known as Boxers had been for some time engaged in agitating against foreigners, but did not become generally known to the English-speaking race until the murder of an Anglican missionary, the Rev. Sidney Brooks, on December 31st, 1899. They first, about a century ago, selected the title of I Ho Ch'uan, or Fist of Righteous Harmony; the society was, however, suppressed by Imperial edict, but, like most associations which have a strong religious or political basis, it was not exterminated, and as the political interference of certain missionary enthusiasts became more daring their activities again burst forth. They determined to destroy the foreigner and all his works, so that their countrymen might pursue their own ways, and work out their own salvation. All very laudable, in a way, but the patriots degenerated into bloodthirsty fanatics, and as the Chinese authorities were unable to deal with the rebels against Imperial authority—or indeed, as some say, appeared to make covert cause with them—it became necessary for the principal Powers to send troops to suppress the disturbances and protect the lives and property of their subjects. Great Britain, France, Germany, America, Japan, and Russia sent troops, which took part in the military and naval operations in the North of China and the Yangtse Valley from June 10th to December 31st, 1900. The most stirring events in the campaign were the defence of the Legations in Pekin from June 10th to August 14th, and the relief of Pekin between the same dates.

Defence of Legations.—On Sunday, June 10th, 1900, the Summer Legation at the Hills was burnt down, and from that time onward no one dared leave the city, while inside the Europeans lived in constant fear of the Chinese inhabitants, for within the very walls the German Minister actually took prisoner a fully dressed and armed Boxer. The Boxers and Imperial troops attacked the city again and again, until it became necessary for the Europeans and their servants to take refuge in the various Legations, and then on June 20th the siege began in earnest. Eighteen officers and 389 men, assisted by 100 volunteers, formed the combined force, and they had with them four guns of uncertain quality. It has been said by Dr. Morrison that the Japanese had only 100 rounds of ammunition per man, the Italians 120, the Russians 145, and the rest at most 300 rounds per man. On the first day the Austrians, for some unexplained reason, abandoned their Legation, and thus weakened the general defence by leaving a corner of the square open to the enemy. Then the Belgian Legation was burnt, and on the morning of June 22nd the Americans, Austrians, Germans, Italians, Russians, and Japanese began to retire on the British Legation, but the masterly activity of Sir Claude MacDonald induced the French and Austrians to return to the French Legation, and the others to resume their posts, thus preventing a catastrophe. The fear of fire was ever present with the besieged, for again and again buildings were ignited, and but for the well-organised fire committee would have helped the investing Boxers to make short work of the defence. Meanwhile the enemy vigorously attacked the Legations, and bombarded the city until a somewhat peaceable period ensued during an armistice which, however, hardly raised the hopes of the besieged, for news had reached them that the European powers believed them to have been massacred.

On August 10th the spirits of the defenders were greatly raised by the receipt of messages from the relief force, and the knowledge gained that the Japanese and American troops had defeated the enemy near Pei Ts'ang on the 5th, and had occupied Yang Ts'un next day, while the allied forces of America, Britain, and Russia were marching on from that place, and hoped to reach Pekin by the 13th or 14th. This news was conveyed in a letter from the Japanese Colonel Shiba, and it was indeed prophetic, for at 2 o'clock on August 14th the Sikhs marched on to the lawn of the British Legation, followed by General Gaselee and his staff, and then, the work of clearing the enemy out of the environs having been completed by the Sikhs, the relief of the Legations was effected.

Medal for China, 1900.—Queen Victoria had decided that those who had taken part in the campaign should be awarded a medal to commemorate their labours. Her decease, however, delayed the consummation of her desire, but in January 1902 the late King Edward gave the necessary instructions for the striking of the medal. It was decided to use the same reverse as on the China medal of 1842, and to revert to the dating of the exergue, so that 1900 appears thereon. On the obverse is the same bust of Queen Victoria as appears on the South African medal. Instead of the strange suspender issued with the second China medal, a straight clasp was used, and in place of the old fishtail bar the straight type was utilised. Three bars were issued with the medal: one with TAKU FORTS to those who were engaged in the Peiho River in the capture of the Forts on June 17th, 1900; one for the DEFENCE OF LEGATIONS to those who took part in the defence; and one for the RELIEF OF PEKIN to those who took part in the operations at or beyond Taku between June 10th and August 14th inclusive, which resulted in the relief of the city and Legations. The names of the recipients were mostly impressed upon the edge of the medal in light Roman capitals, and the same type of ribbon as supplied for the 1842 medals was used for suspension.

The following regiments were engaged: 2nd Batt. Royal Welsh Fusiliers; 12th Field Battery R.A.; 1st and 14th Sikhs; 3rd Madras N.I.; 4th Goorkas; 2nd, 7th, and 26th Bengal Infantry; 22nd and 30th Bombay N.I.; 24th Punjab Infantry; 1st Madras Pioneers; No. 2 Company Bombay Sappers; No. 3 Company Madras Sappers; No. 4 Company Bengal Sappers, and 1st Bengal Lancers. A Naval Brigade from thirty-six of H.M. ships also participated in the campaign, but most naval medals were awarded without bars, consequently those with bars are scarce, particularly those awarded to men of the Naval Brigade who had previously fought in the Boer War, scaled the Tugela Heights, and taken part in the Relief of Ladysmith.

THE BOER WAR

Little over a decade has passed since the South African, or Boer War, was brought to a close, so that it is hardly necessary for me to enter into details of the fighting, since so many will readily call to mind the events which led up to the ultimatum from the Boer Government in October 1899, and the necessity for Britain to place in the field the largest army ever sent out of England—about 200,000 men, which makes the army of 30,000 that fought under Viscount Wolseley in Egypt look very insignificant. During the first seven months of the war the battles of Talana Hill, Elandslaagte, Belmont, Modder River, Tugela Heights, Paardeberg and Driefontein, and many another battle had been fought and won; the sad disaster at Magersfontein had taken place, and the oft-wounded Wauchope killed. Ladysmith and Kimberley had been gallantly defended and relieved, and the siege of Wepener raised. Two hundred and eighteen British officers and 2,062 soldiers had been killed in action; 53 officers and 492 men had died of their wounds; 64 officers and 2,028 men had died of disease, and 664 officers and 9,225 men had been wounded. Indeed, a grand total of 24,253 officers and men had been killed or placed *hors de combat*. During the war Great Britain and her colonies lost in action 5,774 officers and men, and 16,168 by disease, while of those sent home 508 succumbed to their wounds or disease, 22,829 were wounded, and 5,879 were invalided out of the service.

Cape Colony.—There is a mistaken idea that the single bar for CAPE COLONY is not worth consideration as an engagement bar, but as a matter of fact many a brave fellow found a billet for a Boer bullet in the north of Cape Colony during the early days of the war, as the casualty records of certain regiments show. It is true, however, that many a man never got within miles of the enemy, and had to be content with the "donkey work" of looking after things for those at the front, while some whose bars include that for Cape Colony only landed for a day or two, and were then transported by sea to the fighting area. NATAL on a bar means, as every one knows, service in the zone where Buller did so much hard fighting—indeed, it was the Boer invasion of Natal on October 12th, 1899, that necessitated the defence of Ladysmith, and occupied the efforts of the Natal Field Force during 1900.

The Queen's Medal.—Twenty-six bars were issued in connection with the Queen's medal. Many think this constitutes a record, and it does for one campaign, but twenty-eight were issued with the Military General Service medal which was awarded in 1847; this, however, covered a series of campaigns. No less than 230 bars were issued with the Naval General Service medal also authorised in 1847, but these were obviously to cover a series of operations and campaigns. The bars for the Boer War are as follows, ranged in chronological order: CAPE COLONY, NATAL, RHODESIA, RELIEF OF MAFEKING, DEFENCE OF KIMBERLEY, TALANA, ELANDSLAAGTE, DEFENCE OF LADYSMITH, BELMONT, MODDER

RIVER, TUGELA HEIGHTS, RELIEF OF KIMBERLEY, PAARDEBERG, ORANGE FREE STATE, RELIEF OF LADYSMITH, DRIEFONTEIN, WEPENER, DEFENCE OF MAFEKING, TRANSVAAL, JOHANNESBURG, LAING'S NEK, DIAMOND HILL, WITTEBERGEN, BELFAST, SOUTH AFRICA 1901, SOUTH AFRICA 1902. The latter two were issued to those who, although engaged during the period for which the King's Medal was awarded, were nevertheless not entitled to it by the terms of the grant. Talana, where General Penn Symonds was killed on October 20th, 1899, was the first general engagement, and Belfast, fought on August 26th and 27th, 1900, the last. The bars should read upwards from the medal in the order arranged above.

This medal by G. W. De Saulles is $1\frac{2}{5}$ in. in diameter, and has on the obverse the bust of Her Majesty Queen Victoria (as shown in the illustration facing page 176). On the reverse is a spirited figure of Britannia, grasping the Union Jack in her left hand, while with her right she offers the laurel wreath to the army which marches past:

> "A varied host, from kindred realms they come,
> Brethren in arms, but rivals in renown."

Behind her in the distance is a man-of-war in Table Bay, and in the foreground, lying to the left, her shield and trident; above is SOUTH AFRICA. It is noteworthy that the Army Order specifically mentions not only the Colonial and Indian forces, but nurses and nursing sisters. Bronze medals without bars were given to non-enlisted men of whatever nationality who drew military pay, and to authorised camp followers. The ribbon has a broad centre stripe of orange, with dark blue and red stripes at the sides. The names, etc., were mostly indented in skeleton block letters, some tall and some square, but many were engraved in slanting Roman capitals. The first medals awarded, including those issued to the Canadian contingent, had the date 1899–1900 printed on the field to the right of Britannia, but this, of course, quickly rubbed off when the medal was worn. Medals, however, in mint condition, are sometimes found with the date still clearly showing.

The King's Medal.—In October 1902 His Majesty King Edward VII authorised the issue of a medal bearing his effigy, with the legend EDWARDVS VII REX IMPERATOR, to all those who had served for eighteen months at the front, and were still serving on January 1st, 1902, or had completed such term before June 1st, 1902—that is, just one month after the terms of surrender were signed. The reverse of this medal is the same as the Queen's, but on the obverse is the bust of King Edward in Field-Marshal's uniform. Two bars were given; SOUTH AFRICA 1901 and SOUTH AFRICA 1902. The ribbon is composed of three equal stripes of green, white, and orange, and the names are mostly indented or engraved in the same manner as the Queen's medal.

The Mediterranean Medal.—A medal bearing the legend MEDITERRANEAN to the right of Britannia, but otherwise exactly the same as the Queen's South African medal, was awarded to those who garrisoned the Mediterranean forts during the South African War; no bars were issued with this. It is worn with the Queen's ribbon, and the names, etc., are generally indented. Those engaged in this work were mostly Militia units, who volunteered for service abroad during the war.

Kimberley Star.—The Mayor of Kimberley presented to all those who took part in its defence a silver six-pointed star, but it should be particularly noticed that only those bearing the hall-mark, including the date-mark **a**, are genuine. The ribbon is red, white, and blue in the centre, edged with black on one side and yellow on the other. The stars were issued unnamed, and generally depended from an ornamental suspender, but the one illustrated facing page 192 was issued as depicted.

The following troops were engaged during the campaign:

Cavalry.—One squadron each of the 1st and 2nd

Life Guards and Horse Guards; 1st, 2nd, 3rd, 4th, 5th, 6th, and 7th Dragoon Guards; 1st Dragoons; Scots Greys; 3rd and 4th Hussars; 5th Lancers; 6th Dragoons; 7th and 8th Hussars; 9th Lancers; 10th and 11th Hussars; 12th Lancers; 13th, 14th, and 15th Hussars; 16th and 17th Lancers; 18th, 19th, and 20th Hussars; and the 21st Lancers.

Infantry.—Grenadier, Scots, and Coldstream Guards, and the following Regiments of the Line: Bedfordshire, Berkshire, Border, Cheshire, Cornwall Light Infantry, Derbyshire, Devonshire, Dorsetshire, Durham Light Infantry, Essex, Royal Fusiliers, Gloucestershire, Hampshire, East Kent, West Kent, King's Royal Rifle Corps, Lancashire Fusiliers, East Lancashire, South Lancashire, Loyal North Lancastrian, Royal Lancashire, Liverpool, Leicestershire, Lincoln, Manchester, Middlesex, Norfolk, Northampton, Northumberland Fusiliers, Oxford Light Infantry, Rifle Brigade, Shropshire Light Infantry, Somersetshire Light Infantry, North and South Staffordshire, Suffolk, Sussex, East and West Surrey Regiments, Warwickshire, Wiltshire, Worcestershire, Yorkshire, Yorkshire Light Infantry, East also West Yorkshire, York and Lancaster, and West Riding Regiments, Argyll and Sutherland Highlanders, Royal Highlanders ("Black Watch"), Cameron, Seaforth and Gordon Highlanders, King's Own Scottish Rifles, Royal Scots Fusiliers, Scottish Rifles, Highland Light Infantry, Royal Irish Fusiliers, Royal Dublin Fusiliers, Inniskilling also Munster Fusiliers, Royal Irish, Royal Irish Rifles, Leinster, Connaught Rangers, South Wales Borderers, Royal Welsh Fusiliers.

Volunteers.—Thirty-two Battalions of Imperial Yeomanry, City Volunteer Battalions, Companies, and Corps.

Royal Garrison Artillery.—Eastern Division: 5th, 6th, and 10th Companies; Southern Division: 14th 15th, 16th, and 36th Companies; Western Division: 2nd, 6th, 10th, 14th, 15th, 17th, and 23rd Companies.

Royal Horse Artillery.—A, G, J, M, O, P, Q, R, T, U, and V Batteries.

VICTORIAN MILITARY LONG SERVICE MEDAL.

NAVAL DISTINGUISHED SERVICE MEDAL INSTITUTED BY HIS MAJESTY KING GEORGE V OCTOBER 1914.

This medal is illustrated from lead squeezes specially taken for illustration in this book. The medal depends from a straight suspender as used with the Naval Long Service Medals.

Royal Field Artillery.—2nd, 4th, 5th, 7th, 8th, 9th, 13th, 14th, 17th, 18th, 19th, 20th, 21st, 28th, 37th, 38th, 39th, 42nd, 43rd, 44th, 53rd, 61st, 62nd, 63rd, 64th, 65th, 66th, 67th, 68th, 69th, 73rd, 74th, 75th, 76th, 77th, 78th, 79th, 81st, 82nd, 83rd, 84th, 85th, 86th, 87th, and 88th Batteries.

Royal Engineers, Army Service and Army Ordnance Corps, Army Veterinary and Army Pay Departments, and R.A. Medical Corps.

Irregular Corps.—Imperial Light Horse, South African Light Horse, Cape Mounted Rifles, Kitchener's Fighting Scouts, Thorneycroft's Horse, Brabant's Horse, Bethune's Horse, British South African Police, South African Constabulary, National Scouts (Boers), Scottish Horse, Lumsden's Horse and Strathcona's Horse, New South Wales Military Forces, Imperial Bushmen, New Zealand Mounted Rifles and Rough Riders, Queensland Mounted Infantry, South Australian Mounted Infantry and Bushmen's Contingent, Tasmanian Infantry, Artillery, and Bushmen, Victorian Infantry, Victorian Mounted Infantry, Cameron's Scouts, West Australian Contingent, Royal Canadian Dragoons and Batteries of Field Artillery, Canadian Mounted Rifles, Canadian Scouts, Ceylon Mounted Infantry, Bechuanaland Rifles, Border Horse, and Mounted Rifles, also Scouts, Brabant's Scouts, British South Africa Police, Cape Cavalry Brigade, Cape Colony Cyclist Corps, Cape Garrison Artillery, Cape Medical Staff Corps, Cape Mounted Rifle Club, Cape Police, 1st City (Grahamstown) Volunteers, Colonial Defence Force, Commander-in-Chief's Body-guard, Dennison's Scouts, Diamond Field Artillery, Diamond Field Horse, District Mounted Rifles, Driscoll's Scouts, Duke of Edinburgh's Own Volunteer Rifles, Durban Light Infantry, East Griqualand Mounted Rifle Volunteers, Eastern Province Horse, French's Scouts, Frontier Mounted Rifles, Gatacre's Scouts, Griqualand East Mounted Rifle Volunteers, Herbert District Mounted Rifles, Herschell Mounted Volunteers, Imperial Light Horse and Light Infantry, Johannesburg

Mounted Rifles, Kaffrarian Rifles, Kenny's Scouts, Kimberley Regiment, Kimberley Mounted Corps and Light Horse, Kimberley Rifles, Kitchener's Horse, Kuysna Rangers, Komgha Mounted Volunteers, Loch's Horse, Lovat's Scouts, Maritzani Mounted Irregulars, Marshal's Horse, Merre's Scouts, Military Foot Police, Modder River District Mounted Rifles, Namaqualand Border Scouts, Natal Volunteers, Natal Mounted Infantry, Nesbitt's Horse, New England Mounted Rifles, Orpen's Horse, Pioneer Railway Regiment, Prince Alfred's Own Cape Artillery, Prince Alfred's Volunteer Guard, Prince of Wales's Light Horse, Queenstown Rifle Volunteers, Rand Rifles, Rimington's Guides, Roberts's Light Horse, Rundle's Colonial Scouts, Rhodesian Regiment, Scottish Horse, South African Constabulary, South Rhodesian Volunteers, Steinaeker's Horse, Stellenbosch Mounted Infantry, Tembuland Mounted Rifle Corps, Transkei Mounted Rifles, Uitenhage Volunteer Rifles, and the Umvoti, Utrecht, Victoria, and Vryburg Mounted Rifles, Warwick's Scouts, Western Light Horse, and Western Province Mounted Rifles.

Town Guards.—The Aliwal North, Barkly East, Barkly West, Boshof, Burgherdorn, Campbell Town, Colesburg, Cradock, Dordrecht District, Douglas, East London, Grahamstown, Griquatown, Hopetown, Hoppesia, Indwe, Jamestown, Kimberley, King Williamstown, Klipdam, Kokstad, Kuruman, Lady Grey, Molteno, Naauwpoort, Port Elizabeth, Queenstown, Qumbu, Starkstroom, Steynsburg, Stormburg, T'somo, Uitenhage, Vryburg, and Warrenton.

Africa General Service Medal.—In June 1902 it was decided to strike a medal to take the place of the two which had hitherto been awarded for service in Central, also East and West, Africa. It was to be known as the Africa General Service Medal. The obverse is the same as the King's South Africa medal, but on the reverse is the figure of Britannia, with a lion beside her, as on the East and West African medal illustrated facing page 296, but with AFRICA in the exergue. The ribbon is yellow, with two narrow green stripes touching the broad black edging. Bronze medals were issued to camp followers. Numerous bars have been issued with this medal: NIGERIA for operations carried on in the early part of 1900, and in December of the same year, also in August and September 1901; S. NIGERIA for operations in March, April, and May, 1901, and for JUBALAND to those who were in the small force sent against the truculent Ogaden Somalis between November 1900 and the end of April 1901. Only 465 silver and 26 bronze medals were issued, and these mostly to the 16th Bombay Cavalry. The other recipients of the medal and bar were sailors and marines from three of H.M. ships. The bar for SOMALILAND 1901 was issued to those who were engaged there during May, June, and July 1901, and a bar for UGANDA 1900 to those who were engaged in the Uandi country during the period between July and October 1900. Only 5 officers and 1 British non-commissioned officer received the bar, and only 373 other bars were issued—268 to the 4th King's African Rifles, and 105 to Indian soldiers. To the 24th Punjab Infantry, which was engaged in different districts between August 1899 and December 1900, a bar inscribed B. C. A. 1899–1900—meaning British Central Africa 1899–1900—was awarded. The bar for GAMBIA was granted to a detachment of the 2nd King's African Rifles, four companies of the 3rd W.I.R., and the crews of three of H.M. ships who were engaged in the operations between January and March 1901. The bar for ARO 1901–1902 was issued to those who were engaged against the Aro tribe between November 1901 and March 1902. Fourteen British officers and the crew of H.M. "Thrush," with 1,830 native soldiers, composed the Aro Field Force. The bar for LANGO 1901 was awarded to those members of the 4th King's African Rifles who were engaged against the Sudanese mutineers and the Langos between April and August 1901. To those engaged in the Bornu expedition, from February to May 1902, and in the Kontagora expedition in February 1902, the bar for N. NIGERIA 1902 was issued (the same bar was given to those who were serving at Argungu and on convoy duty), and to those who were engaged in Nigeria between July 1902 and June 1903. In April 1905 a bar was issued inscribed N. NIGERIA 1903 to those who had taken part in the Kano-Sokoto campaign. To this medal it was decided to add the bar SOMALILAND 1902–04, and one for JIDBALLI to those engaged in the operations between January 1902 and May 1904.

The only fairly complete regiment engaged in the campaign was the 1st Hampshire, and this, together with detachments of the Norfolks, Yorkshires, and Rifle Brigade, and units of the Royal Engineers and Army Service Corps, formed the only English Corps engaged. The rest of the expedition included the 1st, 2nd, 3rd, and 5th King's African Rifles; Bikanir Camel Corps; Kajawa Corps; Gudabursi Horse; Rajputs; Sikhs; Punjabees and Indian Mounted Infantry; also the crews of fourteen H.M. ships.

Nigeria, 1902–6.—At the battle of Jidballi the Mullah's army of 5,000 men were defeated by this little force, over 1,000 of the enemy being killed. The loss in British officers was great; Captain Lister of the 10th Hussars, heir to Lord Ribbledale, and 2 others being killed, and 9 wounded. In 1905 the bar for S. NIGERIA 1902–03 was issued to those who were engaged in South Nigeria between July 1902 and June 1903, against the Iris and the Omonoha and Ebima tribesmen, and in the operations against the Chief Adukukaiku of Igarra. In 1906 it was decided to give the same bar to those who took part in the expedition in the Afikpo district between December 1902 and January 1903. In April 1905 a bar was issued inscribed S. NIGERIA 1903 to those who had taken part in the operations between April and July 27th, 1903. In 1906 it was decided that the war should be issued to those who had taken part in the expeditions in the Ebegga country west of Anan. In April 1905 an Army Order set forth

(Reverse.)
EGYPTIAN MEDAL FOR BRAVERY.

THE MILITARY CROSS.

(Obverse.)
EGYPTIAN MEDAL FOR BRAVERY.

that a bar was to be issued to N. NIGERIA 1903, and granted to those who took part in the Sokoto-Burmi operations between April 15th and July 27th, 1903. In January 1906 a bar for N. NIGERIA 1903–04 was authorised for distribution among the 700 native soldiers and carriers engaged in the operations in the Bassa Province between December 1903 and March 1904. A bar for N. NIGERIA 1904 was granted to those who took part in four little expeditions, and S. NIGERIA 1903–04 to those employed in the expedition between December 1903 and January 1904 against the towns of Osca, Oriri, and N'doto. The bar for S. NIGERIA 1904 was added for a series of five expeditions, and a further addition to the Nigerian clasps was made in June 1906 by the approval of one inscribed S. NIGERIA 1904–05, given for operations between November 1894 and February 1895 in the Ibibio and Kwa countries. The bar for S. NIGERIA 1905 was awarded to members of the force which operated in the Kwale district between October 10th and 18th, 1905, both dates inclusive. The bar for S. NIGERIA 1905–06 was issued to those who took part in the Bende-Onitsha Hinterland expedition, and the bar for N. NIGERIA 1906 was given to those who took part in the operations against the Satiru rebels and the Emir of Hadeija. The bar for KISSI 1905 was awarded to those engaged in that district from March to June 1905, and for NANDI 1905–06 to those employed in that neighbourhood between October 1905 and July 1906, making a total of twenty-seven bars to this medal, and indicating the constant vigilance and discipline which is necessary to maintain the outposts of our Empire.

East Africa, 1902.—The medal with bar for EAST AFRICA 1902 was awarded to all officers and men composing the Maruka patrol, which reached the Maruka district on September 4th, 1902, and operated in the district until October 25th, 1902, both dates inclusive.

East Africa, 1904.—The medal with bar EAST

AFRICA 1904 was issued to all officers and men composing the Iraini patrol, which entered the Iraini country on February 13th, 1904, and patrolled it between that date and March 17th, both dates inclusive.

East Africa, 1905.—The medal with bar for EAST AFRICA 1905 was issued to all officers and men who took part in the operations in and near Sotik, from May 31st to July 12th, 1905, both dates inclusive, also to the officers and men composing the Kissi patrol, which entered the Kissi country on September 1st, 1905, and patrolled it until October 9th, 1905, both dates inclusive.

East Africa, 1906.—The medal with bar EAST AFRICA 1906 was awarded to all officers and men composing the Embo patrol, which entered the Embo country on June 18th, 1906, and operated there until July 19th, 1906, both dates inclusive.

West Africa, 1906.—The medal with bar WEST AFRICA 1906 was issued to all officers and men composing the Owa column, which left Asaba on June 9th, 1906, and operated against the people of Owa until the restoration of peace and the breaking-up of the column on August 3rd, 1906, both dates inclusive.

The medal with bar WEST AFRICA 1906 was also awarded to all officers and men who operated in the Chibuk country between November 12th and December 4th, 1906, both dates inclusive.

West Africa, 1908.—An Army Order issued in July 1910, with the approval of His Majesty King George V, granted a bar inscribed W. AFRICA 1908 to the forces who took part in the fighting in and near the Soukwala Valley on December 11th, 1908, and between December 24th and 31st, 1908, both dates inclusive.

I.G.S., 1903.—This medal, precisely similar on the reverse, except for the deletion of the date, to the India General Service Medal 1895, bears on the obverse the same bust of King Edward as on the South Africa medal. The ribbon is the same, also the suspender. With this new issue the bar for WAZIRISTAN 1901–2 was given to those who took an active part in the Mahsud Waziri blockade between November 1901 and November 1902. To those holding the 1895 medal the bar only was issued. Bronze medals were given to authorised camp followers.

Only three English soldiers received the medal—men of the Cheshires who were employed as signallers. The troops engaged were: 1st, 3rd, and 4th Sikhs; 2nd, 4th, 5th, and 22nd Punjabis; 1st, 3rd, and 5th Punjab Cavalry; Sappers and Miners.

Natal Native Rebellion.—In 1908 the Natal Government awarded a silver medal to those who took part in suppressing the native rebellion in 1906. The medal, which was supplied by the Goldsmiths & Silversmiths Company, bears on the obverse a portrait bust of King Edward VII facing to the right, encircled by the legend EDWARDVS VII REX IMPERATOR. On the reverse Natali is represented holding the sword of justice in her right hand, and a palm branch in her left, supported and protected by Britannia, who holds in her left hand the orb of empire. In the exergue is the word NATAL. A broad plain bar is inscribed ·1906· with a large dot on either side. The names, etc., are impressed in light square block capitals, and the ribbon for suspension is a deep red, with black edges. The medal illustrated facing page 198 was issued to the local white forces, including the Natal Horse Corps and about 350 chiefs and leading natives.

Tibet, 1903–4.—Owing to the non-fulfilment of treaty obligations on the part of the Tibetans, Colonel Younghusband was sent by the Indian Government with a military escort to the seat of Tibetan authority. For some little time he made peaceful progress, until, on arriving at Hot Springs, General Macdonald's force of 1,600 men, with 4 field and 2 Maxim guns, found it necessary to attack the Tibetans, and drove them out of their position. Then, when they had gathered round their Lhassa General, the political agent, Colonel Younghusband, together with Brigadier Macdonald, their staffs, and some Sikhs, went forward to parley with them, and despite the attitude of the Tibetans, the officers dismounted; some began to eat sandwiches, others to take photographs. Suddenly the Tibetans turned upon the ring of Sikhs which surrounded them; all was soon confusion, but the discipline of the Sikhs, and the coolness of the officers, saved the situation, and within ten minutes the flower of the Tibetan army was demolished, despite the grim determination of those composing it, for they disdained to run away, but solemnly marched off as the fight became too hot for them.

The medal for the Tibet Mission 1903–4 bears on the reversé the fortress-capped hill of Potala a Lhassa, with TIBET 1903–4 beneath it, and to those engaged in the mission about Gyantse a bar inscribed GYANTSE was awarded. The obverse is the same as the I.G.S. 1903 medal. The ribbon is dark red edged with green, with two white stripes. It is noteworthy that with this and the medal I shall next describe the old curled suspenders, as on the Sutlej medals, are used.

The Tibet Force comprised the Royal Fusiliers; a detachment of the Norfolks; a Mountain Battery of the R.A.; 8th Goorkas; 40th Pathans; Queen's Own Sappers and Miners, and the 6th Mule Corps.

I.G.S. 1908.—A new medal in place of the 1903 I.G.S., and destined to be known as the I.G.S. Medal 1908, was awarded by an Army Order in December 1908. It has a pictorial design on the reverse, somewhat similar to the above-mentioned medal, and reminding one of the old Ghuznee Medal. It was issued in 1910 to those who had been engaged on the North-West Frontier between February 14th and May 31st, 1908. The obverse bears the bust of King Edward VII in Field-Marshal's uniform, encircled with the legend EDWARDVS VII KAISAR-I-HIND. On the reverse is a fort on a hill-top backed by mountains, and beneath, forming the exergue, an ornamental tablet overlaid by olive and oak branches bearing the word INDIA. The bar bears NORTH-WEST FRONTIER 1908. This medal was designed by Mr. Richard Garbe, R.B.S. It was awarded to both the troops and followers, and was the last issued during the reign of Edward the Peacemaker. (See facing page 204.)

The regiments engaged in the campaign were: the Gordon and Seaforth Highlanders; Royal Irish Rifles and Royal Munster Fusiliers; the Northumberland Fusiliers and Warwickshire Regiment; 10th Hussars; 71st Company R.G.A.; 6 guns of the 18th and 80th Batteries R.F.A.; 62nd and 75th Batteries R.F.A.; 2nd, 3rd, 8th, 21st, 22nd, 23rd, and 28th Mountain Batteries, and Nos. 1 and 7 British Field Hospital; 1st, 4th, 5th, and 6th Goorkas; 19th, 20th, 21st, 22nd, 25th, 28th, 29th, 30th, and 33rd Punjabees; 15th, 23rd, 34th, 45th, 53rd, 54th, 55th, 57th, and 59 Sikhs; 40th Pathans; Queen's Own Corps of Guides Infantry; Cavalry of the Q.O. Guides; 21st Cavalry; 19th and 37th Lancers; Sappers and Miners, and 5 Native Field Hospitals.

Abor, 1911–12.—The massacre of Mr. Williamson and his party necessitated the dispatch of a punitive force, which under Major-General Hamilton Bower, C.B., entered the Abor country in October 1911, and for seven months, under the most trying physical difficulties, operations were pursued. "The paths were quite unfit for use by laden carriers," and progress was exceedingly slow, but as usual the regiments pressed on to their destination, and, after inflicting punishment on the culpable villagers who had been party to the massacre, captured and brought to trial those immediately responsible for the murders. An important result of the operations was the breaking down of the power of the Kebang-Rotung group of villages, and the freeing of the Lakhimpur districts north of the Brahmaputra from Abor aggression. For his services in this campaign General Bower was made K.C.B., and the following officers Companions of the D.S.O.: Majors James Davidson, M.D., James Alban Wilson,

Edward G. Vaughan, Ernest H. Scott Cullen, M.V.O., and Lieutenant Miles A. Claude Kennedy.

Among the regiments taking part in the expedition were the 1st Batt. 8th Goorka Rifles, on whom the brunt of the fighting fell—medals of this regiment have realised from £2 10*s.* to £3 in the sale-room; 1st Co. (King George's Own) Sappers and Miners, to whose skill and energy "the success of the expedition was largely due"; 1st Batt. King Edward's Own Goorka Rifles; 32nd Sikh Pioneers, who did excellent work on the line of communications; dismounted detachment Assam Valley Light Horse; Supply and Transport Corps; Lakhimpur Military Police; 5 Nagar Carrier Corps.

The medal, although described as the India General Service Medal 1908, has on the reverse the crowned bust of King George V truncated by the edge of the medal, a spray of laurel covering the truncation, and the legend GEORGIVS V KAISAR-I-HIND on a raised band surrounding the effigy. This was given to those who served at or beyond Kobo between October 6th, 1911, and April 20th, 1912, both dates inclusive. The medals are engraved in a coarse kind of script —each letter being separate—or in a loose running hand. Officers and men already in possession of the India General Service Medal 1908 received the clasp only.

SUDAN, 1910

By a Special Army Order, dated from Headquarters, Khartoum, June 12th, 1911, His Highness the Khedive approved of a new medal being struck to commemorate military operations in the Atwot district and granted to the troops (including Government police and Jehadia) who took part in the operations under the command of El Kaimakam W. J. St. J. Harvey Bey, against the Atwot Dinkas in the Atwot district of the Bahr el-Ghazal Province. The medal in silver was granted to all troops who formed part of the columns operating in the district between February 9th and March 17th, 1910, both dates inclusive, or who were disembarked at Sheikh Tombe, Aliah district, Mongalla province, between March 29th and April 4th, 1910, both dates inclusive.

Atwot.—The bar inscribed ATWOT in English and Arabic was granted to all troops taking part, and the following non-combatants were entitled to the medal and bar in silver, who participated under the same conditions as the troops: Government Civil employees; engineers of steamers; Raises of steamers; Sheikhs; Armed Guides; whilst non-combatants, *i.e.* Civilian Syces of Officers, Civilian Servants of Officers, and Transport Drivers, were granted the medal in bronze without bar.

South Kordofan.—To those who took part in the operations in Southern Kordofan under the command of El Lewa Asser Pasha, against Jebel Tagoi, between November 10th and 19th, 1910, inclusive, or who formed part of the column operating against Jebels Katla and Tima, under the command of El Kaimakam Conry Bey, D.S.O., between November 27th and December 19th, 1910, both dates inclusive, the silver medal with bar inscribed in English and Arabic S. KORDFAN 1910 was granted, and to those already in possession of the medal, the bar. The Government police were included in the troops, and therefore received the medal and bar. The following non-combatants were also entitled to the medal and bar in silver: Government Civil employees; Sheikhs, and Armed Guides, whilst the bronze medal without bar was granted to Civilian Syces of Officers, Civilian Servants of Officers, and Transport Drivers.

Sudan, 1912.—To the troops, including Sudan Government police, who took part in the operations against the Beir and Anuak tribes, the medal with bar inscribed SUDAN 1912 in English and Arabic was awarded. Those entitled had taken part in the Pibor reconnaissance under the command of El Bimbashi Dickinson between October 12th and December 25th, 1911, both dates inclusive, or had formed part of the columns operating against the Beir tribe under the command of El Miralai Drake Bey, between the dates stated below, both dates inclusive:

I. Northern Column under the command of El Bimbashi Dickinson, December 26th, 1911, to March 15th, 1912.

II. Central Column under the command of El Miralai Drake Bey, December 31st, 1911, to March 15th, 1912.

III. Southern Column under the command of El Kaimakam Arden Bey, January 1st to February 3rd, 1912.

The column operating against the Anuak tribe under the command of El Miralai Leveson Bey, D.S.O., between March 4th and April 12th, 1912, both dates inclusive. The medal and bar in silver was awarded to Government Civil employees, Sheikhs, and Armed Guides, although classed as non-combatants; whilst the medal in bronze without bar was again awarded to Civilian Syces of Officers, Civilian Servants of Officers, and Transport Drivers.

Sudan Medal, 1910.—The medal, $1\frac{7}{8}$ in. in diameter, bears on the obverse the cypher of the Khedive Abbas Hilmi El Thani (Abbas II) and the date Hegira 1328. On the reverse, modelled by Richard Garbe, R.B.S., is a lion standing in the attitude of attention, with its fore paws resting upon a small panel bearing the record SUDAN above the exergue, in which is a Sudanese shield and spears. Forming the background to the lion is the Nile within its banks, with the sun rising and casting its rays over the sky and sparkling the water of the river. The suspender is like that employed on the South African medal, and the bars, $\frac{1}{8}$ in. wide and riveted the same distance apart, are of the same character. The photograph (facing page 204) illustrating the medal does not properly depict the disposition of the bars, as they are unfixed, the medal with loose bars having been kindly placed at my disposal for the purpose of illustration herein by the Military Secretary of the Egyptian Army. The ribbon of watered silk, $1\frac{1}{4}$ in. wide, is black with a narrow edge of green and an under, outer edging of red.

NAVAL WAR MEDALS

I HAVE given military war medals preference over those awarded for naval service, not because I do not recognise the first line of defence as worthy of priority, for, as I have pointed out, the first medals of which we have any record were probably given as mementoes of, or as rewards for, the defeat of the Armada, but because the variety given for land service is considerably greater, and, in many instances, as a naval contingent also participated, it enabled me to deal with the historical aspect out of hand, a necessity which the concise nature of the book demands.

The Armada Medals.—The story of gallant Drake, Frobisher, and Hawkins is perennially fresh in the minds of every Briton, and how the wind came to their aid and practically destroyed the 120 mighty galleons of Spain, while the flower of her nobility, with 30,000 men, were killed in battle or drowned. I have already referred *en passant* to the medals issued in the latter part of the sixteenth century. The "Ark-in-Flood," a very handsome oval medal, is one of the most striking of the Elizabethan series (see facing page 270), and a particularly fine one is in the possession of W. R. Parker-Jervis, Esq., who inherited it from his aunt, Lady Forester. The medal was given to Lord Uppingham after the defeat of the Armada in 1588, and it is on record, in a book at Woburn Abbey, that this medal was awarded to Admirals and Commanders who took part in the battle, and was worn by them as a badge. Pinkerton states that it was given, in gold and silver, to Marine Commanders as a mark of royal approbation, and I think we have sufficient evidence that it was

COMMONWEALTH MEDAL FOR DUTCH WARS, 1653.

ELIZABETHAN NAVAL MEDAL.

given to the leading spirits of the Armada. This medal differs in several respects from the one described on page 2, for while on the one referred to above there are rays behind the head of Elizabeth and it is framed in laurel leaves with a twist of rope as suspender, that referred to in the beginning of the book has only the simple beaded border, the only embellishment being a fancy scroll on either side of a double ring suspender, and the motto reads ELIZABETH · D · G · ANGLIE · F · ET · HI · REG. It will be noted on comparison that the stops between the letters in the Uppingham medal are round instead of diamond-shape as on the simpler medal. James I, as I have previously described, issued a medal very similar. These were, it may be presumed, suspended from the neck, but medallions were sometimes worn in the hat.

Drake's Medal.—This handsome medal, given to Sir Francis Drake by Queen Elizabeth after his voyage round the world, now in the possession of the descendants of the famous commander at Nutwell Court, Devonshire, is a characteristic example of the Elizabethan jeweller's art. The frame, set with diamonds and rubies, and enamelled in various colours, forms a handsome setting for the fine cameo cut in onyx, and attributed to Valerio Vincenteno. Two heads are carved thereon, one representing Europe cut in the lower strata of white, while out of the upper strata of black the head of a negro has been fashioned. Set in the reverse is a beautiful miniature of Queen Elizabeth by the famous painter Nicholas Hilliard, with the date Anno Dom : 1575 Regni 20. From the badge depends a cluster of baroque pearls connecting a pear-shaped drop with the main body of the badge. The virgin Queen also presented the intrepid mariner with a jewelled star of twelve points; rubies set in the rays, and diamonds and opals in the circular centre surrounding an orb indicative of sovereignty. Loops are attached for fixing on to the coat.

Charles I Naval Medals.—Charles I issued a medal

which, because of its size, could not have been intended as a personal decoration, but rather to commemorate the launching of the famous three-decker, the "Royal Sovereign," in 1637. On the obverse is a portrait in profile of the King looking left: it varies, however, and on one the monarch is represented with a ruff round the neck, and a jewel depending from stars on the shoulder; and on the other he is represented in armour, with long curling hair over a turn-down collar. Around the bust is the motto CAROLVS ✦ I ✦ D ✦ G ✦ MAG ✦ BRITAIN ✦ FRAN ✦ ET ✦ HIB ✦ REX. On the truncation is 1639. The medal is in silver 2·35 in. in diameter, and is by Nicholas, Bristol. On the reverse is the "Royal Sovereign" under full sail, and to the left tiny promontories with forts thereon; around the ship is the inscription NEC ✦ META ✦ MIHI ✦ QVÆ ✦ TERMINVS ✦ ORBI. A smaller medal, probably for naval service, was likewise issued with the same inscription. It was not, however, a small replica of the medal just described, although it bore the same effigy of King Charles I on the obverse, and a ship in full sail on the reverse, for the actual design and modelling differ.

Commonwealth Naval Medals.—Although I have stated that the Dunbar medal was the first campaign medal to be distributed to military officers and men of all ranks, the Navy holds the distinction of being the first to be honoured in this way, for in June 1649 it was decided to issue rewards to the officers and men who had "done good service at sea." The medal, designed by Thomas Simon, has on the obverse the Parliament in session, and on the reverse two distinctive cartouches, depending from the stock of an anchor. The one to the left bears the St. George's Cross for England, and that on the right the Harp for Ireland; a rope attached to the anchor is disposed so as to form a decorative surrounding by the arrangement of three twists; above is the word MERUISTI. On the stock of the anchor are the medallist's initials T. S. An order of the Council of State records that Simon was granted the use of the press in the

Tower of London on condition that he did not use it for any other purpose than that of striking these medals, and he had to enter into a surety for £500 that he would " make no unlawful use of the presse."

Wyard Medal.—Robert Wyard of the " Adventure," with 22 guns, was on the night of July 31st, 1650, bold enough to engage six royalist frigates, and after fighting a whole day made them sheer off. Wyard received a gold medal, which I illustrate, valued at £50, and his officers and men medals varying in value from £5 to 10*s*. The obverse of the medal is as that above described. The reverse shows the " Adventure " engaging two of the royalist frigates with the other ships in the distance. Above is the inscription SERVICE ✦ DON ✦ AGAINST ✦ SIX ✦ SHIPS ✦ IVLY ✦ Y ✦ XXXI ✦ & AVGVST ✦ Y ✦ I ✦ 1650. The medal is oval, 1·6 in. by 1·35 in., and was struck in gold and silver. (See facing page 270.)

THE DUTCH WARS

For the victories over the Dutch some fine medals were struck. Probably the most interesting is the gold medal awarded to the Admirals who participated in the war and in the decisive battle fought on July 31st, 1653.

On February 28th, 1653, Generals Blake, Deane, and Monk defeated the Dutch fleet under Admirals Van Tromp and De Rutzer, and after a three-days' fight defeated them off Portland. In this engagement soldiers served for the first time on board ship, and the Marines came into existence. At the beginning of June the Dutch admirals were again defeated by the English Generals Deane, Penn, and Monk; but although the English did not lose a ship, General Deane was killed in the action. The English then followed the Dutch to their own coasts, and on July 31st the enemy was badly beaten after a terrific fight in which they lost 26 ships, their Admiral, Van Tromp, and about 6,000 men. The English losses were 2 ships and 1,300 men killed and wounded. The English Parliament was not slow to reward the victors, for on August 8th it was resolved to award Generals Blake and Monk gold chains valued at £300 apiece; to Vice-Admiral Penn and Rear-Admiral Lawson chains of the value of £100, and to the four staff officers chains worth £40 each for their brilliant services. The money was ordered to be deducted from the £2,000 voted, and the balance spent in the issue of medals among the officers of the fleet.

Types of Medal.—Four types of medal were issued. One with a broad border of naval trophies having on the obverse to left and right cartouches bearing the Arms of Holland and Zealand, and on the reverse in place thereof side drums. The obverse bears an anchor from the stock of which are suspended three elaborate shields bearing St. George's Cross, St. Andrew's Cross, the Irish Harp, and "the Armes of the Comon wealth." A cable attached to the anchor encircles the whole in a decorative manner. On the reverse is depicted a naval battle; a ship sinking in the foreground has on the stern the medallist's surname, and on the prow of another is T.S., while on the lower wing of the anchor on the obverse is the monogram T.S., so that Thomas Simon made quite sure that we should know who was responsible for the medal, which, with the elaborate border, is 2·2 in. by 2 in. with a ring for suspension. Three of the larger medals are known to be in existence. The medal given to the four staff officers was 2 in. by 1·8 in., and the obverse and reverse are the same; but the border on both sides is of laurel as the one illustrated. One of these medals, presented to Captain William Haddock, who commanded the "America," was purchased by Messrs. Spink at auction for £430 in May 1908; it had realised £105 in 1879.

Gold Medals for Officers.—Gold medals with a plain border, but with the same obverse and reverse as those described, were given to officers of the fleet; and I might here remark that if a copy of this rare medal

(Reverse.)
THE WYARD MEDAL, 1650.

ARMADA MEDAL AWARDED TO LORD UPPINGHAM.

(Obverse.)
THE WYARD MEDAL, 1650.

comes under the notice of collectors, they should look for the surname and initials, which forgers have generally overlooked.

Seamen's Medals.—To the seamen was awarded a small oval medal 0·95 in. by 0·85 in. also by Simon. Like the other medals it has a ring for suspension, and bears on the obverse an anchor from the stock of which depends two shields bearing respectively the Cross of St. George and the Irish Harp, encircled by a cable which runs round the whole. Above the stock is MERUISTI (Thou hast merited), and on the reverse the House of Commons as on the Dunbar medal.

The Triumph Medal.—During the fight on July 31st Admiral Robert Blake's old flagship (he was ashore owing to a wound) caught fire, and many of the men jumped overboard; but those who remained extinguished the fire and saved the ship. For this service the officers and those who had stuck to the ship were awarded a special medal with the obverse and reverse as on the other medals, but with the inscription engraved above the battle scene: FOR EMINENT SERVICE IN SAVING Y TRIUMPH FIRED IN FIGHT W^{h} Y DVCH IN IVLY 1653. This medal is 1·6 in. by 1·4 in.

Blake's Jewel.—On May 28th, 1657, the House of Commons voted £5,000 for a jewel to be "bestowed on General Blake" for his service in destroying the Spanish fleet off Teneriffe on April 20th. The jewel, supposed to have been a ring, was dispatched to Blake, but whether he received it or not is uncertain, although, as he died within sight of Plymouth on August 7th, 1657, it is quite probable that he did.

Charles II Medals.—In October 1665 Charles II had proclaimed that a definite percentage of the value of prizes should be paid to those who captured them from the Dutch, and that a portion of the proceeds should be set aside to help the widows of those who died in battle, to assist the sick and wounded, and to provide medals for those who performed special service.

War was again declared against the Dutch on February 22nd, 1665, and on June 3rd the Dutch fleet under Admirals Evertzen and Opdam was engaged and defeated off Lowestoft by the English fleet under Prince Rupert, the Duke of York, and Admirals Lawson and Penn. Opdam was killed when his ship was blown up, and two other Dutch admirals were killed. Twenty ships were sunk or taken by the British ere the Dutch made off, and through carelessness were permitted to escape. In the engagement the capable Vice-Admiral Lawson was killed, also Vice-Admiral Sansome, likewise the Earls of Marlborough, Portland, and Falmouth, and Lord Muskerry. To commemorate the victory, medals were issued in gold and silver. One bore on the obverse the bust of the King with his titles, and on the reverse the Island of Great Britain and the legend QUATUOR MARIA VINDICO. A smaller medal has on the obverse a triumphal chariot drawn by sea-horses, with the King seated therein, and the legend ET PONTUS SERVIET. These medals were not apparently issued as decorations; but the fine oval medal by Thomas Rawlins was probably destined for that purpose; it was struck in copper and 1·6 in. by 1·5 in. in size. It was also struck in gold and in silver. It bears on the obverse the truncated bust of Charles II in armour, looking to the left, and around the effigy is the inscription CAROLVS · II · D : G : M : BR : FR : ET · H : REX., and on the reverse a warship with a flag on the mainmast bearing the King's initials C R. This side bears the legend NOS · PENSES · IMPERIVM.

In 1665 Charles II caused to be struck a large circular silver medal 2 in. in diameter with the motto PRO TALIBVS AVSIS (For such enterprises) in the exergue of the reverse, whereon is depicted a battle at sea with a wreck in the foreground and Charles II habited as a Roman general looking on at the engagement. The medal by Rœttier is a fine example of his workmanship, and a few very fine specimens are still extant. This medal, illustrated facing page 272, was probably intended to be commemorative of the victories of either

CHARLES II SILVER NAVAL MEDAL.

the first or second Dutch wars; but as it is undated it may have been issued to commemorate either or both. It is 2 in. in diameter. Another medal by Rœttier was struck during the reign of Charles II to commemorate the service of James Duke of York; but this, again, is commemorative rather than decorative.

Several battles were fought ere in 1667 peace was declared with the Dutch. It was not, however, of long duration, for in March 1672 war again broke out, and several battles were fought before peace was finally signed on February 9th, 1674.

La Hogue, 1692.—For this brilliant victory over the French by the combined English and Dutch fleets, under Admiral Russell, following a battle which waged for five days, several medals designed by Rœttier were ordered to be struck by Queen Mary, who expressed her satisfaction in the result of the conflict—which secured the throne of England to her and William, and shattered the hopes of James II—by distributing £30,000 among the soldiers and sailors who had been engaged in the battle, and ordering medals for officers. The most important medal, 1·95 in. in diameter, was struck in gold and silver; it bears on the obverse the busts of William III and Mary II, the king habited in Roman armour and wearing long hair, the Queen simply draped. Around the conjoined busts is the inscription GVL : ET . MAR : D : G : B : F : ET · H : REX · REGINA ; and on the reverse the representation of the French Admiral's flagship, "Le Soliel Royal," in flames. This ship was, with three others, driven ashore and set on fire by English fireships. Above the scene is the legend NOX · NVLLA · SECVTA · EST (No night followed), and in the exergue PVGN : NAV : INT : ANG : ET · FR : 21 · MAY · 1692. One of the medals in gold, together with a chain of the value of £50, was presented to Captain John Tupper "for the good services performed by him" when in a dense fog he sailed through the French fleet and brought the news of their presence to Admiral Russell at Spithead.

After this glorious victory Mary founded Greenwich

Hospital as a home for seamen who had been disabled in the service of their country.

Special Gold Medals.—A number of gold medals and chains were awarded during this reign for service at sea, and, although there is no data to go upon, it may be assumed that the obverse of the medals was struck from the die used for the La Hogue medal described; the reverse being left plain for an inscription, as is the case with the gold medal awarded to Captain Peter Jolliffe, Master of the "Sea Adventure" hoy, which bears on the reverse "His Maties' Gift as a Reward to Peter Jollif, of Poole, for his good service ag^{t} the enemy in retaking a Ketch of Weymouth from a French Privateer, and chaceing the said Privateer on Shoar near Lulworth in ye Isle of Purbeck, where shee was broken in pieces, 1694."

War Medals for Fishermen.—During the reign of William and Mary medals were not only given to those actually engaged in the profession of soldier or seaman, but to fishermen and merchant sailors who performed gallant deeds. Among the recipients was William Thompson, the master of a fishing-smack, who, with one man and a boy armed with a couple of small guns and a few muskets, attacked, on May 30th, 1695, a French privateer armed with two guns and manned by sixteen seamen, which he defeated and captured after two hours' fighting. For this service he was awarded a gold chain and medal of the value of £50, and allowed to retain the vessel he had conquered.

As far as can be ascertained, no medals for land service were issued during the reigns of William and Mary, 1689–94, and William III 1694–1702.

Queen Anne.—For the brilliant attack on Vigo, October 12th, 1702, Queen Anne caused medals to be struck for distribution among the officers of the fleet who had taken part in the victory. Vice-Admiral Hopson was made a knight, and granted a pension for his particular service. The medal bears on the obverse the bust of the Queen, as on the medal described hereafter, and on the reverse a representation of the engagement, with the date.

Lampriere Gold Medal.—Captain James Lampriere was, in 1703, awarded a gold medal, and probably a chain also, as there is a ring for suspension, as

> "Her Maj^{ties} reward to Capt. James Lampriere for his Zeal to her Service, and his Successful Conducting ye Squadron commanded by Rear Admiral Dilkes, who destroyed a considerable number of ye Enemy's Merchant Ships under convey of 3 Men of War on their own coast."

This inscription is engraved in writing characters upon the reverse of the gold medal, with the Arms of Lampriere below, with the motto TRUE · TO · MY · TRUST · on a ribbon. On the obverse is the bust of Queen Anne, facing left, wearing the Royal Crown with a row of pearls in her hair, and round her neck the Collar and Star of the order of the Garter, surrounded by the legend ANNA · DEI · GRATIA · MAG : BRITAN : FRA : ET · HIB : REGINA.

Granville, 1703.—It is said that gold medals were presented to Rear-Admiral Dilkes and his officers for their success near Granville in Normandy in July 1703; but there is no official record of their being granted. But to Captain Thomas Legge, who distinguished himself under Rear-Admiral Dilkes in the same year a gold medal was specially struck to commemorate his gallantry, and Campbell states in his "Naval History" that "The Queen . . . ordered gold medals to be struck on this occasion and delivered to the Rear-Admiral and all his officers," and Lampriere was one of the officers.

Capture of Gibraltar.—Medals were struck in this reign to celebrate the capture of Gibraltar by Sir G. Rooke in 1704; but they were not, as far as I can ascertain, issued as decorations.

A few medals were struck and given to pilots and fishermen; five to the crew of the "Leonora" in consideration of their courage and resolution in driving the Frenchmen (13 who had been placed as a prize crew on the captured ship to pilot her to France) then aboard her "off the Quarter-deck, and bringing the said shipp into England."

Boy's Medal for Gallantry.—In Dr. Payne's collection is a unique specimen of a Queen Anne medal awarded to a boy for conspicuous gallantry. On the obverse is the laureated bust of Queen Anne, facing left, draped, with a row of pearls about the shoulders, with the legend ANNA D: G: MAG: BRI: FR: ET HIB: REG:. The medal, which is 2·75 in. in diameter, has a raised border of laurel leaves and roses, and a ring for suspension. On the reverse is the inscription engraved in writing characters:

> "Her Majties reward to Robert Taylor Boy of ye Mary Galley, for his Zeal and Courage at ye taking of ye French Privateer Jacques La Blonde of Dunkirk."

This medal, illustrated facing page 276, is one of the few struck during the reign of Queen Anne "to give all due Encouragement to the Valour and Fidelity of Her Subjects serving aboard any of Her Majesties Ships of War or Privateers." The cost of providing the medals, it was further ordered, was to be paid out of Her Majesty's share of prizes. This declaration, made on June 1st, 1702, in the first year of her reign, differs from the Act of 1692, which directed that they were to be provided out of a tenth part set aside for that special purpose.

George I.—During the reign of George I there is on record the award in 1715 of a gold enamelled jewel and £1,000 sterling to Captain Matthew Martin, Commander of the "Marlborough," for his gallant defence of his ship against three French ships-of-war in the Bay of Bengal, and getting her safely to Fort St. George in 1712. The medal, it will be observed, was awarded for an action in the reign of George's predecessor. It

SILVER MEDAL AWARDED BY QUEEN ANNE TO A BOY FOR CONSPICUOUS GALLANTRY.

bore on the obverse the Arms of the East India Company, and was enamelled and set with diamonds; on the reverse was an inscription recording the services for which it was awarded. No official decoration was awarded for the victory of Sir George Byng over the Spanish fleet off Cape Passaro in the Mediterranean on July 31st, 1718, although a silver medal was struck to commemorate the event.

George II.—Commemorative medals were also struck in connection with the taking of Puerto Bello, in the Caribbean Sea, by Admiral Vernon with six ships in 1739. The city had about 10,000 inhabitants, and Vernon got possession of the place within forty-eight hours and destroyed the fortifications. There are several "Vernon" medals in existence, varying in size and shape, in white metal, bronze, and silver; but most have on the reverse "He took Porto Bello with Six Ships only." It is by no means certain that any of these medals were issued as official decorations.

Callis Gold Medal.—In 1742 a gold medal and chain of the value of £100 was awarded to Captain Smith Callis, R.N., for his gallant service in taking the "Duke" fireship into the Port of St. Tropez in Provence, on June 14th, and burning five Royal Spanish galleys which had fired upon the blockading fleet. On the obverse of the medal, which is 2·1 in. in diameter, George II, laureated and attired as a Roman Emperor, is represented in the act of presenting a medal to an officer. Under the group, at the top of the plain exergue, is the medallist's name T. PINGO F:. Above all on a ribbon is PRO TALIBVS AVSIS (For such enterprise), and on the reverse five galleys in shore with a squadron of ships-of-war preparing to attack, and in the exergue OB. V. TRIREM. HISPAN. A. S. CALLIS. COMBVST. V. IVLII. MDCCXLII. The date on the medal is the day on which the information was given of the successful venture; but the actual driving into port of the galleys, and the successful attack by the "Duke," was on June 14th. It is believed that the medal in silver was awarded to the other officers who took part in the general affair.

Hornby Medal and Chain.—By a decision of the King given at the Court held at Kensington on September 18th, 1744, a gold medal and chain of the value of £100 was granted to Captain Richard Hornby, of the "Wrightson and Isabella" of Sunderland, and a bounty of £5 to each of his five men, and 40*s.* to each of his three boys, for having engaged a French privateer armed with ten carriage and eight swivel guns, and manned by 75 seamen, and, after a contest renewed again and again over a period of five hours, sinking the French vessel. A gold medal of the value of one hundred guineas was also awarded to Captain Phillips of the "Alexander" privateer for the capture of a large French vessel in St. Martin's Road off the Isle of Rhe. There were 230 men on board the ship of 22 guns, and Phillips's crew numbered 150, but he successfully boarded the vessel and brought her into port, when it was discovered that she was H.M.S. "Solebay," which the enemy had captured a couple of years previously.

Louisbourg, 1758.—For the taking of Louisbourg on July 26th, 1753, a medal was awarded to those who distinguished themselves upon land under Generals Amherst and Wolfe, and at sea under Admiral Boscawen. The Navy's participation in the affair was to attack and cut out the two ships, the "Bienfaisant" of 64 guns, and "La Prudente" of 74 guns, which had been left for fighting purposes after some others had been sunk to obstruct the entrance to the harbour. The boats from the British fleet set out on their task at midnight on July 25th, cut the cable under a heavy fire, and having boarded the ships proceeded to tow them out. "La Prudente" grounded and had to be burned; but the smaller vessel was successfully got out and added to the British fleet. The medals, by T. Pingo, with rings for suspension, were struck in gold, silver, and bronze. Four specimens of the gold medal are known to exist—one which was in the Montague collection, another which was sold at Sotheby's in 1895, and those awarded to Sir Alexander Schomberg, great-grandfather of Lieutenant-Colonel Schomberg, R.M.L.I., and to Senior Midshipman (afterwards Sir George) Young.

"The Glorious" 1st of June.—The reign of George III produced a long record of brilliant achievements by the British Navy. When he ascended the throne, in October 1760, Britain was still at war with France and continued to be so almost the whole time he reigned; but it was not until "The Glorious" 1st of June 1794, when Admiral Lord Howe gained his great victory off Ushant, that medals were awarded, and a regulation medal instituted for naval services. Hearing that the French fleet had sailed for Brest, Howe put to sea to meet it, and having sighted the enemy on the morning of May 28th, gave chase. A smart action ensued which was renewed next day; but heavy fog put an end to fighting until the 31st. On June 1st both fleets prepared for battle: the engagement commenced, and the French line was soon broken in several places; but the French seamen fought with great courage. The furious nature of the British attack, however, was too determined, and the enemy sailed away, such of the ships as could, having 10 dismasted and 7 taken by the English fleet, the "Vengeur" going down with 200 of her crew as she was being towed away. The British Admiral's flagship, the "Queen Charlotte," was badly injured, also the "Brunswick," "Defence," and "Marlborough." It is understood that the 2nd and detachments of the 25th and 69th Foot served as Marines on board the ships which fought on that glorious 1st of June.

The vessels engaged were H.M. ships "Queen Charlotte," "Royal Sovereign," "Royal George," "Barfleur," "Bellerophon," "Impregnable," "Queen," "Cæsar," "Culloden," "Defence," "Glory," "Gibraltar," "Invincible," "Majestic," "Leviathan,"

"Marlborough," "Montague," "Ramillies," "Russell," "Orion," "Thunderer," "Tremendous," "Audacious," "Alfred," "Brunswick," and "Valiant," the frigates "Aquilon," "Latona," "Phæton," "Niger," "Southampton," "Venus," "Charon," and "Pegasus," the sloops "Comet" and "Incendiary," and the cutters "Ranger" and "Rattler."

Naval Gold Medal instituted.—On June 13th Lord Howe reached Spithead, and the King and Queen at a levee held on the Admiral's flagship presented him with a sword set with diamonds in the hilt and a gold chain with a gold medal attached thereto. At the same time gold chains were presented by His Majesty to the two vice-admirals, three rear-admirals, and the captain of the fleet with the intimation that gold medals for suspension would be distributed to the principal officers when they had been struck. These were distributed on November 9th, 1796. The obverses of both medals are the same as illustrated facing this page; but whereas on the larger medal the name of the recipient and the event for which the medal was awarded are encircled by a struck wreath of laurel and oak, the reverse of the smaller one is plain except for the inscription. The larger medal was worn by the admirals suspended from the neck by the gold chain referred to, and the other flag officers hung theirs by means of a white ribbon with dark-blue edges, which became the ribbon for the N.G.S. medal when issued in 1847. The captains' medals, the smaller size, were dependent from a ring and bar made from wire and suspended by the white, blue-edged ribbon through the third or fourth button-hole on the left side of their coats. Although Lord Howe's fleet consisted of twenty-five ships of the line, as well as frigates, only fourteen medals were awarded to his captains, *i.e.* to those whom he had particularly mentioned in dispatches.

This medal bears on the obverse an antique galley on the prow of which Victory is represented in the act of alighting and placing a wreath upon the head of

(Obverse.) (Reverse.)

WILLIAM III SILVER MEDAL FOR LA HOGUE, 1692.

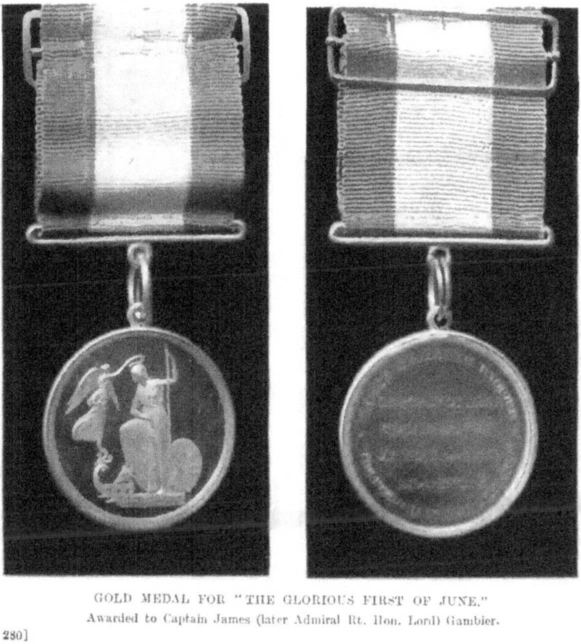

GOLD MEDAL FOR "THE GLORIOUS FIRST OF JUNE."
Awarded to Captain James (later Admiral Rt. Hon. Lord) Gambier.

Britannia, who stands with her right foot on a Greek helmet; she holds in her left hand a spear; beside her is an oval shield whereon is depicted the Union Jack. The larger medal is 2 in. in diameter and the smaller one 1·3 in. Both sizes were struck in gold, and the medal became a general award to naval officers who had distinguished themselves in great naval victories until 1815, and it is noteworthy that the only officers below the rank of post captain who earned the captains' medal were Lieutenant Pinfold of the "Ajax" and Lieutenant Stockham of the "Thunderer," who in the absence of their captains commanded their ships at Trafalgar; but they were not given the medal until they had reached the rank of post captain. It was ultimately awarded to Captain Mounsey of the "Bonne Citoyenne" for the capture of "La Furieuse" on July 16th, 1809. The total number of gold medals issued was 140, of which eight were admirals' medals awarded with chains; flag officers' were issued with broad ribbons for suspension from the neck, and 117 of the captains' size, as illustrated, for suspension between the third and fourth buttonholes of their coats. Gold chains were only given with the admirals' medals for June 1st.

The vessels engaged were H.M. ships "Queen Charlotte," "Royal Sovereign," "Royal George," "Barfleur," "Bellerophon," "Impregnable," "Cæsar," "Culloden," "Defence," "Gibraltar," "Glory," "Invincible," "Leviathan," "Marlborough," "Majestic," "Montague," "Orion," "Ramillies," "Russell," "Tremendous," "Thunderer," "Valiant," "Alfred," "Audacious," "Brunswick," and the frigates "Aquilon," "Charon," "Latona," "Phæton," "Southampton," "Niger," "Pegasus," the sloops "Incendiary" and "Comet," and the cutters "Rattler" and "Ranger."

St. Vincent, 1797.—On St. Valentine's Day, 1797, the battle of St. Vincent was fought between the Spanish fleet of 27 ships of the line and 8 frigates under Admiral Don Josef de Cordova, and 20 British ships of the line, 2 sloops, and a cutter which had been cruising off the coasts of Spain and Portugal to prevent the fleets of France, Spain, and Holland—which countries had entered into an alliance against Great Britain—from combining. Sir John Jervis commanded; his flagship was the "Victory." Passing through the two divisions of the Spanish fleet, he cut off nine of the enemies' vessels: the Spaniards then attempted to break the British line; but the effort was frustrated, two of the enemy's ships striking their colours. This action, which only lasted four hours, resulted in the loss of four ships to the enemy, Commodore Horatio Nelson of the "Captain," 74 guns, having taken the "San Josef" of 112 guns, and the "San Nicolas" of 80 guns.

In this battle Nelson by his intrepid action placed himself in the front rank of commanders, and his bravery and dash were rewarded by the Order of the Bath, and his advancement to the rank of Rear-Admiral. In the fight off Cape St. Vincent he not only attacked at close quarters the "Santissima Trinidad," a great galleon of 112 guns, but with the aid of the "Culloden" (Captain Trowbridge), which gallantly began the fight, he fought for nearly an hour this great ship and four other galleons which had come to her assistance, until the "Blenheim" (Captain Frederick) and the "Excellent" (Captain Collingwood) came to assist the English, when Nelson ran the "Captain" alongside the "San Nicolas," quickly sprang through one of the stern cabin windows, and leading his men across the Spaniard's deck, boldly boarded the "San Josef," on the quarter-deck of which he soon received the swords of the conquered officers. By 5 o'clock the Spaniards were in retreat, having lost to their enemy four of their finest vessels, and, besides receiving a blow to their naval power, had considerably minimised their value to the French in their effort to command the Channel, and upset the calculations of the Dutch. For the victory Sir John Jervis was created Earl of St. Vincent, and awarded a pension of £3,000 per annum, while the unfortunate Spanish Admiral was dismissed in disgrace.

The vessels engaged were: H.M. ships "Victory," "Britannia," "Barfleur," "Blenheim," "Prince George," "Captain," "Goliath," "Excellent," "Egmont," "Culloden," "Colossus," "Diadem," "Namur," "Orion," "Irresistible," and the frigates "Dido," "Lively," "Minerva," "Niger," and "Southampton," sloops "Bon Citoyenne" and "Raven," and the cutter "Fox."

Camperdown.—When the battle of Camperdown was fought on October 11th, 1797, Great Britain was in the unfortunate position of standing isolated among the nations of Europe, and still concerned with the series of mutinies which had broken out in her fleet owing to the poor pay and bad food meted out to the seamen who had served their country so well. As a consequence Britain was, had her enemies but realised it, in a very precarious position. Fortunately Admiral Duncan had "a way with him," which enabled him, by a combination of firmness and reasoning power, to get his men into such a frame of mind as to state in writing that there was not a man on board the "Venerable" "but would lose the last drop of blood in his body before they (the enemy) should obtain any victory over us, therefore, honoured sir, we once more implore your gracious pardon." Fortunately Duncan's ability to deal with the mutineers enabled him, at least, to have a fleet of some pretensions wherewith to meet the enemy. Meanwhile the mutiny at the Nore had been suppressed, "Admiral" Richard Parker, the leader, hanged, and some effort made to remedy the grievances of the seamen while Duncan had been hoodwinking the Dutch.

On October 9th, 1817, Duncan set sail from Yarmouth Roads with 20 ships, 4 cutters, and a lugger, to meet the Dutch fleet which had at last left the Texel. Just before noon on the 11th, Duncan signalled to his ships "clear for action and close," and by half-past 12

Vice-Admiral Onslow, in the "Monarch," had got through the enemy's line and engaged Vice-Admiral Reyntier. Admiral Duncan engaged the "Vyrheid," the Dutch Admiral's flagship, and after a hard fight lasting until 2 o'clock, De Winter only struck his ship to the "Venerable" when he was the only unwounded man on the quarter-deck of his mastless ship. It is significant of Duncan's character and manliness that when his opponent offered his sword he refused to take it, saying "rather a brave man's hand than his sword." On October 16th the victorious Admiral anchored off the Nore with the eleven ships he had taken, and next day King George raised him to the Peerage with the titles of Baron Duncan of Lundie and Viscount Duncan of Camperdown.

The vessels engaged were: H.M. ships "Venerable," "Ardent," "Adamant," "Agincourt," "Bedford," "Belliqueux," "Brackel," "Lancaster," "Monarch," "Montague," "Monmouth," "Powerful," "Russell," "Director," "Isis," "Veteran," "Triumph," the frigates "Beaulieu," "Circe," and "Martin," the cutters "Active," "Diligent," "King George," "Rose," and the lugger "Speculator."

The Nile.—For over a year hardly a British man-of-war had passed into the Mediterranean until Rear-Admiral Sir Horatio Nelson, with Captain Trowbridge, entered with his fleet to frustrate the pretensions of Napoleon. There were 13 74-gun ships and 1 of 50 guns; but before the action commenced on August 1st, 1798, the former had been reduced to 12 by the grounding of the "Culloden." The French fleet was of equal proportions; but included several heavier ships: "L'Orient," the flagship, a magnificent three-decker carrying 120 guns, and three others with 80 guns apiece. The French fleet was moored on the edge of a shoal in a compact line running north-west by south-east. Captain Foley, with the "Goliath," led the way between the front ships and Aboukir Island, thus executing a manœuvre which the French Admiral had

NAVAL GENERAL SERVICE MEDAL AND KNIGHT (COMMANDER'S) HANOVER BADGE AND STAR.

Awarded to Rear-Admiral Sir Thomas Ussher, C.B., K.C.H.

thought impracticable. Five ships thus engaged the enemy on the port or land side, while the "Vanguard," followed up by other ships, made for the centre of the line. At half-past 6 in the evening the battle commenced, and by 8 o'clock 8 British ships and an equal number of French were engaged in one of the most deadly struggles in naval history.

Early in the contest the French Admiral, de Brueys, had been badly wounded, but, gallant sailor that he was, refused to go below, and met his death about this time by a shot that almost cut him in two: even then his resolute spirit refused to yield, and death only gave his comrades the power to remove his maimed body. At 9 o'clock his ship, "L'Orient," was seen to blaze out; but her determined crew stuck to their guns and fired them until they were literally scorched out of their places—and then the great ship blew up. So awful was the scene that the gunners of both fleets were apparently too dazed to fire again for about a quarter of an hour. Meantime a French lad won undying fame by his devotion to his father, the Commodore of "L'Orient," and his name has been immortalised by a British poetess, Mrs. Hemans, who has so well portrayed the story of the little ten-year-old lad who refused to leave his mortally wounded father, and despite the fact that, as a French writer states, "the English seamen made the most strenuous efforts to save the young Casabianca," he went down to his death with the father he loved so well.

Just before "L'Orient" blew up, Nelson received a severe wound on the forehead which was for a time thought likely to prove mortal; but on hearing that the great vessel had blown up, he gave orders for the launching of the only boat so that the unfortunate survivors of the explosion might be rescued. The night with its horrors had exhausted the fleets, and firing gradually ceased until the day broke at 4 o'clock: by then 6 French ships had hauled down their colours and two were aground, while the "Tonnant," commanded by the bravest of brave seamen, Captain Dupetit-Thouars, showed fight to the last, for her seamen were filled with the spirit of their dead captain, who had his colours nailed to the mast, and when wounded unto death refused to yield. Such courage as his must be infectious. With both arms and a leg shot away, seated in a tub of bran, this defiant and gallant Frenchman commanded his ship until the conqueror of all—Death—claimed the victory.

Only 2 French ships-of-war and 2 frigates escaped. The flagship and the "Arteruise" had blown up, 2 were aground, a frigate had been sunk when Rear-Admiral Villeneuve endeavoured to get away with the badly damaged "Tonnant" and 3 others. Of these the "Timeleon" grounded, and after her crew had got ashore, her captain burnt her; the "Tonnant" was captured; and only the "Guillaume Tell" and the "Genereux," which had hardly taken any part in the conflict, escaped with the frigates "Justice" and "Diane." Not a British vessel went down, but all showed severe evidence of the fight that had been waged on that August day in 1798. The victory was due not only to the excellent seamanship of Nelson, who was raised to the peerage as Baron Nelson of the Nile, and of Burnham Thorpe, for what Lord St. Vincent described as "the greatest achievement the history of the world can produce," but to the superb gunnery of his men, who served their guns with such rapidity and precision that their opponents were absolutely outclassed.

The ships engaged in the battle were the "Vanguard," "Bellerophon," "Orion," "Audacious," "Culloden," "Defiance," "Zealous," "Minotaur," "Goliath," "Alexander," "Majestic," "Leander," "Theseus," "Swiftsure," and the sloop "Mutine."

Davison's Medal for the Nile.—As a souvenir of this historic sea-fight Mr. Alexander Davison, Lord Nelson's prize agent, presented to every officer and seaman of Nelson's fleet a medal; gold to captains and lieutenants, silver to warrant officers, bronze gilt to petty officers, bronze to seamen and marines. The medal, which is $1\frac{17}{20}$ in. in diameter, is illustrated facing page 288. Upon the edge is impressed FROM ALEXANDER DAVISON ESQ., ST. JAMES SQUARE, A TRIBUTE OF REGARD. The medals were issued unnamed, but many had their names and the name of their ship also engraved above the sky-line on the reverse. The medals, modelled by C. H. Kuchler, were highly prized by all who received them, from Nelson—who wrote very gratefully to Davison—to the seamen, many of whom had the medals mounted in gold or silver frames and wore them suspended from a broad blue ribbon. This is the first instance of a medal given by a private individual being accepted and worn in the Service.

Copenhagen.—This battle, fought on April 2nd, 1801, will long be remembered for the gallant conduct of Nelson, who, when Admiral Sir Hyde Parker signalled "leave off action," placed his glass to his blind eye and saying, "I really do not see the signal!" ordered his own signal for close action to be kept flying. Had it not have been for this daring disregard of orders, Copenhagen would have been among the failures of our navy, instead of one of its most glorious victories, for it was one of the most stubborn fights on record. For five hours Britons and Danes fought with equal courage and determination. For five hours they fought with equal success, and it was only through the pertinacity of Nelson that the fortune of war fell to Britain. This battle has a double significance, for not only did Nelson prove his great capacity as a seaman, but also gave further evidence of his humane character; for when the stubborn spirit of the Danes was likely to lead to useless bloodshed he appealed "to the brothers of Englishmen—the Danes," to recognise the fact that they were beaten, and so prevent unnecessary carnage.

It was an action which he knew would be misunderstood and misrepresented, but he did not falter. Fortunately his message was received by the Crown Prince in the spirit in which it was sent, and the battle of Copenhagen, "the most terrible of all" his engagements, was added to the laurels of the immortal and gentle Nelson, although because Nelson requested the cessation of hostilities, the Danes regarded it as a drawn battle.

The following ships were engaged: "Elephant," (Vice-Admiral Nelson's flagship), "Defiance," "Edgar," "Monarch," "Bellona," "Ganges," "Russell," "Agamemnon," "Ardent," "Polyphemus," "Glatton," "Isis," "Amazon," "Alcméne," "Blanche," "Désirée," "Jamaica," sloops "Arrow," "Dart," "Cruiser," and "Harpy," bomb vessels "Discovery," "Explosion," "Hecla," "Sulphur," "Terror," "Volcano," and "Zebra," and the brigs "Otter" and "Zephyr."

Trafalgar.—Napoleon having placed the crown of Italy upon his head (May 6th, 1805), within a fortnight declared Genoa and the Ligurian Republic part of the French Empire. He then had hopes that Admiral De Villeneuve would successfully escort the "Army of England," then encamped at Boulogne, across the Channel. His plans were, however, frustrated by the engagement forced upon Villeneuve by Admiral Sir Robert Calder off Ferrol, while Napoleon, charging his Admiral with cowardice, caused him to leave the harbour of Cadiz on October 19th, 1805, and the epoch-making battle off Cape Trafalgar was the result. Owing to light winds, Villeneuve did not make the Atlantic until the next day, and then his combined fleet of 33 French and Spanish ships of the line, 5 French frigates, and 2 brigs, were sighted off Cape Trafalgar by Nelson's look-out ships. It was not until next day, however, that the fleets formed in battle array. At 6.30 a.m. the British ships cleared for action. The English fleet comprised 27 sail of the line, 4 frigates, a cutter and a schooner. Lord Nelson led the Weather or Northern Division in his flagship the "Victory,"

DAVISON'S MEDAL FOR THE VICTORY OF THE NILE, 1798.

while Vice-Admiral Collingwood led the Lee or Southern Division. Just before noon the famous signal "England expects every man to do his duty" was run up, and the battle began. "Engage the enemy more closely" was the only other signal given by the British Admiral, and that was when the enemy's ship "Fougueux" had opened fire upon the "Royal Sovereign"; then followed a series of contests between the French and British ships, which reflect the greatest credit upon the seamen of both nations.

Nelson, being desirous of engaging the French Admiral's flagship, came upon the "Bucentaure" at 12.30 p.m., when she fired upon the British flagship, but, unheedful of the shot from this and seven other ships, the "Victory" ploughed her way ahead in order to break the enemy's line. For three-quarters of an hour her guns refused to bark, though her men were falling fast from the shots which broke through her sides, and tore her sails to shreds; but at 1 o'clock the "Victory" closely engaged the "Bucentaure," and so well were the British shots placed that 20 of the enemy's guns were soon dismounted and, according to a French estimate, 400 seamen were killed. Leaving the "Bucentaure" the "Victory" engaged the "Redoubtable" at ten minutes past 1, and the rigging of the two ships fouling, they were locked in a deadly embrace. Then the crews of the two vessels fought with exceeding gallantry, for both were led by brave men, and the "Redoubtable's" captain cheered his men on to board the "Victory"; indeed, he lowered his mainyard on to the "Victory's" deck, and his men made a brave effort to capture her, but those daring Frenchmen who gained her decks paid for their effort with their lives; not, however, before they had placed 30 of the "Victory's" crew *hors de combat.*

Death of Nelson.—Meanwhile, one of the sharpshooters in the mizzen-top of the "Redoubtable" had given Viscount Nelson his death wound. At half-past one the fatal bullet passed through his shoulder and spine, but Britain's greatest admiral lived long enough to learn that the victory was his, and he died at 4.30 p.m. satisfied that he had done his duty. By 5.30 p.m. 18 of the enemy's ships had struck their colours, and the survivors of the fight were making the best of their way to Cadiz, but a gale coming on, the British fleet with its prizes stood out to sea, where some of the prizes foundered with their crews; others were driven ashore and wrecked, while four were retaken by the enemy, so that by the time the triumphant fleet sailed into Gibraltar Bay it only retained 4 prizes. The victory of Trafalgar had, however, made Britain indisputable mistress of the seas, destroyed the sea power of France and Spain, and, while it added to the laurels of British seamen, gave to Nelson a glorious death and secured for him a resting-place in Westminster Abbey.

The following ships were engaged in the battle: "Victory," "Royal Sovereign," "Temeraire," "Britannia," "Conqueror," "Neptune," "Agamemnon," "Leviathan," "Ajax," "Africa," "Minotaur," "Orion," "Belleisle," "Mars," "Thunderer," "Spartiate," "Bellerophon," "Achille," "Colossus," "Polyphemus," "Revenge," "Dreadnought," "Swiftsure," "Defence," "Defiance," "Prince" and "Tonnant." "Naiad," "Sirius," "Euryalus" and "Phœbe" frigates. The cutter "Entreprenaute" and the schooner "Pickle."

Gold Medal for Trafalgar.—As already stated, the gold medal instituted by George III to reward the officers of Lord Howe's fleet who took part in the action on "The Glorious" June 1st, 1794, was afterwards awarded for all great naval victories, and among those who received them were Nelson's officers—from post-captains upwards—who took part in the battle of Trafalgar. The gold medal was awarded for 18 different actions, the last award being for the fight between the "Endymion" frigate of 48 guns and the American 50-gun frigate "President" on January 15th, 1815. See facing page 280 for smaller type of medal as awarded to James (afterwards Right Hon. Admiral Lord James) Gambier of the "Defence" for June 1st, 1794.

Boulton's Trafalgar Medal.—To commemorate the last and greatest of Nelson's victories, Matthew Boulton, the partner of James Watt of the famous Soho Works near Birmingham, decided to strike medals and present one to each participant in the fight. The medal, $1\frac{9}{10}$ in. in diameter, bears on the obverse a fine bust of Nelson, with the inscription surrounding it HORATIO VISCOUNT NELSON. K.B. DUKE OF BRONTE. On the reverse the battle is represented *en cameo*, with the famous signal

on a ribbon running with the line of the medal, and in the exergue TRAFALGAR OCT. 21 1805. On the edge of the medal is the inscription TO THE HEROES OF TRAFALGAR FROM M. BOULTON. The medal was struck in silver for the senior officers, and in pewter for distribution among the junior officers and seamen. The intrinsic value of the medal did not, however, appeal to many of the recipients of the pewter variety, and they either refused them or threw them overboard. Those who retained them wore them suspended from a blue ribbon. A few were struck in bronze as proofs.

Davison's Trafalgar Medal.—This medal, generally supposed to have been struck at the instance of Mr.

Alexander Davison for distribution among the surviving members of the crew of the "Victory," has on the obverse a shield bearing the arms of Nelson, encircled by a garter inscribed TRIA · JUNCTO · IN · UNO · ensigned by a bust of Viscount Nelson, with a laurel branch to the left, and a palm branch to the right, and on a scroll beneath the shield PALMAM QUI MERUIT FERAT and the double inscription ADMIRAL LORD NELSON D. OF BRONTE NATUS SEP. 29TH 1758. HOSTE DEVICTO REQUIEVIT OCT. 21ST 1805 and ENGLAND EXPECTS EVERY MAN WILL DO HIS DUTY. On the reverse is a man-of-war with sails furled, and above THE LORD IS A MAN OF WAR. EXODUS C 15 V 3, whilst below is VICTORY OFF TRAFALGAR OVER THE COMBINED FLEETS OF FRANCE & SPAIN OCT 21 1805. Beneath the man-of-war is HALLIDAY FECIT. The medals, 2 in. in diameter, were struck in pewter or white metal, and were sometimes framed by the recipients in gold, silver, or gilt metal rims, with a loop for suspension from a blue ribbon.

The Official Medal.—It was not until June 1st, 1847, that it was made known by a General Order that Her Majesty Queen Victoria had commanded that a medal should be struck, not only to commemorate the battle of Trafalgar, but to recognise the services rendered by her fleets and armies from 1793 to 1815. The medal was ready for distribution in January 1849, and later the naval services for which it might be awarded were extended to 1840. The admitted claims totalled 20,900.

The Naval General Service Medal.—Two hundred and thirty bars were issued with the medal, which bore upon the obverse the bust of Queen Victoria, as on the Military General Service medal, and the legend VICTORIA REGINA, with the date 1848 below, and on the reverse a figure of Britannia seated upon a sea-horse holding in her right hand a trident, and in the left an olive branch. The medal, by Wyon, is suspended by a straight clasp from a white ribbon with blue edges, and the recipient's name is impressed upon the edge of the medal in Roman capitals of the same character as those

(Reverse.)
NAVAL GENERAL SERVICE MEDAL.

SULTAN'S MEDAL FOR ACRE.

(Obverse.)
NAVAL GENERAL SERVICE MEDAL.

used for the M.G.S. medal. Officers and warrant officers only had their rank described, and in the arrangement of the bars the first action was placed nearest the medal. Although so many bars were issued with the medal, six is the most awarded to any one man. Dr. Payne has in his collection two with five bars; one awarded to Thos. Hewitt, midshipman, for 1 JUNE 1794—12 OCTR 1798—4 FEB BOAT SERVICE 1804—CENTAUR 26 AUGT 180'—ALGIERS, and another awarded to Capt. (afterwards Rear-Admiral Sir Thomas) Ussher, Kt., C.B., K.C.H., illustrated facing page 284, which also includes 1 JUNE 1794—REDWING 7 MAY 1808—REDWING 31ST MAY 1808—MAGALA 29TH APRIL 1812—2 MAY BOAT SERVICE 1813. This is a particularly rare set of bars, for only seven of the second were issued, five of the third, seven of the fourth, and forty-nine of the fifth.

Bars granted with the Naval General Service Medal with the Names of the Actions thereon

NYMPH, 18 June 1793. 4 issued. Capture of the French frigate "Cleopatre." Captain E. Pellew, who commanded the "Nymph," was knighted.

CRESCENT, 20 Oct. 1793. 12 issued. Capture of the French frigate "Reunion," which was added to the British Navy under the same name.

ZEBRA, 7 March 1794. 2 issued. Running the "Zebra" sloop alongside the bastion of Fort Royal at Martinique, and storming and capturing the fort.

CARYSFORT, 29 May 1794. None issued. Recapture of the "Caster."

1 June 1794 ("The Glorious First of June"). 576 issued. Defeat of the French fleet by Lord Howe, capture of 6 sail of the line and 1 sunk. 58 vessels (for list of principal vessels engaged see page 279).

ROMENY, 17 June 1794. 2 issued. Capture of French frigate "Sybille."

BLANCHE, 4 Jan. 1795. 5 issued. At the end of the

conflict she had a crew of 190. Capture of the French frigate "Pique."

LIVELY, 13 March 1795. 3 issued. Capture of the French frigate "Tourterelle."

14 March 1795. 111 issued. Action with French fleet, and capture of 2 sail of the line. 22 vessels under Vice-Admiral Hotham were engaged.

ASTRÆA, 10 April 1795. 2 issued. Capture of the French frigate "Gloire."

THETIS, 17 May 1795. 3 issued.
HUSSAR, 17 May 1795. 1 issued.
} Action with four French ships, and capture of "La Raison" and "Prevoyante."

MOSQUITO, 9 June 1795. None issued. This was not due to delay in issuing medals, but to the loss of the captain and crew of the "Mosquito" soon after. Action with, and capture of, privateer.

17 June 1795. 38 issued. Brilliant repulse by Vice-Admiral Cornwallis, of a fleet four times superior in force. 7 vessels engaged.

23 June 1795. 201 issued. Action under Admiral Lord Bridport with French fleet, and capture of 3 sail of the line; the "Formidable" became the British ship "Belleisle" and the "Tigre" was also added under its own name. The "Alexander" was a recapture. 29 vessels engaged.

DIDO, 23 June 1795. 1 only issued.
LOWESTOFFE, 24 June 1795. 6 issued.
} Action with the French frigates "Minerve" and "Artemise," and capture of the former.

SPIDER, 25 Aug. 1795. 1 only issued. Action with 2 French brigs, and capture of 1. This medal realised £33 at the Gaskill sale.

PORT SPERGUI, 17 March 1796. 4 issued to survivors of the "Diamond," "Liberty," and "Aristocrat." Destroying the batteries at Port Spergui, and destroying the corvette "Etourdie," 4 brigs, 2 sloops, and 1 lugger.

INDEFATIGABLE, 20 April 1796. 6 issued. Capture of French frigate "Virginia," in the English Channel.

UNICORN, 8 June 1796. 4 issued. SANTA MARGARITA, 8 June 1796. 3 issued.	Action with the three French frigates "La Tribune," "La Tamise," (formerly the British ship "Thames") and "La Legere," and capture of the two former. Captain William of the "Unicorn" was knighted.

SOUTHAMPTON, 9 June 1796. 4 issued. Capture of French corvette "Utile."

DRYAD, 13 June 1796. 7 issued. Capture of the French frigate "Proserpine," which became H.M.S. "Amelia."

TERPSICHORE, 13 Oct. 1796. 3 issued. Capture of the Spanish frigate "Mahonesa."

LAPWING, 8 Dec. 1796. 2 issued. Action with French ship "Décieux" and brig "Vaillante," and capture of the former. 1 medal sold for £21 on July 24th, 1914.

MINERVA, 19 Dec. 1796. 5 issued. BLANCHE, 19 Dec. 1796. 2 issued.	Capture of Spanish frigate "Santa Sabina" by the "Minerve," and action between "Blanche" and "Ceres."

INDEFATIGABLE, 13 Jan. 1797. 8 issued. AMAZON, 13 Jan. 1797. 6 issued.	Destruction of French "74," "Droits de L'Homme," off the coast of France. The "Amazon" went ashore and her crew were made prisoners.

ST. VINCENT, 14 Feb. 1797. 363 issued. Action with the Spanish fleet, and capture of 4 sail of the line. 23 ships engaged (for list see page 283).

SAN FIORENZO, 8 March 1797. 7 issued. Capture of

"Resistance" and "Constance." The first named became the "Fisgard."

NYMPHE, 8 March 1797. 6 issued.

CAMPERDOWN, 11 Oct. 1797. 332 issued. Battle of Camperdown, when Admiral Duncan defeated the Dutch fleet. 25 vessels engaged (for list see page 283).

PHŒBE, 21 Dec. 1797. 7 issued. Capture of the French frigate "Nereide," which was added to the British Navy under the same name.

MARS, 21 April 1798. 26 issued. Capture of French 74, "L'Hercule." In this severe action the "Mars" lost her captain, Alexander Hood.

ISLE ST. MARCOU, 6 May 1798. 3 issued. Action at the Island of Marcou. 2 boats engaged.

LION, 15 July 1798. 21 issued. Action with 4 Spanish frigates and capture of the "Santa Dorotea," which was added to the British Navy under the same name.

NILE, 1 Aug. 1798. 351 issued. Battle of the Nile. 15 ships engaged (for list see page 286).

ESPOIR, 7 Aug. 1798. 1 issued. Action with and capture of a Genoese pirate, the "La Guria."

12 Oct. 1798. 81 issued. Action under Sir J. B. Warren with a French squadron, and capture of the "Hoche" "74"—which became the "Donegal"—and 2 frigates. 8 ships engaged.

FISGARD II, 20 Oct. 1799. 9 issued. Capture of the French frigate "L'Immortalite."

SYBILLE, 28 Feb. 1799. 12 issued. Capture of the French frigate "La Forte," after a stern fight in which a detachment of the Scotch Brigade which was on board the "Sybille" took part.

TELEGRAPH, 18 March 1799. None issued. Capture of "L'Hirondelle II."

ACRE, 30 May 1799. 42 issued. The record on the bar for this affair is ACRE, 30 MAY 1799; but as the French raised the siege on May 20th, and Bonaparte retreated leaving 23 cannon, the bar is wrongly

BALTIC MEDAL.

NAVAL MEDAL FOR BEST SHOT.

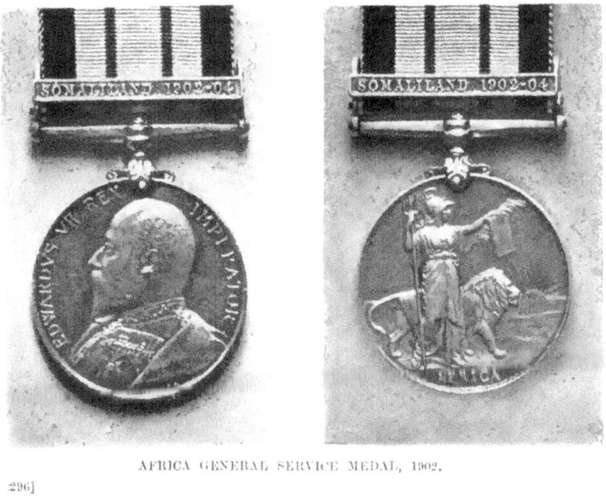

AFRICA GENERAL SERVICE MEDAL, 1902.

inscribed, but it is nevertheless official. The bar was awarded to those who assisted the Turkish fleet to defend "Acre" from the French attack, the ships taking part under Commodore Sir Henry Smith being the "Theseus," "Tigre," and "Alliance."

SCHIERMONIKOOG, 12 Aug. 1799. 10 issued.

ARROW, 13 Sept. 1799. 2 issued. Action and capture of "Draak" and "Gier." 2 vessels. The other boat was the "Wolverine," but no bar was issued for that.

SURPRISE with HERMIONE, 25 Oct., 1799. 7 issued. Boarding and recapturing the "Hermione." For this dashing exploit Captain Hamilton was knighted.

SPEEDY, 6 Nov. 1799. 3 issued. Action with 10 Spanish gunboats and 2 schooners, and successful defence of convoy.

COURIER, 22 Nov. 1799. 3 issued. Action with and capture of the "Guerrier."

VIPER, 26 Dec. 1799. 2 issued. Capture of French brig "Furet."

FAIRY, 5 Feb. 1800. 4 issued.
HARPY, 5 Feb. 1800. 4 issued.
} Action with and capture of the French frigate "Pallas," which became H.M.S. "Pique."

PETEREL, 21 March 1800. 2 issued. Capture of "La Ligurienne."

PENELOPE, 30 March 1800. 11 issued.
VINCIEGO, 30 March 1880. 2 issued.
} Night action with "Guillaume Tell"; the "Lion" and "Foudroyant" were also engaged and actually captured the vessel, which was added to the British Navy as H.M.S. "Malta," but no award was made to the surviving crews of these ships.

DESIRÉE, 8 July 1800. 23 issued. Boarding and capturing the French frigate "Desirée," and other French vessels. 18 vessels engaged.

Seine, 20 Aug. 1800. 9 issued. Capture of French frigate "Vengeance," between Porto Rico and San Domingo.

Phœbe, 19 Feb. 1801. 7 issued. Capture of French frigate "Africaine," which became H.M.S. "Amelia."

Egypt. 626 issued for services on Coast of Egypt. 117 vessels were engaged.

Copenhagen, 1801. 589 issued. 38 vessels engaged.

Speedy, 6 May. 7 issued. Capture of the "Gamo."

Gut of Gibraltar, 12 July 1801. 152 issued. 10 ships engaged. Action with the French squadrons in the "Gut of Gibraltar," and destruction of 2 Spanish ships of 122 guns each, and capture of the "St. Antonio" "74." For his services as commander of the British squadron Sir J. Saumarez was made a K.C.B. and given a pension of £1,200 a year.

Sylph, 28 Sept. 1801. 2 issued. Action with the "Artemise."

Pasley, 28 Oct. 1801. 3 issued. Capture of the Spanish ship "Virgen del Rosatio."

Scorpion, 31 March 1804. 1 issued. Beaver, 31 March 1804. None issued.	Action with and capture of vessels in the Ville road.

Centurion, 18 Sept. 1804. 11 issued. Action with the line-of-battle ship "Marengo," and frigates "Atalante" and "Semillante."

Arrow, 3 Feb. 1805. 8 issued. Acheron, 3 Feb. 1805. 2 issued.	For the protection of 28 British merchant ships, when attacked by 2 French frigates.

San Fiorenzo, 14 Feb. 1805. 11 issued. Capture of the French frigate "Psyche."

Phœnix, 10 Aug. 1805. 25 issued. Capture of French frigate "Didon." The vessel became H.M.S. "Didon."

Trafalgar. 1,710 issued. 33 vessels engaged (see page 290).

4 Nov. 1805. 298 issued. 8 vessels engaged. Capture of 4 sail of French line-of-battle ships.

St. Domingo. 410 issued. Battle of "St. Domingo," and capture and destruction of 4 sail of the line. 11 vessels were engaged.

Amazon, 13 March 1806. 27 issued.
London, 13 March 1806. 28 issued. } Capture of "Marengo" and "Belle Poule."

Pique, 26 March 1806. 7 issued. Capture of the French brigs "Phæton" and "Voltigeur."

Sirius, 17 April 1806. 12 issued. Action with French flotilla at Civita Vecchia, and capture of the "Bergere."

Blanche, 19 July 1806. 22 issued. Capture of the "Guerrière." For this deed Captain Lavie was knighted.

Arethusa, 23 Aug. 1806. 6 issued.
Anson, 23 Aug. 1806. } Capture of Spanish frigate "Pomone."

Curaçoa, 1 Jan. 1801. Taking of Curaçoa. 4 vessels engaged—"Arson," "Arethusa," "Fisgard," and "Latona."

Pickle, 3 Jan. 1807. 1 issued. Capture of the French privateer "La Favorite."

Hydra, 6 Aug. 1807. 10 issued. In the harbour of Bergur, attack of batteries, and capture of "L'Eugene," "Rosario," and "Caroline."

Comus, 15 Aug. 1807. 10 issued. Capture of the Danish frigate "Fredericksooarn."

Louisa, 28 Oct. 1807. 1 issued. Action with and defeat of a French privateer.

Carrier, 4 Nov. 1807. 1 issued. Capture of the French cutter "L'Actif."

Ann, 24 Nov. 1807. None issued. Action with 10 Spanish gunboats.

Sappho, 2 March 1808. 5 issued. Capture of the Danish brig "Admiral Yawl."

San Fiorenzo, 8 March 1808. 16 issued. Capture of the French frigate "Piedmontaise." The British commander, Captain Hardinge, was killed, and a monument to his memory erected in St Paul's Cathedral at the public expense.

EMERALD, 13 March 1808. 12 issued. Destruction of the batteries at Vivero and war vessels.

CHILDERS, 14 March 1808. 4 issued. Captain Wilson and a crew of 65 participated in this affair. Action with the Danish brig "Lougen."

NASSAU, 22 March 1808. 37 issued. STATELY, 22 March 1808. 25 issued.	Destruction of Danish line-of-battle ship "Prince Christian Frederic." 2 ships engaged.

OFF ROTA, 4 April 1808. 20 issued. Fight with gun-boats and convoy. "Alceste," "Mercury," and "Grasshopper" engaged.

GRASSHOPPER, 24 April 1808. 1 issued. RAPID, 24 April 1808. 1 issued.	Action with and destruction of Spanish ships and gun-boats at Faro. 2 vessels engaged.

REDWING, 7 May 1808. 7 issued. Action with Spanish gunboats and destroying them.

VIRGINIE, 19 May 1808. 21 issued. Capture of Dutch frigate "Guelderland."

REDWING, 31 May 1808. 5 issued. Destroying Tarifa Battery near Cape Trafalgar, and taking 2 vessels.

SEAHORSE WH BANDERE LA ZAFFERE, 6 July 1808. 35 issued. Capture of the Turkish frigate "Badere Zaffer."

COMET, 11 Aug. 1808. 5 issued. Action with three French brigs, and capture of the "Sylphe." This became H.M.S. "Seagull."

CENTAUR, 26 Aug. 1808. 38 issued. IMPLACABLE, 26 Aug. 1808. 45 issued.	Fight with Russian fleet and capture of 74-gun ship "Sewolod."

CRUIZER, 1 Nov. 1808. 4 issued. Action with Danish flotilla off Gottenburgh and capture of brig-of-war.

AMETHYST with THETIS, 10 Nov. 37 issued. Capture

of French frigate "Thetis" by Captain Seymour, who was knighted for his prowess.

Off the Pearl Rock, 13 Dec. 1808. Action with batteries and French boats. 6 vessels engaged.

Onyx, 1 Jan. 1809. 6 issued. Recapture from Dutch of the brig "Manly."

Confiance, 14 Jan. 1808. 8 issued. Taking of Cayenne.

Martinique. 523 issued. Capture of Martinique. 43 vessels engaged.

Horatio, 10 Feb. 1809. 14 issued. Supériéure, 10 Feb. 1809. 2 issued.	Capture of the French frigate "Junon," which was added to the British Navy under the same name.

Amethyst, 5 April. 28 issued. Capture of French frigate "Niemen."

12 April. 646 issued. Fight with French squadron and destruction of ships in Basque Road. 35 ships engaged.

Pompée, 17 June 1809. 17 issued. Castor, 17 June 1809. 4 issued. Recruit, 17 June 1809. 3 issued.	Chase and capture of French ship of the line "Hautpoult," which became H.M.S. "Abercromby." 3 vessels engaged. This action took place on April 17 and not as stated on the bars.
Cyane, 25 and 27 June 1809. 5 issued. L'Espoir, 25 and 27 June 1809. 5 issued.	Action with "Ceres," and capture of 18 gunboats, and destruction of 4. Captain Staines, commander of the "Cyane," was knighted.

Bonne Citoyenne with Furieuse. 12 issued. Capture of French frigate "La Furieuse," July 6th, 1809. Captain Mounsey received the Naval Gold Medal for this exploit, as already explained.

Diana, 11 Sept. 1809. 3 issued. Capture of Dutch brig "Zephyr."

ANSE-LE-BARQUE, 18 Dec. 1909. 42 issued. Storming batteries of Anse-le-barque, and capture of " Loire " and " Seine " frigates. 9 vessels engaged.

CHEROKEE, 10 Jan. 1910. 4 issued. Capture of the French lugger " L'Aimable Nelly."

SCORPION, 12 Jan. 1810. 12 issued. Capture of the French brig " L'Oreste."

GUADALOUPE. 509 issued. Capture of Guadaloupe, January and February 1910. 50 vessels engaged.

THISTLE, 10 Feb. 1810. None issued. Capture of the Dutch corvette " Havik."

SURLY, 24 April 1810. 1 issued.
FIRM, 24 April 1810. 1 issued.
} Capture of the French privateer " Alcide."

SYLVIA, 26 April 1810. 1 issued. Capture of the Dutch brig " Echo," in the Straits of Sunda.

SPARTAN, 2 May 1810. 32 issued. Action with the French frigate " Ceres," and consorts, and capture of the corvette " Sparviéve."

ROYALIST, May and June 1810. 3 issued. Action in the Channel with and capture of 6 armed French ships.

AMANTHEA, 25 July. 29 issued. Action with gun-boats, and capture and destruction of a number of transports at Amanthea. " Thames," " Pilot," " Weasel " engaged.

BANDA NEIRA. 69 issued. Capture of the Island of " Banda Neira," 9 Aug. 1810. 3 boats engaged.

BOADICEA, 18 Sept. 1810. 16 issued.
OTTER, 18 Sept. 1910. 8 issued.
STAUNCH, 18 Sept. 1810. 2 issued.
} Action with French squadron and capture of the French frigate " Venus " and re-capture of the British frigate " Ceylon." " Boadicea," " Otter," and " Staunch " engaged. The " Venus " was added to the British Navy as the " Néréide."

Briseis, 14 Oct. 1810. 2 issued. Capture of the French privateer "Sans-Souci."

Lissa. 130 issued. Action with French squadron and capture of frigates off Lissa, March 13th, 1811. 4 vessels engaged.

Anholt, 27 March 1811. 40 issued. Action in defence of Anholt Island when attacked by Danes.

Arrow, 6 April 1811. None issued. Action with "Chasse-marées" and batteries off the French coast.

Off Tamatave, 20 May 1811. 79 issued. Action with French frigates and capture of "Renommée" and "Néréide"; the vessels became H.M.S. "Java" and "Madagascar." "Astree," "Galatea," "Phœbe," and "Racehorse" engaged.

Hawke, 18 Aug. 1811. 6 issued. Capture of the 16-gun French brig "Heron" and convoy.

Java, 715 issued. Capture of Java, August and September 1811. 25 vessels engaged.

Skylark, 11 Nov. 1811. 2 issued. Locust, 11 Nov. 1811. 2 issued.	Action with Boulogne flotilla of 12 gun-brigs and capture of 1.

Pelagosa, 29 Nov. 1811. 64 issued. Action with French frigates "Pauline" and "Pomone" and capture of same. "Alceste," "Active," and "Unitie" engaged.

Victorious with Rivoli. Weasel, 22 Feb. 1812. 6 issued.	Capture of the French "74" "Rivoli," February 22nd, 1812. The "Victorious" lost 27 killed and 99 wounded, but the "Weasel" did not have a man injured. Captain Talbot of the "Victorious" was made K.C.B., and the "Rivoli" became a British war vessel.

Note that the bars are named differently for same engagement.

Rosario, 27 March 1812. 6 issued.
Griffon, 27 March 1812. 3 issued.
Capture of 2 French brigs-of-war off Dieppe. 2 vessels engaged.

Northumberland, 22 May 1812. 62 issued.
Growler, 22 May 1812. 4 issued.
Destruction of the French frigates "Arienne" and "Andromache" and a brig. 2 vessels engaged.

Magala, 29 May 1812. 17 issued. Capture of French privateers "Brave" and "Napoleon," at Malaga. The date of the action was 29 April. 4 vessels engaged.

Off Mardoe, 6 July 1812. 48 issued. Destruction of 2 Danish frigates and 2 brigs. 4 British boats engaged—"Calypso," "Dictator," "Flamer," and "Podargus."

Sealark, 21 July 1812. 4 issued. Capture of the French privateer "Ville de Caen."

Royalist, 29 Dec. 1812. 3 issued. Capture of the French privateer lugger "La Ruse."

Weasel, 22 April 1813. 6 issued. Destruction of 6 French gunboats in the Adriatic.

Shannon with Chesapeake. 49 issued. Capture of American frigate "Chesapeake," on 1 June 1813.

Pelican, 14 Aug. 1813. 4 issued. Capture of the American brig "Argus."

St. Sebastian, Aug. and Sept. 1813. 292 issued. 16 vessels engaged (see page 53).

Thunder, 9 Oct. 1813. 7 issued. Capture of the French privateer "Neptune."

Gluckstadt, 5 Jan. 1814. 45 issued. Capture of the fortress of Gluckstadt. 6 ships and 8 gunboats engaged.

Venerable, 16 Jan. 1814. 31 issued.
Cyane, 16 Jan. 1814. 18 issued.
Capture of the French frigates "Alcméne" and "Iphigénia," which were added to the Navy as H.M.S. "Gloire" and "Dunira."

EARL ST. VINCENT'S MEDAL, 1800.

Presented by Admiral Earl St. Vincent to the petty officers, seamen, and marines of his flagship the "Ville de Paris" for loyalty and good conduct.

EUROTAS, 25 Feb. 1814. 32 issued. Capture of the French frigate "Glorinde," which became H.M.S. "Burma."

HEBRUS with L'ETOILE, 27 March 1814. Capture of French frigate "L'Etoile."

PHŒBE, 28 March 1814. 28 issued.
CHERUB, 28 March 1814. 10 issued.
} Capture of the American frigates "Essex" and "Essex Junior."

THE POTOMAC, 17 Aug. 1814. 107 issued. Daring navigation of the Potomac River and destruction of shipping in the Potomac. 8 vessels engaged.

ENDYMION with PRESIDENT, 15 Jan. 1815. Capture of the American frigate "President."

GAIETA, 24 July. 89 issued. Attack and reduction of Gaieta. "Malta" and "Berwick" engaged.

ALGIERS. 1,362 issued. Battle of Algiers. 27 Aug. 1816. 22 vessels engaged. See page 307.

NAVARINO. 1,137 issued. Battle of "Navarino," Oct. 20th, 1827. 11 ships engaged. See page 309.

SYRIA. 6,877 issued. (Quite a number of these were used for fraudulent purposes, bars being manufactured bearing the records of rare bars and substituted for the common Syria.) Capture of "Acre," Nov. 1840, and operations with it on the coast of Syria. 32 ships engaged. See page 310.

BARS GRANTED FOR BOAT ACTIONS

The boat actions commemorated by these bars cover a number of brilliant actions performed by boats' crews in cutting out, and in some instances actually recovering, British vessels lost to the enemy, or capturing enemies' vessels.

15 March 1793. 1 only issued. 17 March 1794. 30 issued. 29 May 1797. 3 issued. 9 June 1799. 4 issued. 20 Dec. 1799. 3 issued. 29 July 1800. 4 issued. 29 Aug. 1800. 26 issued. 27 Oct. 1800. 5 issued. 21 July 1801. 9 issued. 27 June 1803. 5 issued. 4 Nov. 1803. 1 issued. 4 Feb. 1804. 10 issued. 4 June 1805.

10 issued. 16 July 1806. 51 issued. 2 Jan. 1807. 2 issued. 21 Jan. 1807. 9 issued. 19 April 1807. 1 issued. 13 Feb. 1808. 3 issued. 10 July 1808. 8 issued. 11 Aug. 1808. 12 issued. 28 Nov. 1808. 2 issued. 7 July 1809. 33 issued. 14 July 1809. 8 issued. 25 July 1809. 35 issued. 27 July 1809. 10 issued. 29 July 1809. 11 issued. 28 Aug. 1809. 14 issued. 1 Nov. 1809. 117 issued. 13 Dec. 1809. 10 issued. 13 Feb. 1810. 17 issued. 1 May 1810. 18 issued. 28 June 1810. 24 issued. 27 Sept. 1810. 34 issued. 4 Nov. 1810. 2 issued. 23 Nov. 1810. 66 issued. 24 Dec. 1810. 6 issued. 4 May 1811. 10 issued. 30 July 1811. 4 issued. 2 Aug. 1811. 10 issued. 20 Sept. 1811. 8 issued. 4 Dec. 1811. 18 issued. 4 April 1812. 4 issued. 1 Sept. 1812, 18 Sept. 1812. 24 issued for these 2 clasps. 17 Sept. 1812. 11 issued. 29 Sept. 1812. 26 issued. 6 Jan. 1813. 21 issued. 21 March 1813. 6 issued. 28 April 1813. 2 issued. April and May 1813. 54 issued. 2 May 1813. 49 issued. 8 April 1814. 23 issued. 24 May 1814. 11 issued. 3 and 6 Sept. 1814. 1 only issued. 14 Dec. 1814. 117 issued.

The last two bars are for actions in the war with America.

Algiers.—For this battle, fought on August 27th, 1816, 1,362 medals with a single bar, or combined with others, were issued. The action was brought about by the piratical depredations of the Algerines, and the enslavement of Christians by them. Lord Exmouth sailed for the port with 5 ships of the line and 8 smaller vessels, being joined at Gibraltar by 6 Dutch frigates under Admiral Van de Capellen. They appeared before Algiers on August 26th, and proposed certain terms to the Dey, which were not accepted, so next day they commenced to bombard the city. The Algerine fleet was destroyed, and the Dey accepted the terms he had refused. Over 1,200 Christian slaves were freed, and a promise given that piracy and slavery should cease. For a time the promise was kept, but in 1830 the French had to take strong steps to deal with the Algerines, and in the end Algeria became a French colony.

Exmouth Gold Medal.—As a reward for his services, Admiral Lord Exmouth was created a Viscount, and received from the Prince Regent a large gold medal, bearing on the obverse the bust of the Prince Regent encircled by TO TAME THE PROUD, THE FETTER'D SLAVE TO FREE. THESE ARE IMPERIAL ARTS, AND WORTHY THEE, and in the exergue PRINCE GEORGE REGENT. On the reverse is a representation of the bombardment, and in the exergue ALGIERS BOMBARDED ITS FLEET DESTROYED AND CHRISTIAN SLAVERY EXTINGUISHED, AUGUST 27TH, 1816. Only 4 medals of this kind were struck.

The following ships were engaged in the bombardment: the flagship "Queen Charlotte," "Impregnable," "Leander," "Superb," "Minden," "Albion," "Severn," "Glasgow," "Hebrus," "Granicus," and the sloops "Mutine," "Cordelia," "Heron," "Britomart," and "Jasper," and the following bomb vessels "Fury," "Infernal," "Beelzebub," "Hecla," and three vessels for transport, ordnance, and dispatch. Eighty-four men of the 1st Batt. Royal Sappers and Miners served as marines on the flagship and the "Impregnable."

Ava.—To those officers and seamen who took part in the Burmese War of 1824–6, the Army of India Medal, described on page 133 as having been awarded to officers and soldiers, was also awarded with a bar for AVA. Dr. Payne has in his collection medals with this bar awarded to a junior captain of the Bombay Marine, with his name impressed, likewise one awarded to a man of the H.E.I. Co.'s "Asseeghur," and one to an ordinary seaman of the "Boadicea" with the name engraved.

The following ships' crews took part: H.M.S. "Liffey," "Larne," "Slaney," "Avachne," "Sophie," "Satellite," "Diana," "Tees," "Alligator," and "Mercury," and the H.E.I. Co.'s "Thetis" and "Teignmouth." These were later joined after the capture of Martaban by the crews of the "Boadicea" and "Champion."

Navarino.—This battle was fought on October 20th, 1827. 1,137 Naval General Service Medals bearing the bar for NAVARINO were issued. In 1821 the Greeks had revolted against the Turks, and the latter treated them with ruthless barbarity, murdering in cold blood the inhabitants of Morea and laying waste the country. To put an end to this state of affairs, Great Britain, France, and Russia signed a treaty to ensure the independence of the Greeks, and in September 1827 Vice-Admiral Cordington sailed with the British fleet to Navarino, in the harbour of which the Turco-Egyptian fleet was at anchor. He was joined later by the French and Russian fleets, and an ultimatum was then sent to Ibrahim Pasha to the effect that the barbarities and the ravagings were to cease, and this he agreed to pending communication with the Grand Vizier at Constantinople, but because the allied fleets refused passage to some of his ships from the harbour of Navarino he again resorted to his fiendish practices. The British Admiral thereupon determined to take his ships into the harbour, and anchor near the Turco-Egyptian fleet.

On October 20th, followed by the French and Russian squadrons, Admiral Cordington entered the harbour and anchored close to the hostile fleet, when fire was opened by the Turks upon a ship's boat, killing the lieutenant in charge and several of his men. The "Dartmouth" then opened fire, and the battle commenced. The batteries of the fortress of Navarino then began to fire on the combined fleets, but so bad was the aim of the enemy's gunners that they did as much damage to their friends as to their foes. For four hours the fight continued with great vigour, and had it not been for the splendid co-operation of the French and Russian fleets it is doubtful whether the issue would have been satisfactory to lovers of freedom. It is noteworthy, however, that the French Admiral's flagship was only saved from grappling by a Turkish fireship through the speedy assistance given

WILLIAM IV.

QUEEN VICTORIA. (Reverse.)

EDWARD VII. (Obverse.)

NAVAL LONG SERVICE AND GOOD CONDUCT MEDALS.

by the British brig "Rose" and some boats of the Russian fleet. The British losses were 75 killed, and 197 wounded. The French lost 43 killed, and 144 wounded, and the Russians 59 killed, and 139 wounded, while the enemy—according to his own estimate—had thousands killed.

The following British ships were engaged: the flagship "Asia," "Albion," "Genoa," "Glasgow," "Cambrian," "Dartmouth," and "Talbot," and the brigs "Brisk," "Mosquito," "Philomel," and "Rose." There were 4 French ships of the line, a frigate and a schooner, and 5 Russian ships of the line and 3 frigates.

Hereafter the history of the British Navy and its engagements was, until 1914, confined to the bombardment of ports, or combined with that of the army in land service.

Syria.—Mehemet Ali, Pasha of Egypt, had in 1839 defeated the Turkish armies and made himself master of Syria, and, as has happened often since, Great Britain went to the sick man's aid, co-operating with Austria, Russia, and Prussia, to compel the recalcitrant Pasha to accept the Vice-Royalty and be satisfied with the governorship of Syria for his lifetime. An ultimatum was sent to him and his generals to the effect that Syria was to be restored to the government of the Sultan of Turkey within twenty-one days, but as no notice was taken of the demand, a squadron of British ships blockaded Alexandria, in the harbour of which the Egyptian fleet was at anchor, while the other portion of the fleet, under Admiral Sir Robert Stopford, bombarded Beyrout. Meanwhile, the towns on the coast of Syria had been retaken, Tyre captured on September 24th and Sidon on the 27th. Concentrating at Acre, the Egyptians were worsted there on November 2nd, and the war was brought to a conclusion by Mehemet Ali accepting the terms he had originally refused.

The Sultan's Medal for Acre.—To the officers and men a medal, illustrated facing page 292, was awarded; to the officers ranking as field officers in gold, in silver to quarter-deck and warrant officers, and in bronze to petty officers, seamen, and marines.

The following ships were engaged in the operations upon the coast of Syria in the years 1840–41: "Princess Charlotte," "Powerful," "Revenge," "Bellerophon," "Thunderer," "Ganges," "Benbow," "Edinburgh," "Hastings," "Vanguard," "Rodney," "Asia," "Implacable," "Cambridge," "Pique," "Castor," "Dido," "Carysfort," "Talbot," "Daphne," "Cyclops," "Wasp," "Magicienne," "Vesuvius," "Gorgon," "Hazard," "Phœnix," "Hydra," "Stromboli," "Medea," "Zebra," and "Hecate."

The bar for SYRIA was given with the Naval General Service medal for these operations.

China, 1840–2.—In this war, already described on pages 98 and 99, the following ships' crews were engaged: in 1840 "Melville," "Blenheim," "Wellesley," "Blonde," "Druid," "Conway," "Volage," "Larne," "Alligator," "Pylades," "Modeste," "Cruiser," "Nimrod," "Algerine," "Columbine," "Rattlesnake," and the following ships' crews of the Honourable East India Company: "Atalanta," "Queen," "Enterprise," "Calliope," "Madagascar," "Samarang," "Herald," and "Nemesis"; the latter, a well-armed iron steamer, created great terror amongst the Chinese, who named it the "devil ship." The following ships were added from time to time to the fleet: H.M.S. "Cornwallis," "Endymion," "Vindictive," "North Star," "Cambrian," "Hazard," "Pelican," "Harlequin," "Clio," "Wanderer," "Wolverine," "Hebe," "Serpent," "Royalist," "Plover," "Starling," "Driver," "Vixen," and the hospital ship "Minden," likewise the H.E.I. Co.'s steamers "Auckland," "Akbar," "Memnon," "Hooghley," "Proserpine," "Pluto," "Sesostris," "Medusa," and "Phlegethon."

Scinde, 1843.—The same medal as awarded to the army was awarded to the naval force taking part in the campaign. One hundred and ten medals were issued to the Indus Flotilla, and of these 40 were to

Europeans. For Hyderabad 115 medals were issued to the crews of H.E.I. Co.'s "Comet," "Meteor," and "Nimrod." Medals were given to naval men for Meeanee 1843, Hyderabad 1843, and Meeanee-Hyderabad 1843 (see pages 107–109).

Punjab, 1848–9.—A brigade of seamen, 100 strong with 7 officers, from the Indian Navy under Commander F. T. Powell, served at the Siege of Mooltan, and it is noteworthy that this was the first occasion upon which seamen served so far away from their ships. Medals were awarded with and without bar for Mooltan.

Second China War.—The squadron of British ships engaged under Rear-Admiral Sir Michael Seymour in 1856–60, in the operations already described on pages 98–100, included the flagship "Calcutta," "Sybille," "Pique," "Winchester," "Encounter," "Bittern," "Hornet," "Comus," "Racehorse," "Barracouta," "Sampson," "Coromandel," "Nankin," "Esk," "Elk," "Amethyst," "Sanspareil," "Cruiser," "Acorn," "Niger," "Inflexible," and a number of gunboats. In the attack on the Taku Forts, August 21st, 1860, the only naval co-operation was made by the gunboats "Clown," "Drake," "Woodcock," and "Janus." Bars were awarded for FATSHAN 1857 (navy only), CANTON 1857, TAKU FORTS 1858 (navy only), TAKU FORTS 1860, and PEKIN 1860. These medals were issued to the navy unnamed, and very few seamen appear to have troubled about having their medals engraved. Many issued to the Indian Navy and some of the marines were, however, impressed in capital Roman letters.

Pegu.—To those seamen and marines who took part in the second Burmese War, 1852–3, the India General Service Medal 1854 was awarded with the bar for PEGU, which, by the way, is somewhat smaller than the other bars issued with this medal (see pages 137–138). The following ships' crews were engaged: H.M.S. "Fox," "Rattler," "Serpent," "Sphinx," "Salamander," "Hermes," "Winchester," and a gunboat, together with the H.E.I. Co.'s boats "Berenice,"

"Bhagerutee," "Moofuffer," "Feroze," "Zenobia," "Sesostris," "Medusa," "Pluto," "Proserpine," "Phlegethon," "Soane," "Spy," "Tenasserim," "Mahanuddy," "Fire Queen," "Enterprise," "Indus," "Krishna," "Luckea," "Damoodah," "Sutledge," and "Lord Wm. Bentinck."

South Africa, 1853.—This medal, as previously explained, was awarded to cover all the wars and operations that had taken place in South Africa between 1834 and 1853. No bar was issued with the medal, and the only means of discovering when a recipient was engaged is by looking up the period during which his regiment, or ship, was engaged; but this is not quite so easy in the case of naval medals, as the name and rank only are impressed upon them in Roman capitals in the same type as was used on the Naval General Service medal. The third Kaffir War was the first in which the navy took part, and from the end of 1850 to March 1853 a small naval brigade composed of seamen and marines from the following ships were engaged ashore: "Castor," "Dee," "Grecian," "Hermes," "Penguin," "Pantaloon," "Gladiator," "Orestes," "Rhadamanthus" and "Styx."

CRIMEAN WAR

During the early stages of the war the combined British and French fleets were co-operating in Kavarna Bay, and on April 5th anchored before Odessa, when the Russian commandant firing upon H.M.S. "Furious," which was carrying a flag of truce, the following ships were ordered to open fire on the batteries on April 22nd, 1854. H.M.S. "Furious," "Terrible," "Retribution," "Niger," "Arethusa," and 3 French vessels. The fort was blown up, and the shipping destroyed. Later the "Firebrand" and the "Fury" destroyed the Russian batteries at Sulina, and then with the rest of the fleet took part in the bombardment of Sebastopol.

Sebastopol.—The following ships were engaged in

the bombardment of Sebastopol on October 17th, 1854: "Albion," "Britannia," "London," "Agamemnon," "Queen," "Firebrand," "Niger," "Furious," "Trafalgar," "Retribution," "Vesuvius," "Rodney," "Bellerophon," "Highflyer," "Spitfire," "Arethusa," "Cyclops," "Triton," "Lynx," "Sphinx," "Tribune," "Terrible," "Sampson," and "Sanspareil." In the bombardment 44 were killed and 264 wounded on board the British ships, the "Albion" and the "Agamemnon" suffering most heavily. The French fleet lost about 200 killed and wounded. Prior to the bombardment a naval brigade of 1,050 seamen, under the command of Captain Lushington, landed with 50 heavy guns, and demonstrated their aptitude, and their right to the name of "handymen," by constructing their own batteries and being ready to take the defensive long before the artillerymen had their guns in position. The success of the attack on the Malakoff tower was in no small measure due to the assistance of the naval brigade with their heavy guns. In the assault on the Redan the naval brigade also played a conspicuous and daring part, as the casualty list showed, for out of 120 men 14 were killed and 46 wounded.

Azoff.—The following ships were engaged in the expedition which, together with 3 French men-o'-war and 20 frigates, sailed on May 22nd, 1855, for the Sea of Azoff, blowing up *en route* the works and magazines at Kertch, and destroying Yenikale: H.M.S. "Agamemnon," "Algiers," "Hannibal," "St. Jean d'Acre," "Princess Royal," "Royal Albert," and 27 frigates. The bar for AZOFF was awarded to the officers and crews of the following ships: "Vesuvius," "Stromboli," "Miranda," "Swallow," "Curlew," "Beagle," "Lynx," "Ardent," "Viper," "Snake," "Medina," "Recruit," "Arrow," "Clinker," "Boxer," "Grinder," "Fancy," "Sulina," "Weser," "Jasper," and "Cracker"; also the first and second launches of the "Agamemnon," "Hannibal," "Algiers," "St. Jean d'Acre," "Princess Royal," and the "Royal Albert."

During these operations Lieutenant (afterwards Admiral Sir) E. Commerell of the "Weser" gained the Victoria Cross for his intrepid conduct while engaged in destroying stores on the shore of the Sivash.

The Baltic Medal.—For their services in connection with the blockade of the Baltic, Queen Victoria commanded in 1856 that a medal should be struck and issued to the officers, seamen, and marines who had been engaged in the Baltic from March 1854 to August 1855, during which period they had blockaded Bomarsund, attacked and captured it, and bombarded and destroyed Sveaborg. Two officers and 90 sappers and miners who served aboard the flagship and at Bomarsund were also awarded the medal, which bears on the obverse the same bust of Queen Victoria as on the medals previously described, and on the reverse the seated figure of Britannia, holding a trident in her right hand, with a naval gun and pile of shot, flags, and a naval coronet beside her, and behind in the distance the fort of Bomarsund to the left, and Sveaborg to the right; above all is BALTIC, and in the exergue 1854–55. The medal, by L. C. Wyon, is $1\frac{2}{5}$ in. in diameter, and depends from a yellow ribbon $1\frac{1}{4}$ in. wide with blue edges, by means of a curled suspender similar to the Sutlej type. The medals were issued unnamed, except in the case of the officers and sappers mentioned above, and theirs were issued with their names impressed.

The following ships were employed in the Baltic: H.M.S. "Duke of Wellington,*" "Saint George," "Neptune," "Princess Royal," "Royal George,*" "Saint Jean d'Acre," "Majestic,*" "Nile,*" "James Watt,*" "Prince Regent," "Cæsar,*" "Monarch," "Boscawen," "Cumberland," "Cressy,*" "Blenheim,*" "Hogue,*" "Edinburgh,*" "Amphion,*" "Arrogant,*" "Ajax,*" "Euryalus,*" "Imperieuse,*" "Cruiser," "Odin," "Dauntless," "Archer," "Leopard," "Valorous," "Desperate," "Magicienne," "Vulture," "Dragon." "Porcupine," "Bulldog," "Conflict," "Driver," "Hecla," "Basilisk," "Rosamund," "Lightning," "Alban," and H.M. Hospital Ship "Belleisle."

After the recall of Sir Charles Napier, who was censured for not attacking Sveaborg, a fleet was dispatched to the Baltic under Rear-Admiral Hon. R. Dundas. It included the ships named above marked with an asterisk, and the "Calcutta," "Colossus," "Orion," "Cornwallis," "Exmouth," "Russell," "Hawk," "Pembroke," "Hastings," "Retribution," "Falcon," "Esk," "Tartar," "Cossack," "Archer," "Harrier," and "Pylades," with a number of smaller vessels and gunboats. During the second operations in the Baltic, Lieutenant Dowell of the Marine Artillery gained the V.C. for leading a volunteer crew in a boat from the "Ruby," and taking the cutter of the "Arrogant" in tow when she was swamped after the explosion of her magazine during an attack on some vessels at Viborg.

The following vessels took part in the attack on Sveaborg: H.M.S. "Duke of Wellington," "Arrogant," "Euryalus," "Exmouth," "Cornwallis," "Cossack," "Pembroke," "Merlin," "Vulture," "Dragon," "Locust," "Volcano," "Lightning," "Hastings," "Amphion," "Edinburgh," "Magicienne," "Geyser," "Eolus," "Belleisle," "Cruiser," "Princess Alice," and the gunboats and mortar vessels "Redwing," "Lark," "Magpie," "Starling," "Skylark," "Stork," "Drake," "Redbreast," "Weasel," "Badger," "Mastiff," "Snapper," "Biter," "Growler," "Pincher," "Porpoise," "Snap," "Blazer," "Dapper," "Pelter," "Pickle," "Havock," "Prompt," "Manly," "Sinbad," "Beacon," "Carron," and "Grappler."

Naval Brigade in Crimea.—Some of the naval brigade in the Crimea received medals with three clasps for BALAKLAVA, INKERMANN, and SEBASTOPOL; others two clasps for INKERMANN and SEBASTOPOL, or BALAKLAVA and SEBASTOPOL. Others, showing they had fought *per mare, per terras*, three clasps for AZOFF, BALAKLAVA, and SEBASTOPOL, or AZOFF, INKERMANN, and SEBASTOPOL.

The only single bars are for AZOFF or SEBASTOPOL. Some of the marines who fought ashore received the whole four clasps for land service as awarded to soldiers. The medals issued to the navy were generally unnamed, but I have in my possession a few named with the same impressed lettering as on the medals issued to the army. The sailors and marines who fought in the Crimea were also awarded the Turkish medal described on page 152, while fifty of them received the Sardinian medal described on page 153.

INDIAN MUTINY

The deeds of the "Pearl" and "Shannon" Brigades during the Indian Mutiny have always impressed students of the war, and collectors have invariably been found willing to pay relatively high prices for medals without a bar if awarded to a man of the "Pearl" Brigade of 250 men, because of the splendid work they did in no less than twenty engagements. The "Shannon" Brigade was composed of 410 seamen and marines from that vessel, under the command of Captain W. Peel, who landed with six 68-pounders—which were employed for the first time as field pieces—eight 24-pounders, a battery of 8 rockets, and 2 howitzers, also some of the crew of the "Pearl," together with 120 sailors recruited from merchant ships. At Allahabad part of the "Shannon's" company was left to garrison the place, the rest marching to Cawnpore. Near Futtehgur 100 men of the naval brigade, co-operating with 450 soldiers, defeated a force of 4,000 mutineers and captured 2 guns. In November the brigade was attacking and assisting in the capture of the Martinière, and later breached the walls of the Secundrabagh and Shah Nujeef at Lucknow. In this daring work Lieutenant (afterwards Admiral Sir) M. Salmon and three of his men won the V.C. In January the brigade took part in the battle of Kallee Nudee, and in March 1858 was again at Lucknow storming the Dilkoosha and breaching the Martinière, where Captain Peel was wounded. He died on March 27th from an attack of smallpox, just when his gallantry had earned for him a knighthood. With the capture of Lucknow the work of the "Shannon" brigade was finished, and on September 15th she sailed for home. Five hundred and thirty medals were issued to men of the "Shannon" and 232 to men of the "Pearl." Three hundred Europeans of the Indian Marine also received the medals without bars; this medal was generally inscribed INDIAN NAVAL BRIGADE.

Persia, 1857.—The crews of the following ships of the Indian Navy received the medal with bar for PERSIA: "Ajdaha," "Assyria," "Assaye," "Berenice," "Clive," "Comet," "Constance," "Euphrates," "Falkland," "Ferooz," "Hugh Lindsay," "Lady Falkland," "Napier," "Nitocris," "Planet," "Punjab," "Semiramis," "Victoria."

New Zealand.—I have already described at some length the extended operations in New Zealand between 1845 and 1866. In these operations the navy on many occasions took an important part, particularly the crew of H.M.S. "Hazard" at the beginning of the war in 1845. The ships' crews engaged in 1845–7 were H.M.S. "Hazard," "North Star," "Castor," "Elphinstone," "Racehorse," "Osprey."

New Zealand, 1845–6–7.—The ships engaged during this period were H.M.S. "Castor," "Calliope," "Driver," "Elphinstone," H.E.I. Co.'s "Hazard," "Inflexible," "North Star," "Osprey," and "Racehorse."

New Zealand, 1860–6.—Members of the crews of the following ships received the medal for this period, although after August 1864, following the Gate Pa affair —in which Samuel Mitchell, captain of the foretop of H.M.S. "Harrier," gained the V.C.—the navy was not actively employed: H.M.S. "Brisk," "Cordelia," "Curacoa," "Eclipse," "Esk," "Falcon," "Harrier," "Himalaya," "Iris," "Miranda," "Niger," "Pelorus," and "Victoria" (see page 186 for description of medal).

Abyssinia.—In the Abyssinian campaign (1867–8), already referred to on pages 189–191 a naval brigade consisting of 83 men, with twelve 12-pounder rocket tubes, commanded by Commander Fellowes of H.M.S. "Dryad," took part, forming part of the first division under General Staveley. Men of the "Octavia," "Dryad," "Spiteful," and "Satellite" took part.

Ashantee.—In this campaign (1873–4) seamen and marines from H.M.S. "Argus," "Encounter," "Barracouta," "Druid," "Seagull," "Simoom," "Rattlesnake," and "Beacon" were employed ashore and afloat, while the gunboats "Bittern," "Decoy," and "Merlin" were actually engaged on the coast.

Perak.—The crews of H.M. gunboats "Charybdis" and "Hart" were employed in the operations against the Malays during 1875, and the crews of H.M.S. "Egeria," and the gunboats "Fly," "Philomel," "Ringdove," "Thistle," and the corvette "Modeste" in 1876. The I.G.S. Medal 1854, with bar for PERAK, was issued following a General Order of September 1st, 1879. The officers and crews of H.M.S. "Egeria" and the gunboats "Hart" and "Charybdis" received the medal for their services up the Lakut and Lingie Rivers. The naval medals are named in squat, sloping Roman capitals.

South Africa, 1877–8.—Sailors and marines from the following ships were engaged at different periods in the various wars in South Africa from 1877 to 1879 (see pages 196–202): H.M.S. "Active," "Boadicea," "Euphrates," "Forester," "Himalaya," "Orontes," "Shah," "Tamar," and "Tenedos."

In the campaign against the Galekas and Gaikas in 1877–8 two small naval brigades composed of men from H.M.S. "Active" and "Florence" were engaged, and both did excellent work with their rocket batteries.

Zulu War, 1879.—In this war a naval brigade of 170 seamen and marines from H.M.S. "Active" were attached to the third column under Colonel Pearson. They were under the command of Captain Campbell, and had with them two 7-pounders, a Gatling gun, and two rocket tubes. A brigade from H.M.S. "Boadicea," "Shah," and "Tenedos" marched with the relieving force to Ekowe. At the expiration of the war the naval brigade was warmly thanked by General Sir Garnet (later Viscount) Wolseley for the services they had rendered.

EGYPTIAN WARS

Alexandria, 1882.—In this campaign the navy played an important part—indeed, it was opened by the bombardment of Alexandria on July 11th, 1882, the fortifications of which had been strengthened, and the garrison increased by the rebel Arabi Pasha in anticipation of the allied fleets of Britain and France acting on the offensive. Noticing the preparations which Arabi was making, Admiral Sir Beauchamp Seymour called upon him to desist, and as no notice was taken of his demand, or of his ultimatum that unless work was stopped on the fortifications the place would be bombarded, he prepared his ships for action. Meanwhile the French fleet had left the harbour. The British steamed out, and took up a position facing the outer forts; at half-past 6 on July 11th the British ships cleared for action, and the bombardment began by the "Alexandra" firing a shell into Fort Ada. The fort responded, and then the duel between the ships' guns of the British fleet and those of the fort began. In a couple of hours Fort Marsa-el-Kanat was blown up, and very shortly after, nearly all the guns in Fort Mex were silenced. At half-past 10 an attack was made upon the Lighthouse Fort, and within a couple of hours the Ras-el-Tin batteries were silenced.

The Plucky "Condor."—Meantime Fort Marabout had been the scene of a memorable action on the part of Lord Charles Beresford. Noticing that the rifled guns of the fort were dropping their shot very near to the ships anchored at about 4,500 yards, viz. the "Monarch," "Penelope," and "Temeraire," he steamed the gunboat "Condor" within range, and by his daring action in pounding away at the fort for two hours diverted the enemy's fire, and thus materially assisted in the reduction of Fort Mex. The "Condor" was assisted in its plucky work by the gunboats "Beacon," "Bittern," and "Cygnet," which the Admiral had sent to help in the work, and so enabled her to be the means of silencing the guns of the fort by 2 p.m. Three hours later all firing ceased, and the battle of Alexandria was won with comparatively small loss to the fleet, the "Alexandra" suffering most from the effects of the enemy's guns. On this vessel gunner Israel Harding won the V.C. for throwing water over a live shell, which had pierced the side of the ship, and then picking it up and placing it in a tub of water; this hero rose to be chief gunner before he retired. A force of sailors and marines were landed, and engaged in keeping order. The successful Admiral became Lord Alcester, and received a grant of £25,000.

The ships engaged in the bombardment of Alexandria were the flagship "Alexandra," "Sultan," "Invincible," "Inflexible," "Temeraire," "Monarch," "Penelope," "Superb," and the gunboats "Beacon," "Bittern," "Condor," "Cygnet," "Decoy," and the dispatch vessel "Helicon."

Tel-el-Kebir.—In the operations leading up to the battle of Tel-el-Kebir, a number of seamen were engaged; several strategic points were seized by them, and several minor but vital engagements won by them. Lieutenant Wyatt Rawson, R.N., who guided the Highland Brigade across the desert to Tel-el-Kebir by his correct reading of the stars, was the first to enter the entrenchments—alas! to meet his death shortly after. In the battle of Tel-el-Kebir a naval brigade of 250 men, with 6 Gatling guns, took part; also marine artillery and marines. The latter, attached to General Graham's brigade, were placed in the front line, and suffered somewhat heavily, but the naval brigade did not lose a man.

First type.

Republican Variety.

CROSS OF THE LÉGION D'HONNEUR.

For the bombardment of Alexandria the Egyptian Medal 1882 with a bar inscribed ALEXANDRIA JULY 11 1882 was awarded to the officers, seamen, and marines, and for TEL-EL-KEBIR to those who took part in that decisive battle (see facing page 160 for medal, also the Khedive's Star which was issued by Prince Tewfik); the naval medals were named.

El-Teb.—When the Mahdi incited the Sudanese to rise against the Khedive in 1883, and practically gained control over the entire Sudan, it became necessary, as I have already explained, to send a British force to deal with the Mahdi and his principal lieutenant Osman Digna. In the war in the Sudan the men of the navy again distinguished themselves by their cheery willingness and ability. At Suakin, Admiral Sir William Hewitt's brigade of sailors and marines was landed from the "Decoy," "Euryalus," "Ranger," and "Sphinx" to defend the place, which they did quite successfully against Osman Digna. On February 1st, 1884, Osman Digna's force practically annihilated the army of 5,000 Egyptian troops at El-Teb under Valentine Baker Pasha (at one time Colonel of the 10th Hussars).

The work this undisciplined force set out to accomplish was then taken in hand by Major-General Graham, and the army with which he proceeded towards Tokar included 13 officers and 150 sailors, with 6 machine guns, under the command of Commander Rolfe, likewise about 400 marines and marine artillerymen under Colonel Tuson. At El-Teb on the morning of February 29th, 1884, the British force advanced in square, over the sickening area where the previous battle had taken place, to attack the Mahdists, who had entrenched themselves upon a narrow ridge about a mile in length. The Arabs opened fire, but the guns of the seamen and artillerymen proving too warm for them, they came forward to meet the advancing square, upon which they fell in their fanatical manner, which was as mad as it was grand. The fire of the naval machine guns stopped their rush, and then their position was taken, but the brunt of the fighting had fallen upon the naval brigade—as well as the " Black Watch " and the " York and Lancasters "—" who contributed materially to the success of the action," in which they lost an officer and 3 men killed, but one naval officer gained the V.C. when the Mahdists almost penetrated the corner of the square and attempted to spear some of the seamen ; Captain A. K. Wilson's sword was broken in his gallant effort, and himself would have fallen but for the timely action of some of the 65th (York and Lancasters). The British lost 4 officers and 26 men killed, and 160 officers and men wounded, ere the village of El-Teb had been taken, and the enemy made off in the direction of Tokar, which was taken next day. Those who took part in this action had the bar for EL-TEB added to their Egyptian medal, or, if this was their first experience of Egyptian warfare, the medal without the date, but with the bar for El-Teb, was awarded.

Tamaai.—This battle, which took place on March 13th, 1884, again afforded the " handymen " an opportunity of demonstrating their worth. Undismayed by what should have been a salutary lesson at El-Teb, the Mahdists began to preach a strenuous crusade against the infidel. It became necessary, therefore, to again set in motion the force which had returned to Suakin, and to march on Tamaai, where the Arabs had concentrated their forces. On the evening of the 12th the army, in two squares, halted a couple of miles from Tamaai, and it was thought there was little possibility of a night attack, but at 1 a.m. the enemy began to fire upon the British camp, and continued to do so all through the night, until at 8 o'clock the army, arranged in two brigades, advanced—the first under General (later Sir) Redvers Buller, and the second under General Davis, in which the naval brigade and marines were included. The latter, after the skirmishers had fallen back, met the enemy with a rattling fire from their machine guns and rifles, and then the "Black Watch," charging the enemy with considerable impetuosity, left a gap in the formation through which the tribesmen rushed, and all was soon confusion—friend and foe being so mixed, and all so closely packed together, that they could hardly use their weapons; and when they had extricated themselves, hand-to-hand fighting of a most ferocious character took place. In this unfortunate episode the naval guns had to be abandoned for a time, with the loss of 3 lieutenants and several men. Buller's brigade coming up turned the tables, and the guns were retaken and turned upon the fleeing Mahdists. The troops then marched upon Tamaai, took possession of the place next day, and destroyed it together with Osman Digna's camp. In this battle 5 officers and 101 men were killed, and 8 officers and 103 men wounded, but the Mahdists had lost over 3,000 killed.

For this engagement a bar inscribed TAMAAI was issued, and to those who had also taken part in the previous battle of El-Teb one inscribed EL-TEB—TAMAAI.

Suakin, 1884.—Undaunted Osman Digna still managed to preach his holy war, and secure thousands of adherents to his banner. Within four miles of the disastrous defeat which he had suffered he successfully got together, at Tamanieb, about 2,000 fighting men, but General Graham did not give him much rope, and immediately set out to meet him. Leaving Suakin on March 25th, he marched with a force which included marines to meet the Mahdists. To guard Suakin in the absence of General Graham and his troops, a force of 200 seamen was landed; the enemy did not, however, come to close grips, and after driving them off on March 27th, and destroying their camp, the troops returned to Suakin, after a very trying time, in which owing to the intense heat many fell victims to sunstroke. Suakin was then garrisoned with a small body of troops, including a battalion of marines, and after the rest of the troops had retired to Cairo, or returned to England, they had to defend the place for several months, during which time the officers and crews of the "Albacore," "Briton," "Coquette," "Myrmidon," "Sphinx," "Tyne," and "Woodlark" were kept busy in assisting the defenders. For these services a clasp inscribed SUAKIN 1884 was issued.

The Nile, 1884–5.—The valiant defence of Khartoum by General Gordon ultimately forced upon the British Government the task of attempting to relieve him, and his few European comrades. The expedition was organised to take the field under General Wolseley; and as special efforts were to be made to use the Nile as a water road for the troops, 300 boats of a special type were built, and 500 Canadian boatmen (Voyageurs) and 300 Kroomen were recruited to navigate them. It is said of the Voyageurs that some of them "hardly knew the stern of a boat from the bow," and "when they saw the Cataracts went sick."

It will be remembered that Lord Wolseley offered a prize of £100 to the first regiment to reach Debbeh in boats, and the 1st Royal Irish won it.

In this expedition a naval brigade composed of men from the Mediterranean fleet commanded by Captain (afterwards Admiral) Lord Charles Beresford was prominently engaged. Of the difficulties that had to be encountered in navigating the Nile much could be written, for the work was not accomplished without considerable loss of life, and before the expedition reached Dongola 10 of the Canadians and 40 of the troops had been drowned.

Abu-Klea.—At the beginning of January 1885 Lord Wolseley, becoming apprehensive for the safety of General Gordon, ordered General Sir Herbert Stewart to press forward across the desert with about 1,600 men to Metammeh on the Nile, and from there to make for Khartoum—about 100 miles up the river—in Gordon's steamers. After a march of 100 miles the force rested on January 12th at Jadkul; refreshed, the troops marched on until on January 16th they fell in with the

SECOND CLASS PRUSSIAN ORDER OF MERIT.

HESSIAN MEDAL FOR 1814–15.

AUSTRIAN CROSS FOR 1813–14.

enemy about 4 miles from Abu-Klea. A zareba was formed, and the men lay down to rest, which intermittent firing made almost impossible. At 9 o'clock on the morning of January 17th, leaving the camels and wounded inside the zareba, the force advanced towards the enemy's position; the naval brigade under Lord Charles Beresford formed the centre of the near face of the square: they had with them a Gardner gun. After a march of 2 miles the skirmishers were forced back on the square by the rush of thousands of Madhists, horse and foot, who came up to within 400 feet of the partly formed square. "The terrible rain of bullets poured into them by the mounted infantry, and the Guards, stayed them not," for, as Lord Charles Beresford relates, "they were tearing down upon us with a roar like the roar of the sea, an immense, surging wave of white-slashed black forms brandishing bright spears and long flashing swords; and all were chanting, as they leaped and ran, the war song of their faith. . . . These things we heard and saw in a flash, as the formidable wave swept steadily nearer."

Meanwhile Lord Charles had ordered the Gardner gun to be run outside the square to the left flank, and he laid the gun himself to make sure. As he fired, he states, "I saw the enemy mown down in rows, dropping like ninepins. . . . I was putting in most effective work on the leading ranks, and had fired about thirty rounds, when the gun jammed." Then Lord Charles, with the captain of the gun, William Rhodes, chief boatswain's mate, "set to work to try and unscrew the feed-plate in order to clear the barrel, or to take out its lock. The next moment the enemy were on top of us. The feed plate dropped on my head, knocking me under the gun and across its trail." Rhodes was instantly killed by a spear thrust, as was also the armourer, Walter Miller. On struggling to his feet, Lord Charles was "carried bodily backwards by the tremendous impact of the rush, right upon the front rank of the men of No. 4 Company, who stood like rocks." So great was the pressure that the front ranks could not use rifle or bayonet for a few moments; then the pressure forcing the white men back enabled the rear rank to fire over their heads into the mass of the Arabs. "The Arabs fell in heaps, whereupon the front rank, the pressure upon them relaxing, fired, and fought hand to hand with the bayonet, cursing as the rifles jammed and the shoddy bayonets twisted like tin." The men of the naval brigade joined in the furious hand-to-hand fighting—indeed, all round this was the feature of the fight at Abu-Klea, the heavy Camel Regiment, Scots Greys, Royals, and 5th Dragoon Guards desperately fighting in the square, while the Royal Sussex kept the enemy at bay with their steady fire. Every Arab who penetrated the square was killed; outside their comrades wavered, and then sullenly retired, until at half-past 2 the battle was won.

Of the naval brigade 8 were killed and 7 wounded out of the 40 who went into action. Among the killed were Lieutenants Alfred Piggott and R. E. de Lisle. Every man handling the Gardner gun outside the square was killed excepting Lord Charles Beresford, whose providential escape has enabled us to learn the graphic details of the desperate battle of Abu-Klea, which, as Colonel the Hon. Reginald Talbot (who commanded the heavy Camel Regiment at the battle) wrote, "was an Inkermann on a small scale." I have already described the fight towards the river, and the fact that the expedition had been undertaken in vain—for Khartoum had fallen. The bar for ABU-KLEA was issued to all the survivors.

Suakin, 1885.—In the fight at Hasheen, 14 miles from Suakin, on March 20th, 1885, a naval brigade and marines from the fleet in the Red Sea took part and were awarded the bar for SUAKIN.

Tofrek.—At daybreak on March 22nd, 1885, a force again marched out from Suakin under Major-General Sir John McNeill, V.C.; included was a naval brigade made up of men from the "Condor," "Carysfort,"

"Coquette," "Dolphin," and "Sphinx," with two Gardner guns, under Commander Domville, also a battalion of marines. The object of the force was to construct zarebas, which were to be garrisoned by the 49th Regiment. The troops, having made good progress with their work, were taking things rather easily—some were breakfasting—when the sudden appearance of the enemy caused great confusion. Roaring camels, screaming mules, and frightened camp followers helped to make the confusion worse confounded as the Arabs hewed their way through the terrified animals and the helpless camp followers. The soldiers hastily rushed for their rifles, and endeavoured to form squares, into one of which, formed by the marines, about 60 Arabs forced their way, but never got out again. Again the Gardner guns jammed, and the unfortunate seamen had a bad time, as the howling Dervishes rushed into their zareba. In half an hour, however, the fight subsided, but not until a naval lieutenant and 6 men had been killed, and a number wounded. After this battle the men of the naval brigade returned to their ships, and the troops evacuated Suakin.

The Egyptian medal, described on page 211, with the bar for TOFREK, was awarded to the sailors and marines who took part in the battle.

Bars for Second Sudan War.—To those who had taken part in the war in the vicinity of Suakin in the Eastern Sudan the Egyptian medal was awarded with five bars. Those already in possession of the medal received the bars to which their services entitled them. All who served south of Assouan on or before March 7th, 1885, received the medal with a bar for THE NILE 1884–5, if not already in possession of one; likewise those who served at Suakin from March 27th, 1884, to May 14th, 1885. A bar inscribed ABU-KLEA was issued to all who took part in the fight on January 17th, 1885. A bar inscribed KIRBEKAN was awarded to those who took part in the action on February 10th, 1885, in which a small naval brigade was present with a Maxim gun.

A bar with the record SUAKIN 1885 was issued to those who participated in the operations round that place between March 1st and May 14th, 1885, and a bar with TOFREK to those who took active part in the action on March 22nd, 1885. It should be particularly noted that no medals bearing single bars for ABU-KLEA, KIRBEKAN, or TOFREK were issued, and that these should be in conjunction with either SUAKIN 1885, or in the case of ABU-KLEA or KIRBEKAN with THE NILE 1884–5. Medals without bars were issued to the troops who were located in Egypt but did not participate in any of the engagements. The medals are boldly named in Roman capitals.

To those who were engaged at or to the south of Wady Halfa between November 30th, 1885, and January 11th, 1886, the medal without bar was given to those who had not already earned one, and, although the battle of Ginnis was fought during this period, no bar was issued to those who were not presented with a medal. The Khedive's star was also given to those who took part in these operations.

Gemaizah.—In 1888 Suakin was once again the centre of warfare, but the Mahdists were acting on the offensive by endeavouring to invest the place. General Grenfell therefore decided to march out and attack them. In the action, which took place on December 20th, 1888, the navy co-operated by firing from the Nile, H.M.S. "Racer" shelling the enemy's trenches, and H.M.S. "Starling" the wells; the crews of these boats were entitled to the bar for GEMAIZAH.

Tokar, 1890.—To the officers and men of H.M.S. "Dolphin" and "Sandfly," engaged in transport duty, and to the British officers who commanded the Egyptian army, the Khedive's bronze star without date, but with the record in Arabic characters **TOKAR 1890** on the suspender, was awarded. This is an interesting star, and more valued than the other Khedive's, because it is the decoration for the services rendered, no bars being issued for attachment to the silver Egyptian medal, or medals awarded.

Burma, 1885–7.—To the crews of H.M.S. "Bacchante," "Woodlark," "Osprey," "Ranger," "Mariner," "Turquoise," and "Sphinx," which took part in the third Burma War, that resulted in the disposition of Theebaw and the annexation of Burma, the medal with the bar for BURMA 1885–7, as described on page 138, was granted. Those awarded to the navy were generally impressed in tall upright Roman capitals.

West Africa Medal.—The boats' crews of H.M.S. "Acorn," "Icarus," and "Rifleman" co-operated with the 1st W.I.R. in the operations from November 13th, 1887, to January 2nd, 1888, against the Yonnie tribes who had raided British territory in the vicinity of Sierra Leone. The medal, with bar 1887–8, was granted for this expedition.

Witu, 1890.—The crews of H.M.S. "Boadicea," "Brisk," "Conquest," "Cossack," "Humber," "Kingfisher," "Pigeon," "Redbreast," and "Turquoise," together with marines who took part in the expedition under Rear-Admiral D. Freemantle, C.B., to punish the Sultan of Witu for the murder of nine Germans in his territory, received the medal with the bar inscribed WITU 1890.

1891–2.—The crews of H.M.S. "Alecto," "Racer," "Sparrow," "Thrush," and "Widgeon" received the medal with bar inscribed 1891–2 for their expedition against the robber chief Fodi Cabba, up the Gambia River, from December 29th, 1891, to February 5th, 1892.

1892.—The bar with 1892 was awarded to those bluejackets and marines who co-operated with the 1st W.I.R. in the expedition against Tambi from March 8th, 1892, to April 11th, 1892. The naval brigade consisted of men from H.M.S. "Alecto," "Racer," "Sparrow," "Thrush," and "Widgeon." The bar inscribed 1892 was also awarded with the medal to the naval brigade from H.M.S. "Alecto," "Racer," and "Sparrow," who co-operated with the 1st W.I.R. in the expedition which ended in the capture and destruction of the town and fort of Toniataba. The operations extended from March 12th to April 30th, 1892.

Liwondi, 1893.—A bar inscribed LIWONDI 1893 was awarded to 34 men of H.M.S. "Herald" and "Mosquito," who with a Nordenfelt gun marched under Lieutenant-Commander Carr of the latter vessel to the relief of H.M. Commissioner, Mr. H. Johnston, on the Shire River, Central Africa, between February and March 1893, when they routed the chief Liwondi. Only 29 medals were issued.

Witu August, 1893.—This bar, with the West Africa medal, was granted to about 200 men from H.M.S. "Blanche," "Swallow," and "Sparrow," who were landed under the command of Captain Geo. Robert Lindley, and with the assistance of Zanzibar and Sudanese troops punished the chief Fumo Omari. Two hundred and fifty medals were issued.

Juba River, 1893.—The bar for JUBA RIVER 1893 was awarded to about 40 men of H.M.S. "Blanche," who volunteered to proceed up the River Juba to release two Englishmen who were besieged by a force of Somalis, August 25th, 1893. They were commanded by Lieutenant Price Vaughan Lewis, R.N. Forty bars were issued for this expedition.

Lake Nyassa, 1893.—The bar for LAKE NYASSA 1893 was awarded to 36 men of the "Adventurer" and "Pioneer," two screw steamers of 35 tons, who were engaged against Makaujira and other chiefs, and in the bombardment of Okuirka in Nyassaland in 1893. One hundred and one Sikhs, who were also engaged, received the Central Africa medal, with clip and ring as described on page 228. Thirty-six medals with this bar were issued.

Gambia, 1894.—The bar for GAMBIA 1894 was awarded to the naval brigade from H.M.S. "Alecto," "Magpie," "Raleigh," "Satellite," and "Widgeon," which landed early in February at Bathurst on the River Gambia, West Coast of Africa, to operate against Fodi Selah, a slave-raiding chief. The expedition, which lasted from February 22nd to March 11th, 1894, only achieved its object of punishing the chief after surmounting considerable difficulties, and surviving much hard fighting, during which 2 lieutenants and 17 men were killed, and nearly 50 officers and men—mostly of the "Magpie" and "Raleigh"—had been wounded. Surgeon Bowden won the D.S.O. during this expedition.

1893–4.—It is noteworthy that those men of the 1st W.I.R. who were also engaged received the medal with bar for 1893–4, instead of the bar for GAMBIA 1894; a very definite distinction is thus made between the military and the naval medal.

Benin River, 1894.—The bar for BENIN RIVER 1894 was issued to those sailors and marines of H.M.S. "Alecto," "Philomel," "Phœbe," and "Widgeon" who, assisted by Houssas, fought against the chief Nana, the terror of the inhabitants of the Benin River, whom he had plundered for a number of years. The steam launch of the "Alecto," while steaming up the narrow creek leading to the stronghold Brohemie, on August 25th, was fired upon by a concealed battery, the steersman shot dead, and every man on board wounded. In this unfortunate affair Chief Petty Officer Crouch and Leading Stoker Joseph Perkins earned the Conspicuous Gallantry Medal for their exceeding bravery. On September 18th Brohemie was attacked and captured, and the expedition proceeded to another fortified town three miles distant, and this also was captured, together with Nana's war canoes, which had helped him so much in his depredations. Brohemie was then placed under the Niger Coast Protectorate. Commander Heugh and Lieutenants Hickley and Gore-Browne gained the D.S.O. in these operations.

Brass River, 1895.—The bar for BRASS RIVER 1895 was awarded to the men from H.M.S. "Barrosa," "Saint George," "Thrush," and "Widgeon" who took part in the attack on Nimbi, the stronghold of King

Koko and other native chiefs, from February 17th to March 26th, 1895.

Benin River, 1897.—Those sailors and marines who landed from the following ships were awarded the bar for BENIN RIVER 1897: H.M.S. "Alecto," "Barrosa," "Forte," "Magpie," "Philomel," "Phœbe," "Saint George," "Theseus" and "Widgeon"; about 1,400 officers and men took part.

Sierra Leone, 1898–9.—The bar for SIERRA LEONE 1898–9 was awarded to those who landed from H.M.S. "Alecto," "Blonde," and "Fox," likewise the Colonial boat "Countess of Derby," and took part in the boat expedition up the river and were actually under fire. The men from H.M.S. "Blonde" were engaged in the Sherboro neighbourhood from May 1st to 15th, and with the men from the "Alecto" on the Boom Kittam River next day. Those belonging to the "Fox" and the "Countess of Derby" were engaged on the Brempé River between May 11th and 14th, 1898. This bar is wider than the others previously issued, as the inscription is in two lines.

Mwele, 1895–6.—By an Admiralty Order dated January 1st, 1897, a further grant of the West Africa medal was sanctioned, and the medal awarded to those who had taken part in the operations against Mwele, the stronghold of the rebel chief Sheik Mbarnok Bin Rashid, which was captured on August 17th, 1895. No bar was issued for this affair, the name and date MWELE 1895 being engraved on the edge of the medal in slanting Roman capitals, together with the name, etc., of the recipient. Those already in possession of the medal had MWELE 1895 engraved after their name. The little force comprised men from H.M.S. "Barrosa," "Phœbe," "Racoon," and "Saint George," and the 24th and 26th Bombay Infantry—their medals had MWELE impressed on one side of the claw of the suspender and 1895 on the other side—who also received the medal, while camp followers were given the medal in bronze. This medal is described on page 228.

THE IRON CROSS.

AUSTRIAN ORDER OF THE IRON CROWN.

PRUSSIAN MEDAL FOR DISTINCTION IN SERVICE, 1825.

AUSTRIAN OFFICER'S GILT CROSS.

Central Africa, 1895.—Dr. Payne has in his collection the only Central African 1895 medal believed to have been issued to the navy. It was awarded to Lieutenant J. S. Brogden, R.M.L.I., who afterwards became Captain R.N.

BOER WAR

During the Boer War a naval contingent was employed, and their assistance in the defence and relief of Ladysmith is one of the outstanding features of that long conflict. The men of H.M.S. "Powerful," who helped in the defence of Ladysmith under Captain (afterwards Admiral Sir) Percy Scott, R.N., whose genius was such a material asset to the beleaguered, did excellent work, while the naval brigade which took part in the Natal campaign and the relief of Ladysmith made splendid efforts. It comprised 39 officers and 403 men of the Royal Navy, and 2 officers and 50 men of the Natal Naval Volunteers, who during the fighting leading up to the relief, on February 28th, 1900, had fired 4,000 rounds of ammunition from the 4·7-inch guns, and 12,000 rounds from the 12-pounder quick-firing guns.

Bars Gained by Navy.—The officers and men of H.M.S. "Powerful" received the bar for the DEFENCE OF LADYSMITH. The bar for the RELIEF OF LADYSMITH was awarded to officers and men from H.M.S. "Terrible" and "Philomel." Men from the latter ships also took part in the battle of the TUGELA HEIGHTS. The sailors from the "Philomel," in addition to the bars already mentioned, received those for ORANGE FREE STATE and LAING'S NEK, while men from the "Monarch" and "Doris" were awarded bars for BELMONT, MODDER RIVER, PAARDEBERG, DRIEFONTEIN, DIAMOND HILL, and BELFAST. Men from H.M.S. "Naiad" fought in, and received bars for, CAPE COLONY, TRANSVAAL, WITTEBERGEN, and SOUTH AFRICA 1902.

Medals with bars to the navy are rare, and those with more than two particularly so. A leading stoker of H.M.S. "Doris" received, in addition to those mentioned above as awarded to men from his ship, the rare one for WEPENER, but though this bar to the navy was afterwards recalled, Dr. Payne has this rare medal with this and six other bars in his collection.

The crews of the following ships received the medal: H.M.S. "Barracouta," "Barrossa," "Beagle," "Blonde," "Doris," "Dwarf," "Fearless," "Forte," "Gibraltar," "Magicienne," "Magpie," "Monarch," "Naiad," "Niobe," "Partridge," "Pearl," "Pelorus," "Philomel," "Powerful," "Racoon," "Rambler," "Rattler," "Redbreast," "Sappho," "Sybille," "Tartar," "Terpsichore," "Terrible," "Thesis," "Thrush," and "Widgeon."

Africa General Service Medal.—This medal, described on page 254, was awarded to the army and navy. It was instituted by King Edward VII to commemorate and recognise the naval and military operations in East Central and West Africa, and on its institution the East and West and Central Africa medals were no longer issued.

Jubaland.—The medal with bar for JUBALAND for the operations against the Ogaden Somalis, from November 16th, 1900, to April 30th, 1901, was awarded to the officers and men of the Royal Navy and marines from H.M.S. "Magicienne," "Terpsichore," and "Scout."

Gambia, 1901.—Those members of the crews of H.M. ships who had received the Africa General Service medal with the bar for JUBALAND were not entitled to this bar.

Aro, 1901–2.—The crew of H.M.S. "Thrush" took part in the operations against the Aro tribe from November 15th, 1901, to March 23rd, 1902 (see also page 255). The naval recipients of the medal with the bar for ARO 1901–2 were 53 of the crew of the "Thrush," and 3 officers and 27 sailors belonging to the Protectorate gunboat "Jackdaw."

Somaliland, 1902–4.—In this campaign men belonging to the following ships participated in the expedition under Colonel Rochfort, C.B., C.M.G., and were entitled to this bar: H.M.S. "Cossack," "Dryad," "Fox," "Highflyer," "Harrier," "Hussar," "Hyacinth," "Mohawk," "Merlin," "Naiad," "Perseus," "Porpoise," "Pomone," and "Redbreast" (see also page 256).

Somaliland, 1908–10.—This bar was awarded with the medal to those who took part in the campaign. Men from H.M.S. "Hyacinth" and "Proserpine" were engaged.

Transport Medal.—King Edward VII, on his birthday, November 9th, 1903, instituted a medal which was granted as a special recognition of the magnificent work done, and for the nautical skill and perfect efficiency shown, by those who were engaged in connection with the Transport Service during the China and South Africa campaigns. On the obverse is the bust of King Edward VII, facing left, in the uniform of an Admiral of the Fleet, wearing his orders and decorations, and the legend EDWARDVS VII REX IMPERATOR. On the reverse, in the foreground, is a large liner ("Ophir") ploughing through the sea; above is a map of the world on Mercator's projection, which embraces the British Empire from the West Indies to New Zealand, but, owing to the circular form of the medal, omits the North American Continent; below is the legend OB PATRIAM MILITIBUS PER MARE TRANSVECTIS ADJUTAM. The medal, $1\frac{2}{5}$ in. in diameter, is suspended by means of a straight swivel bar from a red ribbon, with two blue stripes each $\frac{1}{4}$ in. broad. The names are impressed round the edge of the medal in capital letters. The official announcement of the issue stated that the "Transport Medal shall be granted in future wars to the Officers of the Mercantile Marine serving in the transports" whenever a war medal is granted for the campaign. These medals have realised from £2 to £4 each at public auction.

MERITORIOUS SERVICE MEDALS

UNDER this heading I might have placed several of those earlier medals which were awarded when campaign medals had not been instituted, but the record for our purpose may begin with the institution of the Meritorious Service Medal on December 19th, 1845, when Her Majesty Queen Victoria decreed that a sum of not more than £2,000 per annum should be set aside for the payment of rewards in the form of annuities not exceeding £20, to sergeants recommended by the Commander-in-Chief in recognition of meritorious or distinguished services. The sum for disbursement was increased to £4,000 in June 1853. This medal bears on the obverse the diademed head of Queen Victoria, with the legend VICTORIA REGINA, and generally has the date of institution in the exergue, but in some of the earlier, and in the later issues also, the date is omitted. On the reverse (as illustrated facing page 136) is the inscription FOR MERITORIOUS SERVICE, surmounted by a crown, and encircled by a broad wreath of laurel. The suspender is of the same pattern as that issued with the India 1854 medal, and the decoration depends from a red ribbon 1¼ in. wide when worn by military sergeants, and a blue ribbon by sergeants of marines, who in 1849 were also granted the medal. Until November 1902 this medal could not be worn with the L.S. and G.C. medal. These medals realise from £3 to £4 10*s*. in the sale-room.

H.E.I. Co.'s Meritorious Service Medal.—On May 20th, 1848, the Governor-General of India instituted a medal for meritorious service for distribution among the East India Company's troops. On the obverse is the bust of Queen Victoria as on the war medals of the period,

SILVER CROSS FOR SAN SEBASTIAN, 1836.
Awarded to Staff-Surgeon Chas. Benson Brearey, M.D.

IRON CROSS FOR SAN SEBASTIAN, 1836.
Awarded to Staff-Surgeon John Callender.

with the date 1848 in the exergue, and on the reverse the arms of the Company encircled by the inscription FOR MERITORIOUS SERVICE. The medal is 1¼ in. in diameter, and was suspended by a scroll clasp as used with the Sutlej medal by a red ribbon 1¼ in. wide. The name of the recipient was engraved in a neat running hand upon the edge of the medal. The issue of this medal was discontinued when the English variety was instituted. These medals have realised from £3 10*s*. to £5.

Distinguished Conduct Medal.—The next in chronological order is the medal for Distinguished Conduct in the Field, and was instituted on December 4th, 1854, as an indication of the "Sovereign's sense of the distinguished service and gallant conduct in the field of the army then serving in the Crimea." It was awarded to non-commissioned officers and privates only, and has since been given for gallant service in many other campaigns. By the decree the medal could be awarded after selection by the commanding officer. Originally a gratuity went with the medal, but that was discontinued in 1862. In 1881 it was decided that a bar bearing the full date of the action for which it was awarded should be given to a recipient who had again distinguished himself.

The obverse is like the Queen Victoria Long Service Medal, and on the reverse is the inscription FOR DISTINGUISHED CONDUCT IN THE FIELD. The same kind of clasp as used with the Meritorious Service Medal is employed for suspension by a red ribbon with a broad blue stripe down the centre. The first medals issued had the name, rank, regiment, etc., impressed or engraved upon the edge in Roman capitals, but the modern medals are engraved, and generally bear the date of the action, while those issued during the reign of King Edward VII were impressed in small block capitals.

These medals (see facing page 140) realise from £4 to £7 7*s*. in the saleroom, generally with a war medal representing the campaign in which it was gained. Groups are relatively higher.

Conspicuous Gallantry Medal.—This medal was instituted on August 13th, 1855, as a reward to petty officers, sailors, non-commissioned officers, and privates in the marines who had distinguished themselves in the Crimean War. With it gratuities were given ranging from £5 to £15. In 1874, after the Ashantee War, it was again decided to issue the medal, and to grant annuities to chief and first-class petty officers, and sergeants of marines, whenever the annual grant authorised by the Treasury had not been exceeded. This medal has the same obverse as the Meritorious Service Medal, and the reverse is practically the same, except for the inscription, which is FOR CONSPICUOUS GALLANTRY. The first medals issued were struck from the same die as the Meritorious Service Medal, the second and third lines being erased, and the words CONSPICUOUS GALLANTRY engraved in Roman capitals. The suspender in the first issue was the same as that used with the sister medal for the army, but in the second issue, which commenced in 1874, a straight bar was used for suspension from a blue ribbon with a broad white stripe down the centre. The names on these medals are engraved in capital Roman letters like the naming on the Distinguished Conduct Medal.

The medal issued during the reign of King Edward VII bore upon the obverse the same bust of His Majesty in Admiral's uniform as is used on the Naval Long Service and Good Conduct medals granted by him. These medals realise from £3 10*s.* to £5, but a medal of the first issue, together with the Baltic medal awarded to George Belding of H.M.S. "Firefly," realised £59.

The Victoria Cross.—This coveted decoration was first suggested by the Prince Consort, and Queen Victoria being desirous of taking into her royal consideration a means of adequately rewarding the individual gallant services, either of officers of the lower grades in the naval and military service, or of warrant and petty officers, seamen, and marines, ordained by Royal Warrant on June 29th, 1856, that "the cross shall only be awarded to those officers or men who have served us in the presence of the enemy, and shall have performed some signal act of valour or devotion to their country." With the cross an annuity of £10 is awarded to warrant officers, seamen and marines, non-commissioned officers, and privates, and for each additional bar £5 per annum is added to the annuity. In July 1898, owing to the unfortunate condition to which a recipient had been reduced, and so induced to sell his cross, it was decided that the Home Secretary might, in his discretion, increase the annuity to £50. On April 23rd, 1881, the warrant was revised to enable officers of any grade to receive the medal, and on August 8th, 1902, King Edward sanctioned the posthumous award of the cross, and of its issue to relatives of deceased men who had earned it; prior to this the brave fellows' names had been gazetted only. The new regulation was retrospective, so that the surviving representatives of men who had earned the cross as long ago as the Indian Mutiny received the bronze token of their relative's valour.

The cross (facing page 136) carries with it the right to append V.C. after the recipient's name.

The cross is made from captured cannon, and is not, as is generally and wrongly described, a Maltese cross, but a cross patée. It bears in the centre of the obverse the royal crest of a lion passant gardant upon the British crown, with a ribbon inscribed FOR VALOUR in a semicircle beneath it, the ends tucked under the raised edge. The cross has a pierced semicircular lug, through which a simple link is run to attach the cross to the laureated suspender by a V. The obverse has raised edges like the front, but the centre is circular. The name, rank, regiment or ship to which the recipient was attached is engraved upon the back of the suspender, and the record of the act which gained the decoration inside the circle on the back of the cross. The cross is suspended by a dark-red ribbon, $1\frac{1}{2}$ in. wide, by military recipients, and by a dark-blue ribbon by naval recipients. The cross, owing to the fact that it is cast and chased, has often been copied, and I have had in my possession one which, but for its size, would have deceived experts, but it had the faults of all cast copies, having shrunk in the casting, and this is the one way of telling a fraud, although it generally needs the genuine article for comparison. The V.C. has realised in the sale-room from £43 for a lieutenant's won in the Mutiny to £175 for a private's won before Sebastopol.

The only Victoria Cross awarded for gallant services not in "the presence of the enemy" was given to Private Timothy O'Hea for extinguishing a fire in an ammunition van during the Fenian Raid in Canada, 1866. It sold in August 1900 for £50.

The Albert Medal.—The Albert Medal, commonly called "The Civilian's Victoria Cross," was originally instituted by Queen Victoria under a Royal Warrant dated March 7th, 1866, as a reward for heroic actions performed in saving life at sea. On April 12th, 1867, the Warrant was revoked by a second which instituted two new decorations respectively styled The Albert Medal of the First Class and The Albert Medal of the Second Class, the original Warrant instituting one class only. Ten years later, on April 30th, 1877, Her Majesty Queen Victoria signed another Warrant which extended the decorations to cases of gallantry in saving life on land.

By the Warrant of April 12th, 1867, it was ordained that the Albert Medal of the First and Second Classes, severally inscribed "For Gallantry in Saving Life at Sea," should be made only on a recommendation by the President of the Board of Trade; and in the Warrant dated April 30th, 1877, it was ordained that the award of the Albert Medal of the First and Second Classes, inscribed "For Gallantry in Saving Life on Land," should be made only on a recommendation by the First Lord of the Treasury. On September 13th, 1881, an

BRONZE MEDAL GIVEN BY LEOPOLD, GRAND DUKE OF BADEN, "TO THE BRAVE ARMY OF LIBERATION," 1849.

BRONZE MEDAL GIVEN BY LEOPOLD, GRAND DUKE OF BADEN, "FOR FAITHFUL SERVICE IN WAR."

SILVER MEDAL FOR SAN SEBASTIAN, 1836.

amended Warrant was issued, ordaining that the award of the Albert Medal should be made only on a recommendation by the Principal Secretary of State for the Home Department, "Providing always that the preliminary steps and inquiry concerning the award of the Albert Medals severally inscribed 'For Gallantry in Saving Life at Sea' be, as heretofore, with the President of the Board of Trade." By Warrant dated April 30th, 1877, it was ordained that a register of the names of those who received the Albert Medal for Saving Life at Sea should be kept at the office of the Board of Trade. By Warrant dated Whitehall, March 24th, 1891, the previous Warrants bearing the dates April 30th, 1877, and September 13th, 1881, respectively, were amended, and it was ordained that the award of the Albert Medal for Saving Life at Sea should be made only on a recommendation of the Principal Secretary of State for the Home Department, "Provided always that the preliminary steps and inquiry concerning the award shall, if the award be to any one belonging to the Royal Navy or the Royal Marines, be with the Lords Commissioners of the Admiralty, and in other cases with the President of the Board of Trade." Further, that a register of the names of those persons belonging to the Royal Navy or Royal Marines, upon whom the Albert Medal for Saving Life at Sea shall have been conferred, shall be kept at the offices of the Lords Commissioners of the Admiralty. (See group facing p. 184.)

Distinguished Service Order.—By a Royal Warrant dated September 6th, 1866, Her Majesty Queen Victoria instituted the Distinguished Service Order for the purpose of recognising the special services of commissioned officers of the Army and Navy, likewise the Indian and Colonial Naval and Military Forces.

The badge consists of a gold cross patée, convexed, the ground enamelled white, leaving an edge of gold; on the obverse, in the centre, within two sprigs of laurel, enamelled green, is the Imperial Crown in gold, upon a red enamelled ground. On the reverse is the monogram V.R.I. within two branches of laurel, also upon a red enamelled ground. The badge is worn suspended from the left breast by a red ribbon edged with blue, 1 in. in width, from a gold laureated bar 1⅕ in. wide, fastened to the badge by two gold loops, and having a similar gold laureated bar with brooch attachment above. The award carries with it the right to append the letters D.S.O. after the recipient's name. (See facing page 208.)

The New Zealand Cross.—This was instituted by Order in Council at the Government House, Wellington, March 10th, 1869, and afterwards sanctioned by Her Majesty as "a Decorative Distinction to be conferred on members of the Militia, Volunteers, or Armed Constabulary, who may particularly distinguish themselves by their bravery in action, or devotion to their duty while on service." The conditions are almost exactly similar to those for the Victoria Cross, and the recipient received an annual pension of £10.

It is a silver Maltese cross, with bright silver double borders, having a six-pointed gold star on each arm. In the centre, in a circle, NEW ZEALAND surrounded by a wreath of laurel in gold. The cross is surmounted by a gold crown and attached to a crimson ribbon—1½ in. wide—by a silver bar ornamented with laurel, in gold, connected by a V and ring, in silver, with the top of the crown. The name of the recipient and the date of the action should be engraved on the back of the cross. Clasps may be added for additional acts of bravery or devotion. The clasp to be a silver bar across the ribbon, having a plain surface burnished and inscribed with the date of the occurrence for which the bar is given, and the name of the action—if any. Twenty-one officers and men have received this decoration. These crosses have realised from £15 to £25.

India Distinguished Service Medal.—This medal was instituted on June 28th, 1907, by an Army Order published in Simla as a reward for both commissioned and non-commissioned officers of the regular and other forces in India. It bears on the obverse the bust of King Edward VII, and on the reverse a laurel wreath encircling the words FOR DISTINGUISHED SERVICE. The medal, $1\frac{2}{5}$ in. in diameter, is ordered to be worn immediately to the right of all war medals, suspended by a red ribbon $1\frac{1}{4}$ in. wide, with blue edges $\frac{3}{8}$ in. wide. This medal may be conferred by the Viceroy of India.

Conspicuous Service Cross.—This cross was instituted by King Edward VII on June 28th, 1901, as a reward for "distinguished service before the enemy," for bestowal upon warrant officers and subordinate officers of the fleet who do not hold commissions in the navy. No person can be nominated for the cross unless his services shall have been marked by a special mention of his name in dispatches by the Admiral, or senior naval or military officer commanding the squadron or detached force.

The cross (facing page 208) is of silver, patée in form, and convex. It bears on the obverse the monogram E.R.I. surrounded by a raised border, surmounted by the imperial crown, with ring for suspension from a blue ribbon with a white stripe down the centre like that used for the Conspicuous Gallantry Medal. The reverse is plain. The recipient has the right to append the letters C.S.C. to his name.

Indian Order of Merit.—This reward of valour takes chronological precedence over those decorations already described. It was instituted by the H.E.I. Co. in 1837, to reward personal bravery without any reference to length of service or good conduct, and is the Sepoy's Victoria Cross (facing page 136.)

It is divided into three classes and is awarded to native officers and men for distinguished conduct in the field. On the advancement from one class to another the star is surrendered to the Government, and the superior class substituted, but in the event of the death of the recipient his relatives retain the decoration. The order carries with it an increase of one-third in the pay of the recipient, and in the event of his death the allowance is continued to his widow for three years. The First Class consists of a star of eight points, $1\frac{3}{5}$ in. in diameter, having in the centre a ground of dark-blue enamel bearing crossed swords in gold, within a gold circle, and the inscription REWARD OF VALOUR, the whole being surmounted by two wreaths of laurel in gold. The Second Class star is of silver, with the wreaths of laurel in gold; and the Third Class entirely of silver. The decoration is suspended from a simple loop and bar from a dark-blue ribbon $1\frac{1}{2}$ in. in width with red edges, bearing a gold or silver buckle according to class.

The Indian Order of Merit realises from £2 10*s*. to £8 8*s*. in the sale-room according to class and action for which it was awarded. It is noteworthy, however, that in June 1900 the following prices were paid in the auction-room: First Class, £58; Second Class, £40; and Third Class, £35. When the order is unnamed, as it was issued, and there is no record of the services for which it was earned, prices invariably range lower than these.

Order of British India.—This order was instituted at the same time as the Order of Merit, to reward native commissioned officers for long and faithful service in the Indian Army. Since 1878, however, any person, European or native, holding a commission in a native regiment, became eligible for admission to the Order without reference to creed or colour. The First Class consists of a gold eight-pointed radiated star $1\frac{1}{2}$ in. in diameter. The centre is occupied by a lion statant gardant upon a ground of light-blue enamel, within a dark-blue band inscribed ORDER OF BRITISH INDIA, and encircled by two laurel wreaths of gold. A gold loop and ring are attached to the crown for suspension from a broad ornamental band $\frac{9}{10}$ in. in diameter, through which the ribbon, once blue, now red, is passed for suspension from the neck. The Second Class is $1\frac{7}{10}$ in. in diameter with dark-blue enamelled centre; there is no crown on this class, and the suspender is formed of an ornamental gold loop. The reverse is plain in both classes. The First Class carries with it the title Sirdar Bahadur, and an additional allowance of two rupees a day; and the second the title of Bahadur, and an extra allowance of one rupee per day.

India Meritorious Service Medal.—This was instituted on July 27th, 1888, and on receipt of the medal the order states "a non-commissioned officer must surrender his Long Service and Good Conduct medal"; but on being promoted to a commission he may retain the M.S. medal, but the annuity attached to it will cease. On the obverse is the diademed bust of Queen Victoria facing left, with a veil falling over the crown behind, encircled by the legend VICTORIA KAISAR-I-HIND. On the reverse is a wreath of lotus leaves enclosing a wreath of palm tied at the base, having a star beneath; between the two wreaths is the inscription FOR MERITORIOUS SERVICE. Within the palm wreath is the word INDIA. The medal, $1\frac{2}{5}$ in. in diameter, is suspended from a scroll by means of a red ribbon $1\frac{1}{4}$ in. wide.

The medals issued during the reigns of Queen Victoria's successors bear on the obverse their bust in profile with the legend altered to EDWARDVS or GEORGIVS. The Victoria medals have realised from £3 3*s.* to £4 10*s.*, and those of King Edward £4 to £5.

Egyptian Medal for Bravery.—As a means of rewarding N.C.O.'s and men of the Egyptian Army who distinguish themselves on the field of battle, the Khedive Abbas II instituted in May 1913 the silver medal illustrated. It is $1\frac{7}{8}$ in. in diameter and depends from a pale-blue ribbed ribbon and a suspender of the same pattern as the British Distinguished Conduct Medal. It bears on the obverse the cypher of Abbas Hilmi El Thani (Abbas II) and on the reverse in the upper half of the field FOR BRAVERY, and in the lower half in Arabic characters EL SHAHAMA, which is the equivalent of the English lettering above it.

Distinguished Service Cross.—As already explained

on page 343, His Majesty King Edward VII instituted the Conspicuous Service Cross in 1901 to reward "distinguished service before the enemy" on the part of warrant or acting warrant officers and subordinate officers of His Majesty's Fleet who do not hold naval commissions. On October 14th, 1914, the Lords Commissioners of the Admiralty—being "of opinion that it would be desirable to enable the said Cross to be granted also to commissioned officers" of the Navy below the rank of Lieutenant-Commander "for meritorious or distinguished services in cases where those services may not be considered sufficient to warrant the appointment of such officers to the Distinguished Service Order"—memoralised His Majesty King George V, who graciously approved of the suggestion that the Cross be in future designated the Distinguished Service Cross with the right to the recipient to append the letters D.S.C. to his name. The Cross is identical with the Conspicuous Service Cross, illustrated on page 208, except in the change of the Royal Cypher, G.R.I. replacing E.R.I. on the obverse. The ribbon, from which the decoration is suspended is dark blue with a broad white stripe down the centre. The decoration, facing page 208, is made by Messrs. Garrard & Co.

Naval Distinguished Service Medal.—In response to a memorial from the Lords Commissioners of the Admiralty, His Majesty King George V decided at the Court held at Buckingham Palace on October 14th, 1914, to institute a medal for courageous service in war by chief petty officers, petty officers, and men of His Majesty's Navy, and by non-commissioned officers and men of His Majesty's Corps of Royal Marines, and all other persons holding corresponding positions in His Majesty's Service afloat, in cases where the award of the Conspicuous Service Medal would be inappropriate (Her Majesty Queen Victoria having instituted it as an award for pre-eminent bravery in action with the enemy), the Distinguished Service Medal will be awarded to such men "as may at any time show themselves to the fore in action, and set an example of bravery and resource under fire without performing acts of such pre-eminent bravery as would render them eligible for the Conspicuous Gallantry Medal."

The design of the Naval Distinguished Service Medal is the same as the Conspicuous Gallantry Medal, except that the suspender is straight and the obverse bears the profile of His Majesty King George V and the reverse FOR DISTINGUISHED SERVICE. The decoration is suspended by a dark-blue ribbon, $1\frac{1}{4}$ in. wide, with two broad white stripes down the centre divided by a thin dark-blue stripe. (See page 252 for squeezes, the medal not being struck at time of going to press.)

The Military Cross.—Following the institution of the Cross and medal above described, His Majesty King George V, by a Royal Warrant published as a supplement to the *London Gazette* on December 31st, 1914, announced that he had instituted a new decoration to be known as The Military Cross for distinguished service in time of war. The decoration consists of a silver cross, having on each arm the Imperial Crown, and bearing in the centre the letters G.R.I. The Royal Warrant states that " No person shall be eligible for this decoration unless he is a captain, a commissioned officer of a lower grade, or a warrant officer in Our army, or Our Indian or Colonial Military Forces, and the Cross shall be awarded only to officers of the above ranks on recommendation by the Principal Secretary of State for War," and that " The Cross shall be worn immediately after all Orders and before all decorations and medals (the Victoria Cross alone excepted), and shall be worn on the left breast pendant from a riband $1\frac{3}{8}$ in. in width which shall be in colour white with a purple stripe." The Cross does not carry with it any individual preference or entitle the recipient to use any letters after his name. The list of first recipients of the decoration was published the same day, included in the list being Lieutenant Dimmer, K.R.R.C., a ranker, who has also the distinction of winning the V.C. in the European War now waging.

LONG SERVICE MEDALS

MILITARY LONG SERVICE MEDALS

THE institution of the Long Service and Good Conduct medal is due to an order of King William IV, who notified his intention to establish it in an Order to the Secretary of War on July 30th, 1830. In it His Majesty states that "discharged soldiers receiving a gratuity for meritorious conduct shall be entitled to wear a medal having on one side the words FOR LONG SERVICE AND GOOD CONDUCT, and on the other in relief the King's Arms, with the name and rank of the soldier, and the year, inscribed on the medal." Those entitled to the medal "must have completed twenty-one years of actual service in the Infantry, or twenty years in the Cavalry, never have been convicted by Court Martial, and must have borne an irreproachable character, or have particularly distinguished themselves in the service." The medal, designed by Pistrucci, was suspended from a dark-crimson ribbon by means of a steel bar attached to a steel clip which gripped the medal as on the Waterloo medal. This medal is sometimes found, however, with a ring through which the ribbon is passed. The name, rank, and regiment and the year of discharge were impressed in large capital Roman letters round the edge of the medal. On the accession of Queen Victoria the Arms of Hanover were omitted. About 1851 a silver scroll suspender replaced the steel clip, and in the new die the lettering on the reverse was cut smaller and the date was omitted. By a warrant dated January 16th, 1860, the grant of the medal was extended to non-commissioned officers on the permanent staff of the Militia. Since 1874 the names and rank of the recipients have been engraved on the edge of the medal. The earlier

(Reverse.)

(Obverse.)

FRENCH MEDAL FOR MEXICO, 1862–3.

(Obverse.)

(Reverse.)

FRENCH MEDAL FOR TONKIN, 1883–5.

medals realise from 10*s.* to £1, and the later ones from 5*s.* to 7*s.* 6*d.* each, according to rank and condition.

King Edward Long Service Medal.—On the accession of King Edward VII the obverse of the Military Long Service and Good Conduct medal was altered, and the effigy of King Edward, with the legend EDWARDVS VII REX IMPERATOR substituted for the trophy of arms with the shield of Great Britain in the centre.

King George's Long Service Medal.—The same alteration has been made in this latest issue; the bust of H.M. King George V takes the place of King Edward, and the legend being slightly altered so far as the name is concerned.

NAVAL LONG SERVICE MEDALS

William IV Naval Long Service.—King William IV instituted by Order in Council dated August 24th, 1831, a medal for Long Service and Good Conduct to men of the navy. It was at first granted for not less than twenty-one years' service; but this was subsequently reduced to twenty. It bears on the obverse an anchor surmounted by a crown, encircled by a wreath of oak leaves tied with a ribbon at the base. On the reverse is a blank space within a beaded circle in the centre for the recipient's name, rating, ship, and length of service, and running with the edge of the medal FOR LONG SERVICE AND GOOD CONDUCT. The medal, $1\frac{7}{20}$ in. in diameter, was issued with a silver ring passed through the medal, which was suspended from a narrow blue ribbon; but many of the recipients had a wire suspender made to take a broad ribbon for attachment to the breast. Only about 450 of these medals were issued. (See facing page 308.) They realise from £1 5*s.* to £2 2*s.* according to condition.

Victorian Naval Long Service.—On the accession of Queen Victoria a larger and thicker medal was instituted. It bears on the obverse the bust of Queen Victoria, as on the military medals, and on the reverse a line-of-battle ship at anchor, surrounded by a cable tied at the base in a reef knot, encircled by FOR LONG SERVICE AND GOOD CONDUCT. In the earlier medals the recipient's name, rank, etc., were engraved, and the medal bore on the obverse 1848—very few were issued thus—and the bar for suspension was nearly half as wide again, $1\frac{3}{5}$ in. by $1\frac{7}{10}$ in., as those on the later medals. The later medals are mostly impressed in capital letters, and are suspended from a clasp which takes a $1\frac{1}{4}$ in. ribbon of blue with white edges. The earlier medals realise about 10*s*., and the later ones about 5*s*. to 6*s*.

India, 1859.—The H.E.I. Co. issued a medal in 1859 to men of the old Indian Local European Service for "long service and good conduct"; but none were issued to the H.E.I. Co.'s navy! About 100 were issued to men in the three Presidencies before the mistake was discovered. The medal bears on the obverse the diademed head of Queen Victoria as generally used at this time on all war medals, and on the reverse FOR LONG SERVICE AND GOOD CONDUCT. Above the wording is a crown, and below an anchor surrounded by two oak branches. The medal, $1\frac{3}{5}$ in. in diameter, was suspended by a red ribbon $1\frac{3}{10}$ in. wide by a scroll suspender. The names were impressed on the edges.

Edward VII Naval Long Service.—The obverse of this medal bears the bust of the King in Admiral's uniform, and the legend EDWARDVS VII REX IMPERATOR, and the recipient's name, rank, and ship are impressed upon the edge. (See facing page 308.) The reverse is as that of the Victoria variety. It realises, according to condition and rank, 5*s*. to 10*s*.

George V Naval Long Service.—The medal issued during the reign of our present Sovereign His Majesty King George V is in all respects the same as the above mentioned, except that his bust in Admiral's uniform adorns the obverse, and GEORGIVS takes the place of EDWARDVS. The recipients' names are indented in block capitals.

Volunteer, Militia, and Territorial Medals.—I also illustrate representative oval medals issued by King Edward and King George to the Militia, Territorials, and Special Reserve, for Long Service, Good Conduct and Efficiency. These are named in light skeleton block capitals. The Victoria Volunteer Officer's Long Service and the Edwardian Colonial and Auxiliary Forces Officer's Long Service Medals, illustrated facing page 240, were issued unnamed.

Naval Best Shot.—The fine medal illustrated facing page 296 was instituted by His late Majesty King Edward as a reward for excellence in Naval Gunnery, and although, like Volunteer Long Service Medals, it can hardly be classed among war medals, the importance of encouraging good gunnery has been well demonstrated in the war now in progress, where our gunners have made wonderful hits at a range of eight or more miles. The obverse of this medal bears the bust of King Edward in Admiral's uniform, as does also that of the Naval Reserve Medal facing page 224.

I have not herein considered the large number of regimental medals, or those awarded by officers to their men for prowess in war or excellence of conduct in the ranks; these, together with the old Volunteer medals of the eighteenth century, are too numerous for adequate treatment in a volume such as this, and as I have restricted my pages to war medals of an official nature, only including those private ones which have great historical interest, I must leave the consideration of these to the time when I may find leisure to treat of them as they deserve.

I have endeavoured, while making my book helpful, to make it also interesting, for I am of opinion that those who do not generally take more than a mere handling interest in the things they sell lose more than half the interest which a business life affords, while those who simply take a pleasure in the mere possession of the things they collect fail to understand the true meaning and value of the collector's spirit.

HOW MEDALS ARE NAMED

I APPEND a few sketches and notes which may be helpful, together with the facts I have given, to those who collect or sell the golden medallions, the bright discs of silver, or the intrinsically valueless but precious bronze crosses or stars which have adorned the breasts of the brave.

First let me state that any re-engraved medal is comparatively valueless, and that those who purchase high-priced medals should not trust to their own judgment, but employ the services of those experts and dealers who possess the necessary technical knowledge and have at their command the means of identification.

The names on the Waterloo medals were impressed with Roman capitals like **ROBERTS** or MURPHY, so large that they left little margin between the edge of the medal and the top and bottom of the letters.

The Peninsular, or Military General Service, medal had the names indented in Roman capitals as follows: **STEWARD.** The same type was used on the Naval General Service and the Army of India medals, also the Mutiny medal.

On the Afghan medal of 1842, for Candahar, Ghuznee, and Cabul, the same kind of Roman capitals were used, but sometimes rather larger and narrower, and the down-strokes somewhat lighter: these medals were not always impressed, and many were engraved in this style JOHN and others in a light script.

The Punjab medals for 1849 were indented in the same style as the M.G.S. medal; but on many of the Sutlej medals the rather taller type of Roman capital was used.

The South Africa 1853 medals were named with the same letters as the M.G.S. medals.

MEDAL TO PAPAL BRIGADE
FOR 1860 CAMPAIGN.

PRUSSIAN BRONZE CROSS FOR KÖNIG
GRÄTZ (SADO WA), 1866.

GENEVA CROSS FOR FRANCO-
GERMAN WAR, 1870–1.

On those of the Crimean medals which were returned for naming the same type was used, but the letters were rather more openly spaced. Some were also impressed in large and others in small skeleton Roman, so, MURRAY, many were engraved in square Roman capitals, and others in upper and lower case upright block.

The India General Service medal, 1854, had the names mostly impressed with the same type as the M.G.S. medals up to Bhootan 1864–6; some, particularly naval medals, were named in a smaller kind of Roman capitals stamped closer together. The same type of Roman capital was likewise used on the New Zealand medals, while the China medals for 1857 and 1860, which were only named for the army, had the names indented in the same type. But the naming on the China medal for 1842 was very characteristic, and heavy Roman impressed capitals of a square nature, like DUNCAN, were employed in the stamping, with three long stars to fill the gap on either side between the name and the lug of the suspender.

The Abyssinian medal is distinctive in its naming, the recipient's name and regiment being impressed in relief on the centre of the reverse, or, in the case of the Indian troops, in the same manner or struck incuse, while some are engraved.

The Canada General Service medal has the names indented in square block capitals, as BROCK, or in caps and lower case, as Boston. Some were engraved in square upright Roman capitals.

The home troops who received it had the regimental numbers impressed and not the territorial designations, as on the South African medals. The Naval medals were generally engraved in dwarfed Roman capitals.

The Ashantee medal for 1873–4 was engraved in this style, **BUCHANAN, 42ND** with the date 1873–74, after the regiment, but on the 1892 medals, while the same style was used, it was engraved with thinner down-strokes and was smaller.

When the I.G.S. medals, 1854 type, were first issued the names were impressed similar to the M.G.S. medal; but those medals awarded to the navy and the army for Pegu have very small tall Roman capitals on the edge; for Perak the names are engraved in sloping Roman capitals like the Afghan medals described below; or like the medals for the rest of the campaigns with the exception of Jowaki, 1877–8, which by the way was the first campaign medal to have the date indicated on the bar, the names were generally engraved as *P. Murphy*, and the same style of script, but very lightly engraved, as *Edward*, was used to name the I.G.S. 1895 medal. This light and somewhat scratchy form of script was also employed in a very characteristic manner, to name the Punniar and Maharajpoor stars, and also some of the first Cabul medals, 1842.

The Burma 1885–7 medal is generally engraved in a light running hand; but I have seen several engraved in slightly sloping squat Roman letters, ***2/SCOTS FUS:*** The script kind of naming was used on the 1887–9 and 1889–92 Burma medals, but some of the latter medals were engraved in a neat round hand.

The Naval medals awarded for Burma 1885–7 are mostly named with rather tall thin Roman capitals.

The medals for Jowaki 1877–8 are impressed with Roman capitals after the character of M.G.S. lettering, and the regiments are generally described so, 2/9th Foot.

The Afghan medals for 1878–9–80 were engraved thus, ***RENDELL***, or in the same style of letter but upright, as were also the Egyptian medals of 1882 onward; but most of those issued to the Royal Marine Light Infantry and seamen who took part in the Suakin Expeditio nof 1885 were indented like this, HENRY. This kind of type was also used for naming the King Edward Naval Long Service medals. Some, however, were impressed in tall Roman capitals having thick down-strokes.

The engraving on the Egyptian medals awarded to some of the Indian troops, dooley bearers, and Indian transport department is in a very finely cut, neat script.

Some of the Zulu War medals of 1878–9–80 were engraved in this way, SHEA, the letters being rather badly shaped and spaced, but most were engraved in the same style as the Afghan and Egyptian medals.

The Soudan medals were engraved in this style *CURTIS*, and some **GORE.R.A.**: some of the Queen's and Khedive's medals were impressed in very small Roman letters, thus MACPHERSON. In a few instances the Queen's medals, like the Khedive's, were issued unnamed, and the recipients had to have them engraved.

The medals granted for the Punjab Frontier are named with a rather coarse script.

The engraving on the Hazara medal for 1888 has rather finer-cut strokes, and the naming on the Sikkim medal, 1888, is the same.

King Edward's medal for Waziristan, 1901–2, is engraved in the same style; but the medal for Chin-Lushai 1889–90 is named in a neat round hand.

The Maharajah of Kashmir's bronze medal for Chitral, 1895, is impressed in badly aligned and carelessly spaced block letters.

The Cape of Good Hope medal is engraved in neat, squat, upright Roman capitals, and that for Natal is generally named in very lightly impressed upright skeleton block.

The East and Central Africa medal with swivel ring is named in a light script; but the same medal with bar for Uganda, 1897–8, is engraved rather roughly in light Roman capitals.

But these medals are named in several ways. For Sierra Leone in square block capitals. With bar for 1891–2 the medal is named in slightly sloping Roman capitals, that for 1892 the same. That with bar for

1897–8 named like the tall thin block letters used on some of the Boer War medals.

The Boer War medals were generally indented with block capitals, either square, like H I L L, or tall like this, HILL, or engraved in this style, *F. HILL, 2/Linc Rgt:*

King Edward's medal for Ashanti, 1900, is generally impressed in rather small skeleton block lettering.

The China medal, 1900, is named—for the navy—with a bold but light kind of Roman capital, similar to that described on page 354 as used on the Suakin, 1885, medals.

The Tibet medal is generally named in a combination of large and small Roman and script, as *G. Hand, 1st Bn.*

The Indian medal for 1908 is engraved in script, but so badly that one could not conceive an engraver's apprentice of one year's standing doing so badly.

King George's medal for Abor, 1911–12, is engraved in a rather crude script.

The illustrations of the namings are mostly considerably enlarged to render comparison easier.

GERMAN MEDAL FOR FRANCO-PRUSSIAN WAR, 1870-1.

FRENCH MEDAL GIVEN TO VETERANS OF THE WAR OF 1870-1.

FIRST CHILIAN WAR MEDAL, 1880.

SECOND CHILIAN WAR MEDAL, 1881.

SOME CONTINENTAL AND FOREIGN WAR MEDALS

A CENTURY ago the Prussians were our allies. To-day the position is reversed, and we have hastened to assist the descendants of the brave men who routed the Prussians at Jena and the Austrians at Marengo, and so strenuously fought against our forefathers at Waterloo. I have already described most of the medals awarded for that famous battle, but there are others of an interesting character which were awarded for the campaigns of 1813, 1814, and 1815, when the various continental nations strove against the genius and power of Napoleon.

Prussian Oval Iron Medal for 1815.—This reminds us of the fact that the exchange made by modern German women of their gold wedding rings for those of iron is simply a repetition of a sacrifice which the women of Germany made during the Napoleonic Wars, when they gave their gold jewellery and replaced it by delicately made ornaments of iron, as visitors to the Victoria and Albert Museum may see. The medals awarded to the survivors of the wars were invariably of metal, some, like our Victoria Cross, having been made from captured cannon. This oval medal, however, is of iron and bears on the obverse King Frederick William's initials in German text capitals, surmounted by the Prussian crown and underneath Fur Pflichttreue im Kreige (For faithful service in war); encircling the whole is the legend Gott war mit uns, Ihm sey die Ehre; (God was with us, to Him the honour). On the reverse is a cross patée with rays issuant between the arms and in the centre 1815. The medal is suspended from a white watered-silk ribbon with black and yellow stripes close to the edges. It was awarded to non-combatants.

Prussian Medals for 1813–15.—These were awarded by King Frederick William III to all those of his subjects who took part in the campaigns of 1813–14–15. On the obverse are the initials F. W. surmounted by a crown; below, Preussens tapfern Kriegern (To Prussia's brave warriors), surrounded by the legend, with a border, Gott war mit uns, Ihm sey die Ehre ! (God was with us, to Him the honour). On the reverse, within a wreath of oak and laurel, tied with a ribbon at the base, upon a cross patée with rays, 1815—the date of the campaign. In the medals for 1813 and 1814 the arms of the cross are rounded at the ends and terminate at the medal rim. Indented round the edge, AUS EROBERTEM GESCHUTZ (from the captured guns). See page 52. The ribbon is yellow, flanked by stripes of black and white.

The Austrian Cross for 1813–14.—This is an interesting decoration, for the Cross itself is enamelled green in the *basse-taille* manner, the edge only being left to show the metal, likewise the laurel wreath which connects the arms of the Cross. On the obverse, across the centre arms, is PRINCEPS ET PATRIA, on the upper arm is the word GRATI, and on the lower arm FRANC IMP AUG. On the reverse, across the centre arms, is LIBERTATE ASSERTA, on the upper arm EUROPAE, and on the lower MDCCCXIII / MDCCCXIV. It is suspended from a yellow corded ribbon with very broad black edges.

Hessian Medal for 1814–15.—This is a bronze medal $1\frac{1}{8}$ in. in diameter, bearing on the obverse in German text capitals, K W II reinen tapfern Hessen, *1821*, surmounted by a crown, and encircled by a wreath of conventionalised oak leaves. On the reverse in the centre is a cross patée, the arms of which rest upon a wreath of laurel. In the centre are the dates 1814 / 1815. Two tilting spears are arranged between the arms. Over the top arm, and resting upon a circle enclosing a light wreath, is a helmet, around is the motto in German text, Gott brach des feindes macht und Hessen ward befreit. On the edge in Roman is impressed AUS EROBERTEM GESCHTZ (from captured guns). The medal is suspended from a dark-blue ribbon with wide red edges. (See facing page 324.)

Légion d'Honneur.—The Order of the Légion d'Honneur was proposed on May 15th, 1802, by Napoleon Bonaparte when he was First Consul, as a reward for Military and Civil Services. The Legislative Assembly, however, was not favourably disposed, and it was not until four days later that the institution of the now famous Order was agreed to. Napoleon knew the value of a ribbon to stick in the coat. (See facing page 320.)

The Order was originally divided into three classes—Légionaries, Grand Officers, and Commanders. After the Coronation of Napoleon (July 14th, 1804) the first class of Grand Officers was divided into Knights of the Grand Eagle (as the highest) and Grand Officers. At present there are five classes of the Order: Knights of the Grand Cross, Grand Officers, Commanders, Officers, and Knights.

The original Badge or Cross consists of a white enamelled badge with five double rays, with silver balls on the points, resting on a laurel and oak wreath, tied at the base, surmounted by an Imperial crown attached to the badge by a loop and ring, and has a ring for suspension from a crimson corded ribbon, $1\frac{1}{2}$ in. wide.

Obverse: On a silver-gilt radiated centre, the laureated head of the Emperor Napoleon, facing right, surrounded by a blue enamelled band, with gilt borders, inscribed, in gold letters, NAPOLEON EMPEREUR DES FRANÇAIS. Reverse: Also on a silver-gilt centre of horizontal lines, the French Imperial eagle, surrounded by a blue enamelled band, with gilt borders, inscribed, HONNEUR ET PATRIE, with a sprig of laurel below.

The Cross for the Knight is in silver, and for the other classes in gold and of a larger size.

The Knights and Officers wear it at the button-hole

or on the left breast; the Commanders round the neck. The Grand Officers wear besides, upon the right breast, a silver Star, similar to that of the Grand Crosses; and at the button-hole the Cross in gold. The Knights of the Grand Cross wear a similar golden Cross, but larger, suspended by a ribbon across the right shoulder towards the left hip; and also, on the left breast, a silver Star.

The Star is of silver, similar to the Cross, without the wreath, having rays between the angles, and in the centre, within a band inscribed HONNEUR ET PATRIE, the Imperial French eagle.

The first claim to the Order must begin with the lowest degree of Knights, as no degree can be passed over.

The Republican Cross.—The Badge of the Légion d'Honneur, given by the Republic, bears on the obverse, facing right, a laureated female head symbolic of the Republic, and surrounded by the legend REPUBLIQUE FRANÇAISE, 1870. On the reverse in the centre are the crossed French flags, surrounded by the motto HONNEUR ET PATRIE. In place of the French Imperial crown which was employed as a suspender in the original badge or cross is an enamelled wreath, otherwise the general appearance of the decorations is the same.

Prussian Iron Cross, 1813.—The Order of the Prussian Iron Cross was instituted by King Frederick William III on March 10th, 1813, to reward those, either military or civil, who distinguished themselves in the war then being carried on. It was divided into three classes. The Grand Cross, which was double the size of the Knight's Cross, and was worn round the neck, was given exclusively for the gaining of a decisive battle, the conquest of an important position or place, or the brave defence of a fortress. The first class also wear upon the left breast, instead of a Star, a similar Cross or Badge. In the bestowal of the Cross, neither rank nor condition was regarded. (See facing page 332.)

It was worn by the military with a black ribbon

with two white stripes near the edge; and by civilians with a white ribbon with black borders, and was suspended from a silver loop and ring. At the close of hostilities, the distribution of the Order ceased, but was revived on July 19th, 1870, for the war then about to commence with France. The decoration is a cast-iron Cross, in the form of a cross patée, with silver borders and mountings. There are three classes, both for military and civilians. Obverse: In the centre, within a silver milled border, three oak leaves; above, F. W. surmounted by the Prussian crown; below, 1813. The Cross awarded for the Franco-German War bears on the reverse: In the centre, also within a silver milled border, the initial W; above, is a crown; below, 1870.

For Distinction in Service.—In 1825, on the anniversary of the battle of Waterloo, King Frederick William III instituted a gold Cross for officers who had served twenty-five years, bearing on the obverse the initials, 𝔉. 𝔚. III, surmounted by a crown, and on the reverse the number XXV. It was worn on the left breast, suspended by a blue ribbon. See facing page 332. For sub-officers and privates, silver buckles, $1\frac{13}{20}$ by $\frac{9}{20}$ in., with the initials 𝔉. 𝔚. III in relief, on a rough ground, within a raised double border, worn on the left breast, with brooch attachment, suspended by a blue ribbon with yellow borders for twenty-one years' service; by a blue ribbon with white and blue borders for fifteen years' service; and by a blue ribbon with black borders for nine years' service. Years of war service counted double.

San Sebastian, 1836.—This medal was granted by the Spanish Government to the British Legion in Spain, who served under General Sir de Lacy Evans against the Carlists on the heights of St. Sebastian, on May 5th, 1836. It was given in silver to officers, and in white metal to privates. On the obverse, surrounded by the Collar of the Golden Fleece, within a circle, on a mottled ground, is a Lion, statant gardant. Above,

ESPAÑA; below, AGRADECIDA. On the reverse, in the centre of a cross of four arms, with plain, raised borders, having crowns in the angles, surrounded by a wreath of laurel, from which rays issue, 1836; around which is inscribed SAN SEBASTIAN 5 DE MAYO. Beneath, and to the right of the lower limb of the cross, I.D.

The medal, $1\frac{2}{5}$ in. in diameter, is suspended by a ring passed through the medal, to which is attached a silver bar. The ribbon is $1\frac{1}{2}$ in. wide, dark purple with two yellow stripes near the edges.

Iron Cross for San Sebastian, 1836.—An iron Cross of four arms, with bright borders, with rays in the arms issuing from the centre, having four gold crowns between the arms. On the obverse, in the centre, in high relief, in gold, is a Lion, statant gardant, with a gold border, on which is inscribed in gold letters, HESPEINA AGRADECIDA. The whole within the Collar of the Golden Fleece, in bronze, the pendant Lamb being in gold. On the reverse, also in relief, within a laurel wreath of gold, SAN SEBASTIAN 5 DE MAYO in gold, having in the centre of a gold circle, 1836 in gold. The Cross, similar to that on the circular silver medal, is $1\frac{1}{20}$ in. in diameter, with a gold loop and gold ring for suspension.

Medal for Irun.—For the capture of the fort and town of Irun, in Spain, on May 17th, 1836, a gold medal was issued by the Spanish Government to the officers, and among the recipients was an English surgeon, whose medal is in Dr. Payne's collection. It is $1\frac{1}{16}$ by $\frac{19}{20}$ of an inch and has a plain gold border. On the obverse, encircled by two oak leaves enamelled green, tied at the base with a ribbon enamelled white, is a tower enamelled white; behind which is a sword with a gold hilt, the blade enamelled blue, and above the tower on a dark blue enamelled ribbon the word IRUN standing up in gold. The whole is enamelled champléve. On the reverse on a plain gold ground is enamelled in blue 17 DE MAYO DE 1837, encircled by the name and rank of the recipient, STAFF SURGEON B. A. L. SPAIN, C. B. BREAREY, M.D., M.R.C.S. The ribbon from which the medal is suspended is black with red edges, symbolical of the blood spilt in capturing the fort and town, over which the black flag was waving, indicating that no quarter was being given.

Staff Surgeon Brearey likewise received the Silver Cross for San Sebastian, 1836, and this also is in Dr. Payne's collection.

Silver Cross for San Sebastian, 1846.—This is a silver Cross of four arms, $1\frac{1}{20}$ in. in diameter, with plain, raised borders, with obverse and reverse the same as the medal, the words ESPANA AGRADECIDA being omitted on the obverse. (See facing page 336.)

Russian Medal for Hungary, 1849.—The medal awarded to the Russian soldiers who took part in the Pacification of Hungary at the request of Austria in 1849, bears on the obverse the Russian double-headed eagle, surmounted by a crown, having a shield on its breast, encircled by a collar and badge, containing the figure of St. George and the Dragon. The eagle to the left is holding a baton in its claws, and the one to the right a globe with a cross above it. Above, the radiated Eye of Providence; encircling the whole, in Russian characters, GOD IS WITH US: HEAR, O YE PEOPLE, AND SUBMIT, FOR GOD IS WITH US. On the reverse, inscribed in Russian characters, is FOR THE PACIFICATION OF HUNGARY AND TRANSILVANIA, 1849. This is a small circular medal, $1\frac{3}{20}$ in. in diameter, with plain, raised double borders. It is made in silver and bronze.

For Zeal.—An interesting silver medal $1\frac{1}{10}$ in. in diameter was issued by the Emperor Nicholas I. It bears on the obverse the bust of the young ruler facing right and his titles in Russian characters, and on the reverse, following the line of a ribbed border, the legend FOR ZEAL in Russian, a rope-like circle makes a central shield. The medal is $1\frac{1}{16}$ in. in diameter.

Médaille Militaire.—This, the French equivalent of the British Distinguished Conduct Medal, was founded in 1852 by Louis Napoleon, when President of the

French Republic. It is of silver, 1 in. in diameter, formed of a close band of laurel wreaths, encircling on the obverse the gilt bust of Louis Napoleon, facing left, with his name in gilt letters on a blue enamelled band. On the reverse, the wreath encircles a narrow band of blue enamel which borders the gilt centre bearing the legend VALEUR ET DISCIPLINE. The medal is surmounted by the French eagle, with outspread wings, and is suspended by an orange-coloured ribbon with green edges. After the Crimea, Louis, who had then become the Emperor Napoleon III, presented this medal to 500 British non-commissioned officers and men who had distinguished themselves in the campaign. (See facing page 132.)

The Republican variety bears the female head symbolic of the Republic on the obverse and the legend REPUBLIQUE FRANÇAIS ✵ 1870 ✵; on the reverse is the inscription VALEUR ET DISCIPLINE. The medal is connected with the suspender by means of a trophy of French arms. The medal illustrated was won in France during the war now waging, by the late Sergeant Hunt.

Turkestan.—For the campaign in Turkestan the Emperor Alexander II gave a small silver medal $1\frac{1}{16}$ in. in diameter which bears on the obverse the Emperor's initial A in ornamental cypher, surmounted by the Imperial crown and II arranged between the legs of the initial. The reverse bears in the centre the date 1857–1858–1859 encircled by an inscription in Russian explaining that it was awarded for the success of arms in Turkestan. (See facing page 364.)

Garibaldian Medal, 1860.—This medal was given in 1860 to the Garibaldians in Sicily by the Municipality of Palermo, and distributed to the troops by Garibaldi, November 4th, 1860, in the square fronting the Royal Palace, Naples.

On the obverse, within the words AI PRODI CUI FU DUCE GARIBALDI, is a spread eagle standing on a scroll, on which are the letters S.P.Q.P. Three stars below. All in relief. On the reverse, outside a circle of laurel

SILVER MEDAL "FOR ZEAL," AWARDED BY NICHOLAS I.

SILVER MEDAL FOR CAUCASUS, 1859–64.

BRONZE CROSS FOR CAUCASUS, 1864.

SILVER MEDAL FOR TURKESTAN, 1867–8–9.

BRONZE MEDAL FOR POLISH INSURRECTION, 1863–4.

RUSSIAN WAR DECORATIONS.

leaves are the words MARSALA, CALATAFIMI, PALERMO. Within the circle, IL MUNICIPIO PALERMITANO RIVENDICATO MDCCCLX. A star is in the exergue. This is a small circular medal, $1\frac{1}{5}$ in. in diameter, within a plain, raised double border, suspended by a silver loop and ring from a crimson ribbon, $1\frac{2}{5}$ in. wide, with narrow yellow edges, on which is attached the Arms of Sicily in silver, bearing the words UNO DEI MIL.

Papal States Campaign, 1860.—This medal was given by Pope Pius IX to his troops, including the Irish Brigade of Volunteers, who served under Major O'Reilly against Garibaldi. It was issued in gold to officers, and in silver and white metal to privates.

The obverse: An open ring formed by the body of a serpent—symbolic of eternity—on the head of which rests an inverted cross; around, on a plain band with simple, moulded edge is the motto, PRO PETRI SEDE ▲ PIO · IX · P · M · A · XV ▲. On the reverse, in the band, VICTORIA QVAE VINCIT MVNDVM FIDES NOSTRA ✠.

This (facing page 352) is a circular medal, $1\frac{1}{2}$ in. in diameter, with a plain, raised, double border, and a scroll bar and claw clip suspender, from a crimson ribbon, $1\frac{1}{5}$ in. wide, with two white stripes, edged with yellow.

The Caucasus.—For services in the Caucasus between 1859–1864 the Emperor Alexander II granted a small silver medal $1\frac{1}{16}$ in. in diameter bearing on the obverse his bust in profile facing left, and on the reverse across the centre is the date 1859–1864 encircled by an inscription in Russian characters stating that it is awarded for the subjugation of the Western Caucasus.

A bronze Cross, with crossed swords between the arms, to be attached to the uniform by means of a ring and bar attachment, was also awarded for the same campaign. It bears the explanatory inscription across the horizontal arms and Alexander's initial in Russian character surmounted by the Russian Imperial crown in the upper arm, and the date 1864 in the lower arm. In the circle which occupies the centre is the Russian eagle. It is $1\frac{9}{10}$ in. across the arms. (See facing page 364.)

The Polish Insurrection.—For the suppression of the Polish Insurrection a bronze medal, $1\frac{1}{16}$ in. in diameter, was granted to the Russian troops. It is suspended from a $\frac{3}{4}$ in. ribbon composed of three equal stripes of white, yellow, and black, and bears on the obverse the two-headed Imperial eagle, surmounted by the Imperial crown, holding in its claws the orb and sceptre. On the reverse across the centre is the date 1863–1864 encircled by the inscription, in Russian, " For the suppression of the Polish Rebellion."

Cross for König Grätz.—The Cross for König Grätz was awarded to the Prussian troops who fought against the Austrians during the sanguinary and disastrous seven weeks (" seven days ") war in 1866. It was this short campaign which demonstrated the superiority of the needle gun or breech-loader, on the bolt principle, over the old-fashioned muzzle-loader used by the Austrians, who assert that they were defeated simply because it took them three or four times as long to load and fire their guns as it did the Prussians. In this fight, also called the battle of Sadowa, the Austrians lost 20,000 men placed *hors de combat,* and as many prisoners. The Prussian loss was 10,000.

The Cross, which is $1\frac{2}{5}$ in. in diameter, bears in the centre the Royal Cypher W.R., surrounded by the legend PREUSSEN SIEGRIECHEM HEERE (To Prussia's victorious army); on the upper arm of the Cross is the Prussian crown, on the right arm GOTT MIT IHM, and on the left WAR UNS SEI, on the bottom arm is DIE EHRE. On the reverse is the crowned Prussian eagle, resting upon a cannon; on the upper arm is KONIG-GRATZ, on the left DENZ, on the right, JULI, and on the lower arm, 1866. The ribbon is of black corded silk, 1 in. wide, with narrow stripes of white and orange at the side. (See facing page 352.)

Geneva Cross, 1870–71.—This bears on the obverse a red enamelled Cross of four arms, with a silver indented border in the centre in white enamel, with a plain silver edge. Reverse: On a silver perpendicular lined surface, within a raised silver double border, above, 1870; below, 1871; between, AUX BLESSÉS DES ARMÉES DE TERRE ET DE MER. The cross has a ring and loop for suspension from a white ribbon, 1½ in. wide, with a red cross in the centre. This Cross appears to be like the one worn by King Edward VII.

The Silver Cross, 1½ in. in diameter, bears on the obverse, above, 1870; below, 1871; between, SOCIÉTÉ FRANÇAISE DE SECOURS AUX BLESSÉS DES ARMÉES DE TERRE ET DE MER. The reverse is plain, with a simple border. Ribbon, the same as the above. The Bronze Cross is the same, 1½ in. in diameter, and is also suspended from the white ribbon with the Geneva Cross in the centre. (See facing page 352.)

French Medal for 1870–71.—The bronze medal recently awarded by the French Government to the survivors of the Franco-Prussian War of 1870–71 is what one might expect from an artistic nation. It is somewhat larger, however, than most French military medals. It bears on the obverse the helmeted and armour-clad bust of a woman, symbolising the Republic, with the legend REPUBLIQUE FRANÇAISE and on the reverse a finely arranged trophy of French arms, supporting the French tricolour, above a panel inscribed AUX DEFENSEURS DE LA PATRIE (To the defenders of the country). It is suspended by a green ribbon, with four black stripes of equal width. (See facing page 356.)

German Medal for 1870–71.—After the conclusion of the Franco-Prussian War the German Emperor distributed circular medals of iron or bronze to combatants and non-combatants who rendered service during the campaign: the reverse of the medal explains whether it was awarded to one or the other. On the obverse of the medal granted to combatants is W, the initial of Wilhelm, surmounted by the Prussian crown, and underneath, Dem siegreichem Heere (To the victorious army), encircled by Gott war mit uns, Ihm sei die Ehre (God was with us, to Him the honour). On the reverse, surrounded by a laurel wreath, $\frac{1870}{1871}$. On the steel medals the wreath is of oak (on the medals for 1813, 1814, 1815 the wreath is of laurel and oak), but in each case the wreath overlays a cross patée with rays issuant between the arms. On the edge is sometimes impressed AUS EROBERTEM GESCHUETZ (From captured guns). The same inscription is on the edge of the bronze medals issued to combatants for the campaigns of 1813-14 and 1815. The medal awarded to noncombatants bears the inscription Fur Pflichttreue im Kreige (For faithful service in war); there is no inscription on the edge. The medal is suspended by a ribbed silk ribbon, 1 in. wide, with a red stripe down the centre, flanked by black and white stripes and black edging. (See facing page 356.)

Chili-Peruvian War, 1879-81.—On April 5th, 1879, war was declared against Peru by the Chilian Republic, it being alleged as a *casus belli* that the Government of Peru had made a treaty with Bolivia—Chili having strained relations with the latter country—which was antagonistic to Chili. This, however, as the terms of the treaty show, was merely a pretext for taking by force the province of Tarapaca with its mineral wealth and guano deposits. The battle of Tarapaca was fought on November 17th, 1879, and despite the gallant efforts of the Peruvian Army, the city was taken by the Chilians. The latter, under General Baquedano, landing an army farther north, fought the battle of TACNA on May 26th, 1880, the action commemorated upon the bar attached to the medal illustrated and shown above the bar for SAN FRANCISCO, fought on November 19th, 1879. The Bolivian and Peruvian force of 9,000 men under General Camero was decisively defeated by the Chilians, 14,000 strong, who followed up their success by capturing Arica on June 7th, 1880. Incidentally I might mention that the bars slip over the ribbon and bear on the back the date of the action. The campaign

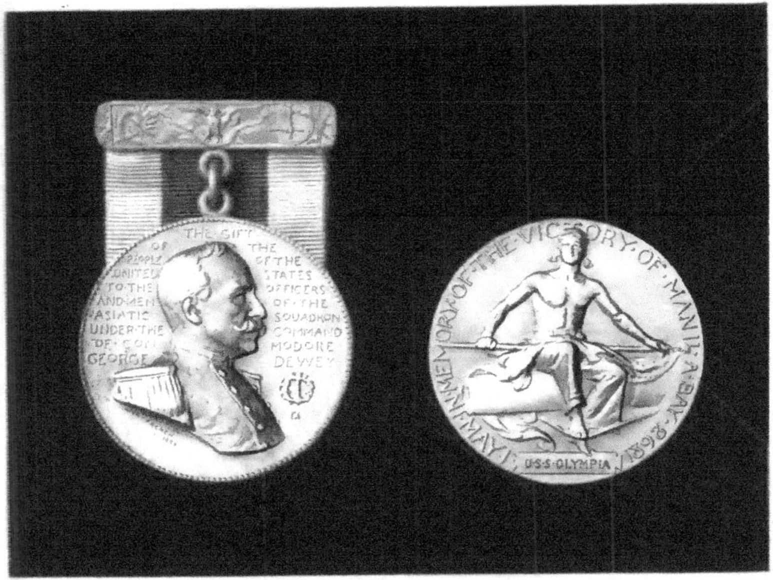

GOLD MEDAL PRESENTED BY THE AMERICAN NATION TO ADMIRAL DEWEY FOR THE VICTORIOUS BATTLE OF MANILA BAY, 1898.

in Lima succeeded the annihilation of the Peruvian Army. In this the raw recruits and volunteers who fought for Peru made desperate struggles at Chorrillos on the 13th and at a final concerted struggle at MIRAFLORES on January 15th, 1881, but they were beaten, losing about 3,000 men killed and wounded (their opponents' losses were over 2,000), and despite the splendid defence of Lima, it was taken on the 17th; but its defender, General Caceres, struggled on with a guerilla warfare until October 1883, when the war was concluded upon terms which pressed very heavily upon the unfortunate Peruvians.

The First Chilian Medal.—This was awarded to the participants in the first campaign, which, as the inscription on the reverse states, covered the inclusive period of DE 14 DE FEBRERO DE 1879 A 7 DE JUNIO DE 1880. This legend is on a blue enamelled ground, in the centre of the silver medal, bordered by a band of red. The five arms of the star are covered with scaling which forms the centre of the panels; the points are tipped with balls; conventional rays are issuant between the arms of the star. On the obverse, in the centre, is a laureated, Grecian head, symbolic of the Republic; it is gilt and encircled by a blue enamelled band bearing the legend CAMPAGNA A BOLIVIA I EL PERU. The decoration depends from a red, ribbed ribbon, and regulation brooches were issued with the medals for suspension.

The Second Chilian Medal.—For the campaign which resulted in the capture of Lima a silver Cross with five arms was awarded. The terminals of the arms are protected by beads and the panelling is scaled as in the other decoration; issuant between the arms, three leaves of laurel. In the centre of the obverse and reverse is a gilt five-pointed star upon a radiated ground. On one side in a blue enamelled band is the record CAMPAGNA DE LIMA 1887, and on the other REPUBLICA DE CHILE. This decoration depends from a blue ribbon, from a regulation silver buckle brooch. (See facing page 356.)

Tonkin, 1883–5.—The French medal for the China War of 1883–5 bears on the obverse, within a wreath of laurel, with ribbon entwined, the head of "the Republic," helmeted and laureated, with the word PATRIE on the forehead. Around, REPUBLIQUE FRANÇAISE. Below truncation is the artist's name, DANIEL DUPUIS. On the reverse, in the centre of a circle, within a similar wreath of laurel, are the battle names, SONTAY, BAC-NINH, FOU-TCHÉOU, FORMOSE, TUYEN-QUAN, PESCADORES. Around, TONKIN CHINE ANNAM, 1883–1885, below, with a dot on either side. This is a small circular silver medal, $1\frac{3}{20}$ in. in diameter, with a ring and loop for suspension from a yellow ribbon, $1\frac{1}{2}$ in. wide, with four light green stripes. (See facing page 348.)

Spanish-American War.—Arising out of the difficulties between the Spanish and the Cubans, came the war between Spain and America and the opportunity for the United States Navy to demonstrate its power, with a squadron of only 6 ships, by destroying the fleet of 11 Spanish battleships in Manila Bay without the loss of a single man. I illustrate the remarkably fine medal which was presented by the American nation to Commodore George Dewey—Admiral Dewey, as he is known in Britain—and to the officers and men who served under him. The medal, designed by the famous American sculptor, Daniel Chester French, was struck by Messrs. Tiffany & Co. It bears on the obverse the bust of Admiral Dewey, and on the ground the following inscription, THE GIFT OF THE PEOPLE OF THE UNITED STATES TO THE OFFICERS AND MEN OF THE OCEATIC SQUADRON, UNDER THE COMMAND OF COMMODORE GEORGE DEWEY. On the reverse is a finely modelled figure of an American sailor, sitting on a naval gun, holding firmly across his knees the American flag, his left foot resting on a small panel upon which the name of the ship was engraved. Encircling all is the record IN MEMORY OF THE VICTORY OF MANILA BAY, MAY 1ST, 1898. (See facing page 368.)

German South-West Africa.—The German medal for

South-West Africa, 1904–6, marks a departure on the part of the Germans in the application of art to the making of war medals. The severity which has usually distinguished them has disappeared, and we have in very low relief the richly helmeted head of Germania, surrounded by the inscription, SUDEWEST AFRICA 1904–06, and on the reverse the Imperial crown with ribbons surmounting the Gothic initial W. II, which is arranged above crossed swords, the whole being encircled by the inscription DEN SIEGREICHEN STREITERN. The medal was given in steel to non-combatants. The ribbon is white with a series of red horizontal lines down the centre, leaving a clear white margin, which is edged with black of an equal width, and to it is attached a gilt bar for services rendered in KALAHARI 1908. Like some of the medals awarded by American States, the bar is not attached to the medal, and is apt to get lost.

The Serbo-Turkish War.—Of this medal, with 1912 in the centre, 400,000 were struck and distributed to the officers and men who took part in the Serbo-Turkish campaign. Five medals were struck in gold for the Court. On the obverse is the Serbian Eagle, encircled by a laurel wreath upon the leaves of which are struck incuse the names of the principal battles, reading downwards on the left: Kumanovo, Ueskūb, Prilip, Adrianople, Welles, N. Pazar, Debar, Istip, Prizren et Medare; and reading downwards on the right: Monastir, Scutari, Ohrid, Dojran, Sjeniza, Tetovo, Ljesch, Elbassan, Durazzo, and Pristina. The names are in Serbian characters.

On the reverse in the foreground is represented a piece of artillery captured from the Turks, with the sun in its splendour above, the date 1912 being inscribed on the orb itself, and running with the line of the medal above is the Serbian inscription, in native characters, signifying "Kossovo avenged," the field of Kossovo being the place where the Serbians lost their independence in 1389. The medal, which bears the

Swiss modeller's name, Hugeunin Frères, is suspended from a silk-woven ribbon, with the national colours, red, blue, and white, repeated perpendicularly six times.

The Serbo-Bulgarian War.—The medal issued for the war with the Bulgarians is in the form of a Cross—41 millimetres in dimension. This is a very plain, bronze, gilt Cross, bearing on the obverse the royal monogram and surmounted by the royal crown, and on the reverse the date 1913. The decoration is suspended from a red ribbon with an edging of black. Five hundred thousand of these Crosses were issued to the officers and men who took part in the campaign, and four Crosses were struck in gold for the Court.

" For Courage."—The third medal is for valour, and was awarded to those officers and men who particularly distinguished themselves in the two campaigns. On the obverse is a strong figure of Obilitch, a warrior revered by the Serbian people. He has been their national hero since 1389, the ideal and the eternal spirit which inspired and encouraged the Serbian people to live for the time when they could throw off the yoke of the Turk. Around the bust of Obilitch is the inscription, in Serbian characters, Miloch Obilitch, National Hero, 1389. It will be noticed that the crest or chimera which surmounts the helmet of the hero helps to give the connection between the shaped ring for suspension and the decoration. On the reverse is the cross of chivalry, and a wreath of laurel encircling the legend, in Serbian characters, " For Courage." Crossed swords fittingly fill the gaps between the arms of the cross. The medal, modelled by M. Henri Hugeunin, is 36 millimetres in diameter, is suspended from a red silk ribbon. Seven of these decorations (see facing page 376) were struck in gold, 101 in silver, 4,000 in silver-plated bronze, and 27,000 in bronze gilt.

The Sanatory Cross.—The " Sanatory Cross," which is bronze gilt and enamelled, and 40 millimetres at its greatest dimension, was distributed to those men and women who rendered service in the medical depart-

FOR SERBO-TURKISH WAR, 1912.

FOR SERBO-BULGARIAN WAR, 1913.

SERBIAN MEDALS.

ment during the two campaigns. On the obverse is the Serbian eagle on an enamelled ground, and on the reverse the inscription in Serbian characters, " For care to the wounded and sick." The inscription also stands out on an enamelled circle. Four thousand five hundred were issued with the sky-blue ribbon arranged in triangular form for men, and 3,500 in tie form for women. (See facing page 376.)

French Veteran's Medal.—An interesting medal is that worn by French veterans who are members of the SOCIÉTÉ DE SECOURS MUTUELS FRANÇAISE DES EX MILITAIRES. It bears this inscription in a circle, and on the reverse a trophy of arms. It is suspended from a wreath of laurel, tied with flying ribbons, by a ribbon into which is woven the colours of various war ribbons.

APPENDIX I

PISTRUCCI'S WATERLOO MEDAL

In the year 1819, when the battle of Waterloo was still fresh in the minds of every one, the Prince Regent, who became George IV, being desirous of fitly commemorating the event, conceived the idea of doing so by the striking of a medal which by reason of its magnificence would do justice to the great achievement of arms. To this end a competition among the great artists of the day was suggested to the Royal Academicians, but they unanimously selected Flaxman to execute the commission. He produced a design which, approved by his peers, was passed on to Benedetto Pistrucci, a Roman who succeeded T. Wyon as chief engraver at the Royal Mint. Signor Pistrucci, however, refused to cut the dies, on the grounds that his abilities and position placed him above that of a mere die cutter. His claims were acceded to; Flaxman's designs were abandoned, and Pistrucci was commissioned to prepare a modelled design, which was "instantly honoured by the fullest and most flattering approbation of" the Prince Regent. He thereupon received instructions to cut the dies for a fee of £3,500, the price being based upon the assumption that the design contained as much work as thirty ordinary-sized medals, and certainly a glance at the reduced photographs facing page 80 —the medal is 5½ in. in diameter—will give some idea of the exceeding amount of work which the famous die cutter and gem carver put into his *chef d'œuvre.* The dies were finished in 1849; and of the great men who were to receive the medal in gold, the Duke of Wellington—the greatest of them all—was the only survivor when a lead impression was taken for submission to the Lords of the Treasury.

It is stated that the reason for the failure to strike from the massive dies was owing to the super difficulties which could not be overcome in any attempt to harden them. The real reason may, however, be found in our altered relationship with the Continental nations, and in consequence thereof the Government of the day had no desire to do anything which might give offence to our ally in 1850. It was felt, however, that Pistrucci's masterpiece should not be kept from the public, and the Lords of the Treasury handed the matrices to a Mr. Johnson of Alexander Terrace, Bayswater, who prepared a number of electrotypes, a pair of which have been kindly lent me for illustration herein by my friend and late student, Cecil Thomas, who, like Pistrucci, is one of the very few craftsmen who are equally at home in the cutting of steel or gem stones.

The Official Description.—"Both sides of the medal are treated allegorically, except the central part of the obverse, which represents the busts of the four allied Sovereigns, the Prince Regent, Emperor of Austria, Emperor of Russia, and King of Prussia, grouped together in profile. Around this group of actual portraits the figures constitute an allegorical and mythological allusion to the treaty of peace which was consequent upon the great triumph on the field of battle. The summit of the surrounding groupings presents Apollo in his car restoring the day. The rainbow zephyr and Iris follow the chariot of the sun in succession, but the zephyr is tending towards the earth, and scattering flowers, as the emblem of peace and tranquillity.

"On the opposite side, the car of Apollo is seen closely approaching the constellation Gemini, personified, as usual, by a pair of graceful youths, indicating the month in which the great contest took place. Castor and Pollux, each armed with spears, are intended to elucidate the apotheosis of Wellington and Blucher. Themis, the goddess of justice, appears on earth, as in the Golden Age. This figure is placed in front of the profile busts of the Sovereigns, to show that Justice is a greater security to government than Power. The goddess is seated on a rock; a palm-tree waves over her head; she is prepared to reward virtue with its branches in one hand, and in the other holds a sword for the ready punishment of crime. Power is personified by a robust man of mature age, bearded, and armed with a club; he is seated under an oak-tree, and forms the corresponding figure, at the back of the group of busts of the Allied Sovereigns, to that of Justice facing it. Beneath Themis the Fates are introduced, to indicate that henceforward human actions will be controlled by Justice alone. These actions and passions are represented by the Furies, which, being placed beneath the emblematical figure of Power, are subjected to its influence, and no longer suffered to quit the infernal regions, or Cimmerian caverns, in which, at the base of this side of the medal, the allegory is completed by the figure of Night, the mother of the Fates, receding into darkness from the ruling daylight of Phœbus' car on the summit.

The central group on the reverse consists of a couple of equestrian figures, classically treated, but having the countenances of Wellington and Blucher. They are full of action; the figure personifying the Hero of Waterloo is galloping in advance, and that of the veteran Blucher is rushing to the aid of his companion in glory, to complete the enemy's destruction. They are guided by a female figure of a flying Victory, placed between them, conducting their horses to the conflict. Quite detached from this central goup, and forming a border round it, a composition of many figures represents the battle of the Giants. They are struck down by the thunder of Jupiter; the youngest ones, being the most daring in the assault of heaven, are the first to receive the Divine punishment. In their descent they tumble over one another in every variety of attitude—symbolical of the confusion of the defeated enemy. The number of the figures of the Giants is nineteen, illustrative of the nineteen years' duration of the war; and in grouping these figures they are represented following each other in succession.

MEDAL FOR COURAGE.

MEDAL FOR SERVICE TO THE WOUNDED.

SERBIAN MEDALS.

APPENDIX II

REGIMENTAL DESIGNATIONS

Cavalry

Name.	Former Title.	Present Title.
Life Guards . .	1st and 2nd . .	Same
Royal Horse Guards	The "Blues" . .	,,
1st Dragoon Guards	The King's . .	,,
2nd ,, ,,	Queen's Bays . .	,,
3rd ,, ,,	Prince of Wales' . .	,,
4th ,, ,,	Royal Irish . .	,,
5th ,, ,,	Princess Charlotte of Wales	,,
6th ,, ,,	Carabineers . . .	,,
7th ,, ,,	Princess Royal's . .	,,
1st Dragoons . .	Royal	,,
2nd ,, . .	Royal North British (Scots Greys) . .	Royal Scots Greys
3rd Hussars . .	Light Dragoons . .	The King's Own
4th ,, . . .	,, ,,	The Queen's Own
5th Lancers . .	Royal Irish . .	Same
6th Dragoons .	Inniskilling . .	,,
7th Hussars . .	Light Dragoons . .	The Queen's Own
8th ,, . . .	,, ,, . .	King's Royal Irish
9th Lancers . .	,, ,, . .	The Queen's Royal
10th Hussars. .	,, ,, . .	Prince of Wales' Own Royal
11th ,, . .	,, ,, . .	Prince Albert's Own
12th Lancers . .	,, ,, . .	Prince of Wales' Royal
13th Hussars . .	,, ,, . .	None
14th ,, . .	The King's . . .	Same
15th ,, . .	Light Dragoons	The King's
16th Lancers . .	The Queen's . .	Same
17th ,, . .	"Death or Glory Boys"	Duke of Cambridge's Own
18th Hussars . .	Light Dragoons . .	Queen Mary's Own
19th ,, . .	1st Bengal European Cavalry . . .	Princess of Wales' Own
20th ,, . .	2nd Bengal European Cavalry . . .	
21st Lancers . .	3rd Bengal European Cavalry . . .	Empress of India's

INFANTRY

Name.	Former Title.	Present Title.
Grenadier Guards .	1st, 2nd, and 3rd Battns.	Same
Coldstream Guards	1st and 2nd Battns.	,,
Scots Fusilier Guards	1st and 2nd Battns.	,,
Irish Guards . .		,,
Welsh Guards .		
1st Regt. . .	The Royal . . .	Royal Scots (Lothian Regt.)
2nd ,, . .	Queen's Royal . .	Royal West Surrey Regt.
3rd ,, . .	East Kent—The Buffs .	East Kent Regt.
4th ,, . .	King's Own Royal .	Royal Lancaster Regt.
5th ,, . .	Northumberland Fusiliers	Same
6th ,, . .	Royal 1st Warwickshire.	Royal Warwickshire Regt.
7th ,, . .	Royal Fusiliers . .	City of London Regt.
8th ,, . .	The King's . . .	Liverpool Regt.
9th ,, . .	East Norfolk . .	Norfolk Regt.
10th ,, . .	North Lincolnshire .	Lincolnshire Regt.
11th ,, . .	North Devonshire . .	Devonshire Regt.
12th ,, . .	East Suffolk . .	Suffolk Regt.
13th ,, . .	Prince Albert's Light Inf.	Prince Albert's (Somersetshire Light Inf.)
14th ,, . .	Buckinghamshire . .	Prince of Wales' Own (West Yorkshire Regt.)
15th ,, . .	Yorkshire, East Riding .	East Yorkshire Regt.
16th ,, . .	The Bedfordshire . .	Bedfordshire Regt.
17th ,, . .	Leicestershire . .	Leicestershire Regt.
18th ,, . .	Royal Irish . . .	Royal Irish Regt.
19th ,, . .	1st Yorkshire, N. Riding .	Alexandra, Princess of Wales' Own (Yorkshire Regt.)
20th ,, . .	East Devonshire . .	The Lancashire Fusiliers
21st ,, . .	Royal North British Fus.	Royal Scots Fusiliers
22nd ,, . .	The Cheshire . .	Cheshire Regt.
23rd ,, . .	Royal Welsh Fusiliers .	Same
24th ,, . .	2nd Warwickshire. .	South Wales Borderers
25th ,, . .	King's Own Borderers .	King's Own Scottish Borderers

N.B.—All the above Infantry Regiments had two Battalions.

26th Regt. . .	The Cameronian . .	1st Cameronians (Scottish Rifles)
27th ,, . .	Inniskilling . . .	1st Royal Inniskilling Fusiliers
28th ,, . .	North Gloucestershire .	1st Gloucestershire Regt.
29th ,, . .	Worcestershire . .	1st Worcestershire Regt.
30th ,, . .	Cambridgeshire . .	1st E. Lancashire Regt.
31st ,, . .	Huntingdonshire . .	1st E. Surrey Regt.
32nd ,, . .	Cornwall Light Inf. .	1st Duke of Cornwall's Light Inf.

Name.	Former Title.	Present Title.
33rd Regt.	Duke of Wellington's	1st West Riding Regt.
34th ,,	Cumberland	1st Border Regt.
35th ,,	Royal Sussex	1st Royal Sussex
36th ,,	Herefordshire	2nd Worcestershire Regt.
37th ,,	North Hampshire	1st Hampshire Regt.
38th ,,	1st Staffordshire	1st S. Staffordshire Regt.
39th ,,	Dorsetshire	1st Dorsetshire Regt.
40th ,,	2nd Somersetshire	Prince of Wales' Volunteers 1st S. Lancashire Regt.
41st ,,	The Welsh	1st Welsh Regt.
42nd ,,	Royal Highland (Black Watch)	1st Black Watch (Royal Highlanders)
43rd ,,	Monmouthshire Light Inf.	1st Oxfordshire Light Inf.
44th ,,	East Essex	1st Essex Regt.
45th ,,	Nottinghamshire (Sherwood Foresters)	1st Sherwood Foresters (Derbyshire Regt.)
46th ,,	South Devonshire	2nd Duke of Cornwall's Light Inf.
47th ,,	Lancashire	1st Loyal N. Lancashire Regt.
48th ,,	Northamptonshire	1st Northamptonshire Regt.
49th ,,	Princess Charlotte of Wales' (Herts)	1st Royal Berks Regt.
50th ,,	The Queen's Own	1st Royal W. Kent Regt.
51st ,,	2nd Yorkshire, West Riding (King's Own Light Infantry)	1st King's Own (Yorkshire Light Inf.)
52nd ,,	Oxfordshire Light Inf.	2nd Oxfordshire Light Inf.
53rd ,,	Shropshire	1st The King's (Shropshire Light Inf.)
54th ,,	West Norfolk	2nd Dorsetshire
55th ,,	Westmoreland	2nd The Border
56th ,,	West Essex	2nd Essex
57th ,,	West Middlesex	1st Duke of Cambridge's Own (Middlesex)
58th ,,	Rutlandshire	2nd Northamptonshire
59th ,,	2nd Nottinghamshire	2nd East Lancashire
60th ,, (4 Battns.)	King's Royal Rifle Corps	Same
61st ,,	South Gloucestershire	2nd Gloucestershire
62nd ,,	Wiltshire	1st Duke of Edin. (Wilts)
63rd ,,	West Suffolk	1st Manchester
64th ,,	2nd Staffordshire	1st Prince of Wales' (N. Staffs)
65th ,,	2nd Yorkshire, North Riding	1st York and Lancaster
66th ,,	Berkshire	2nd Princess Charlotte of Wales' (Royal Berks)

Name.	Former Title.	Present Title.
67th Regt.	South Hampshire	2nd Hampshire
68th ,,	Durham Light Inf.	1st Durham Light Inf.
69th ,,	South Lincolnshire	2nd Welsh
70th ,,	Surrey	2nd East Surrey
71st ,,	Highland Light Inf.	1st Highland Light Inf.
72nd ,,	Duke of Albany's Own Highlanders	1st Seaforth Highlanders, Ross-shire Buffs (Duke of Albany's)
73rd ,,	Perthshire	2nd Black Watch (Royal Highlanders)
74th ,,	Highland Regt.	2nd Highland Light Inf.
75th ,,	Stirlingshire Regt.	1st Gordon Highlanders
76th ,,	None	2nd Duke of Wellington's (W. Riding)
77th ,,	East Middlesex	2nd Duke of Cambridge's Own (Middlesex)
78th ,,	Highland Regt. (Ross-shire Buffs)	2nd Seaforth Highlanders (Ross-shire Buffs), Duke of Albany's
79th ,,	Cameron Highlanders	Queen's Own Cameron Highlanders
80th ,,	Staffordshire Volunteers	2nd South Staffordshire
81st ,,	Loyal Lincoln Volunteers	2nd Loyal N. Lancaster
82nd ,,	Prince of Wales' Volunteers	2nd Prince of Wales' Volunteers (South Lancaster)
83rd ,,	County Dublin	1st Royal Irish Rifles
84th ,,	York and Lancaster	2nd York and Lancaster
85th ,,	Bucks Volunteers (King's Light Inf.)	2nd The King's (Shropshire Light Inf.)
86th ,,	Royal County Down	2nd Royal Irish Rifles
87th ,,	Royal Irish Fusiliers	1st Princess Victoria's (Royal Irish Fusiliers)
88th ,,	Connaught Rangers	1st Connaught Rangers
89th ,,	Princess Victoria's	2nd Princess Victoria's (Royal Irish Fusiliers)
90th ,,	Perthshire Volunteers (Light Inf.)	2nd Cameronians (Scottish Rifles)
91st ,,	Argyllshire Highlanders	1st Princess Louise's (Argyll and Sutherland Highlanders)
92nd ,,	Gordon Highlanders	2nd Gordon Highlanders
93rd ,,	Sutherland Highlanders.	2nd Princess Louise's (Argyll and Sutherland Highlanders)
94th ,,	None	2nd Connaught Rangers
95th ,,	Derbyshire	2nd Sherwood Foresters (Derbyshire)
96th ,,	None	2nd Manchester

GREEK MEDAL FOR GRECO-TURKISH WAR, 1913. (Obverse.)

GERMAN MEDAL FOR SOUTH-WEST AFRICA, 1904-6.

GREEK MEDAL FOR GRECO-TURKISH WAR, 1913. (Reverse.)

Name.	Former Title.	Present Title.
97th Regt. . .	Earl of Ulster's . .	2nd Queen's Own (Royal W. Kent)
98th ,, . .	None	2nd Prince of Wales' (N. Staffordshire)
99th ,, . .	Lanarkshire . . .	2nd Duke of Edinburgh's (Wilts)
100th ,, . .	Prince of Wales' Royal Canadians	1st Prince of Wales' (Leinster Royal Canadians)
101st ,, . .	Royal Bengal Fusiliers .	1st Royal Munster Fusiliers
102nd ,, . .	Royal Madras ,, .	1st Royal Dublin Fusiliers
103rd ,, . .	Royal Bombay ,, .	2nd Royal Dublin Fusiliers
104th ,, . .	2nd Bengal Fusiliers .	2nd Royal Munster Fusiliers
105th ,, . .	2nd Madras Light Inf. .	2nd King's Own (Yorks Light Inf.)
106th ,, . .	2nd Bombay ,, ,, .	2nd Durham Light Inf.
107th ,, . .	3rd Bengal Inf. . .	2nd Royal Sussex
108th ,, . .	3rd Madras Light Inf. .	2nd Royal Inniskilling Fusiliers
109th ,, . .	3rd Bombay ,, ,, .	2nd Prince of Wales' Leinster (Royal Canadians)
Rifle Brigade (4 Battns.) . .	95th (Rifle Corps) Regt.	Prince Consort's Own

[Since the outbreak of war the number of battalions in all Regiments has been considerably increased.]

APPENDIX III

SALE PRICES

The following short record of average values is based upon auction prices realised during the last twelve years. It is difficult to standardise values, as so many factors are involved in the valuation of war medals. Not only rarity, but condition; not only the number issued, but the importance of the engagement in which they were earned, are taken into consideration by most collectors. Then there are periods during which there is a "slump," and prices do not rule so high. High prices are invariably obtained for the first medals placed upon the market even when, as in the case of the Boer War, it was known that thousands would be issued. Medals for South Africa which sold in 1903 for £1 5*s.* have now a fairly regular price of about 10*s.* if in good condition. In 1902 the Khedive's Sudan medal with bar for Khartoum realised £3 15*s.*; it is now valued at about 7*s.* 6*d.* in the sale-room. In the same year the Mayor of Kimberley's siege medal (a star of silver) sold for £6 15*s.*; it now realises from 10*s.* to £1. The last new British war medal to be issued, that for Abor, realised as much as four guineas in the sale-room when it first appeared; it dropped in a few weeks to three guineas. A Military General Service medal with the single bar for "Roleria" has fetched as much as £21—and as little as £3 3*s.*, but this was due to the fact that the latter was faked, *i.e.* some bars had been removed, the medal roll showing that the man whose name was impressed on the edge had served in five engagements, and his medal originally had that record.

On the other hand, because of its exceeding rarity as well as the gallant conditions under which it was earned, the M.G.S. medal with bar for "Benevente" has realised £34. Bars issued with the Army of India medal for "Ava" and "Bhurtpoor" are the most frequently met with, but these realise high prices at auction. At the Gaskell sale a medal with single bar for "Defence of Delhi" sold for £20, a single bar for "Corygaum" £35 (but one of these bars has realised as much as £100), two bars for "Assaye" and "Argaum" £33. Three bars for "Allighur," "Argaum," and "Gawilghur" realised £40, and £60 was paid for one with bars for "Allighur," "Battle of Delhi," and "Laswarree." The "Mutiny" medal has realised as much as £38, the price for which a four-bar medal from the Gaskell Collection sold at auction; but three-bar medals sell at auction for from £2 2*s.* to £3 3*s.*, and single bars for "Defence of Lucknow" for as little as £1 16*s.*; but at the present time there is an inclination to pay more for Mutiny medals, and the conditions under which they were earned warrant it. In the later Indian wars the bar issued for the North-East Frontier 1891, owing to the great interest taken in the Manipur Expedition, realised from £3 to £4, but this bar, which is attached to the I.G.S. 1854 Medal, sells for about a quarter of that sum now. The Waterloo medal, to one of the 42nd (Black Watch), has sold at auction for as much as £15, and £20 has been paid for one awarded to the 27th Regiment (Inniskillings); but a poor specimen has found a purchaser at as low as £1 10*s.*

Great care should be taken to examine the bars and ensure that they have not been tampered with. No medal that has been re-engraved is worth taking into consideration, unless it is one with rare bars which show no signs of faking. Medals with genuine bars, even though the name has been erased from the edge of the medal, realise good prices, and to complete a collection are very useful when named medals fetch very high prices.

Waterloo Medals

Infantry

	£	*s.*	*d.*		£	*s.*	*d.*
1st Guards (Grenadier)	2	2	0	to	3	15	0
2nd Guards (Coldstreams)	3	0	0	to	3	10	0
2nd Battn. 3rd Guards (Scots Guards) and M.G.S. medal. Two bars: Vittoria and St. Sebastian	.	.	.		8	5	0
2nd Battn. 3rd Guards (Scots Guards) . .	2	6	0	to	3	10	0
14th Foot	.	.	.		2	4	0
23rd Foot (Royal Welsh Fusiliers) . . .	.	.	.		3	10	0
27th Foot (Inniskillings)	1	10	0	to	20	0	0

	£	s.	d.		£	s.	d.
27th Foot. Private Sale					25	0	0
30th Foot (Sergeant)					12	0	0
33rd Foot					3	5	0
42nd Foot (Black Watch). (Has sold for £15.)	4	0	0	to	6	6	0
44th Foot (2nd Battn.)					10	0	0
52nd Foot (1st Battn.)					2	10	0
54th Foot					5	5	0
73rd Foot (now 2nd Black Watch)	4	0	0	to	7	15	0
79th Foot (1st Battn.) (Paymaster-Sergeant)					14	0	0
79th Foot (Camerons)					8	5	0
95th Foot (Rifles) (Captain)					12	0	0
Royal Waggon Train (Officer). Private Sale					20	5	0
Royal Horse Guards (Officer). Private Sale					25	0	0
Cavalry							
2nd Life Guards					5	15	0
2nd or R.N. British Dragoons (Scots Greys)	3	0	0	to	6	0	0
6th or Inniskilling Dragoons					5	0	0
7th Hussars	2	0	0	to	6	0	0
10th Royal Hussars					4	5	0
11th Light Dragoons					4	0	0
15th King's Hussars					2	18	0
23rd Light Dragoons					2	17	6
1st Light Dragoons (K.G.L.)	2	14	0	to	3	10	0
King's German Artillery (Gunner)					4	0	0
Hussars, Prinz Regt., Hanoverian Waterloo					1	13	0
Landwehr Battn., Verden, Hanoverian Waterloo					1	16	0
Gold Peninsular Medal for Salamanca							
With bars for Orthes and Toulouse, awarded to Major Beatty, of 7th Regt. (Royal Fusiliers)					115	0	0
Military General Service Medals, 1801 to 1814							
1st Foot Guards. Three bars: St. Sebastian, Nivelle, Nive					2	8	0
1st Foot Guards. One bar: Barossa, 1811					3	12	6
1st Foot Guards. One bar: St. Sebastian, 1813					3	15	0
2nd Coldstream Guards (Captain). Four bars: Roleria, Vimiera, Corunna, Talavera					15	0	0
1st Foot (Royal Scots). One bar: Martinique					12	15	0
3rd Foot (Lieutenant). Five bars: Pyrenees, Nivelle, Nive, Orthes, Toulouse					9	0	0
5th Foot (Sergeant). Nine bars					18	0	0
5th Foot. One bar: Roleria. Private Sale					37	0	0
6th Foot. One bar: Orthes, 1814					6	0	0
6th Foot. One bar: Vittoria, 1813					2	10	0
6th Foot. Three bars					7	0	0
7th Foot. Twelve bars (verified)					36	0	0

THE MÉDAILLE MILITAIRE.

Awarded to Sir John French and non-commissioned officers and men of the British Army.

	£	s.	d.		£	s.	d.
7th Foot (Sergeant). Eight bars					23	10	0
9th Foot (Ensign). Four bars: Vittoria, St. Sebastian, Nivelle, Nive					8	10	0
9th Foot. One bar: Roleria					12	0	0
11th Foot. One bar: Busaco					9	0	0
14th Foot. One bar: Java, 1811					1	4	0
23rd Foot (Sergeant). Ten bars					38	0	0
25th Foot. One bar: Guadaloupe					13	15	0
27th Foot. One bar: Maida, 1806					6	10	0
27th Foot. Two bars: Egypt, 1801; Maida, 1806					7	0	0
29th Foot (Lieutenant). Five bars					16	0	0
30th Foot. One bar: Fuentes d'Onor					13	10	0
31st Foot (Quartermaster). Seven bars: Talavera, Albuhera, Vittoria, Pyrenees, Nive, Orthes, Toulouse					8	10	0
32nd Foot (Sergeant). Two bars: Roleira, Vimiera					3	15	0
36th Foot. One bar: Vimiera, 1808					5	10	0
38th Foot. Three bars: Vimiera, Corunna, Salamanca					3	7	6
39th Foot. Two bars: Vittoria, Toulouse					1	14	0
40th Foot. One bar: Ciudad Rodrigo					18	10	0
40th Foot. Seven bars: Vimiera, Talavera, Albuhera, Toulouse, Ciudad Rodrigo, Vittoria, Pyrenees	5	0	0	to	7	7	0
41st Foot. Two bars: Vittoria, St. Sebastian					2	4	0
42nd Highlanders. One bar: Corunna					4	4	0
42nd Highlanders. Ten bars	10	10	0	to	13	13	0
42nd Highlanders. Six bars					7	7	0
43rd Foot. Two bars: Vimiera, Corunna					3	12	6
43rd Foot. Eight bars					12	10	0
43rd Foot. Nine bars					15	0	0
45th Foot. One bar: Talavera, 1809					4	12	6
47th Foot. Four bars: Barossa, Vittoria, St. Sebastian, Nive					4	10	0
48th Foot. One bar: Talavera, 1809					3	17	6
48th Foot. Ten bars					38	0	0
48th Foot. Thirteen bars					30	0	0
52nd Foot. Eight bars: Fuentes d'Onor, Ciudad Rodrigo, Badajoz, Salamanca, Vittoria, Pyrenees, Orthes, Toulouse	6	0	0	to	12	15	0
59th Foot (Drummer). Four bars					5	0	0
60th Foot (Lieutenant). Two bars: Martinique, Guadaloupe					9	15	0
63rd Foot. One bar: Martinique, 1809					2	17	6
74th Foot. Seven bars					11	11	0
74th Regt. (H.L.I.). Nine bars: Busaco, Fuentes d'Onor, Ciudad Rodrigo, Salamanca, Vittoria, Pyrenees, Nivelle, Orthes, Toulouse	7	0	0	to	12	12	0
77th Foot. One bar: Ciudad Rodrigo					15	0	0
78th Foot. One bar: Java, 1811					4	12	6

	£ s. d.	£ s. d.
83rd Foot. Ten bars: Talavera, Busaco, Fuentes d'Onor, Ciudad Rodrigo, Badajoz, Vittoria, Pyrenees, Nivelle, Orthes, Toulouse		12 0 0
84th Foot. One bar: Nive	2 10 0 to	3 3 0
88th Foot (Lieutenant). Four bars: Talavera, Busaco, Fuentes d'Onor, Ciudad Rodrigo		6 6 0
89th Foot. One bar: Chrystler's Farm		20 0 0
91st Foot. Six bars		7 10 0
91st Foot. One bar: Vimiera		16 0 0
95th Foot (Officer). Six bars		14 10 0
95th Foot. Seven bars		10 10 0
97th Foot. Four bars: Roleria, Vimiera, Busaco, Albuhera		4 17 6
Royal Staff Corps (Sergeant). Three bars: Roleria, Vimiera, Corunna		3 5 0
Staff Surgeon. One bar: Badajoz		17 0 0
Royal Sappers and Miners. One bar: Badajoz, 1812		3 3 0
Artillery Driver. One bar: Fuentes d'Onor, 1811		10 0 0
Artillery Driver. Nine bars		14 10 0
Royal Artillery (Gunner). Three bars: Pyrenees, St. Sebastian, Toulouse		2 10 0
Waggon Train. One bar: Ciudad Rodrigo. Only single bar awarded		23 10 0
Canadian Militia. One bar: Fort Detroit	16 0 0 to	19 10 0
Native Warrior. Three bars: Fort Detroit, Chateauguay, Chrystler's Farm		40 0 0
Royal Garrison Artillery. One bar: Fort Detroit		26 0 0
Native Warrior. One bar: Chateauguay		10 5 0
One bar: Chateauguay		10 10 0
One bar: Fort Detroit		14 10 0
One bar: Chrystler's Farm		13 0 0

Cavalry

	£ s. d.	£ s. d.
4th Light Dragoons. Six bars: Talavera, Albuhera, Busaco, Salamanca, Vittoria, Toulouse		5 5 0
10th Hussars. Four bars		17 0 0
11th Light Dragoons. One bar: Salamanca, 1812		2 17 6
13th Light Dragoons. Four bars		27 0 0
14th Light Dragoons. Ten bars: Talavera, Busaco, Fuentes d'Onor, Badajoz, Salamanca, Vittoria, Pyrenees, Nive, Orthes, Toulouse		8 15 0
14th Light Dragoons. Ten bars: Talavera, Busaco, Fuentes d'Onor, Badajoz, Vittoria, Pyrenees, Nivelle, Orthes, Toulouse		7 0 0
15th Light Dragoons. One bar: Sahagun. Private Sale		35 0 0
16th Light Dragoons. Four bars: Talavera, Busaco, Fuentes d'Onor, Vittoria		3 12 6

	£ s. d.	£ s. d.
18th Light Dragoons. Two bars: Sahagun, Toulouse		16 0 0
20th Light Dragoons. One bar: Vimiera		12 0 0
22nd Light Dragoons. One bar: Egypt, 1801		3 5 0
One bar: Sahagun and Benevente	8 10 0 to	18 0 0
Three bars: Vittoria, Sahagun, Benevente		25 0 0
Ten bars (Officer)		50 0 0

Gold Cross

For the Peninsular, with three clasps for seven battles, awarded to Lieutenant-Colonel Sir Charles Sutton		370 0 0

Gold Medal

For Vimiera		72 0 0

Portuguese Cross

For five engagements in Peninsular. Ensign 5th Foot		30 0 0

Naval General Service Medal

One bar: "Nymphe," June 18th, 1793		35 0 0
Four bars: June 1st, 1794, St. Vincent, St. Domingo, Martinique (a rare combination)		61 0 0
One bar: June 1794		2 0 0
One bar: March 14th, 1795		16 10 0
One bar: "Lapwing," December 3rd, 1796 (two issued)		21 0 0
One bar: San Fiorenzo, March 8th, 1797 (three issued)		50 0 0
One bar: St. Vincent (349 issued)	1 16 0 to	6 15 0
Two bars: St. Vincent, Nile	3 3 0 to	10 5 0
One bar: "Mars," April 21st, 1798		15 15 0
One bar: Camperdown, 1797 (332 issued)		10 0 0
One bar: "Lion," July 15th, 1798 (21 issued)		23 0 0
One bar: "Sybille," February 28th, 1799 (12 issued)	12 0 0 to	40 0 0
One bar: "Harpy," February 5th, 1800		21 0 0
Two bars: "Phœbe," February 19th, 1801, St. Sebastian		25 10 0
One bar: Egypt (626 issued)		2 0 0
One bar: Boat Service, August 20th, 1800		30 0 0
One bar: Gut of Gibraltar, June 12th, 1801, 1st Lieutenant R.N.		3 0 0
One bar: November 4th, 1805		10 0 0
One bar: Curacoa		13 0 0
Two bars: St. Domingo, Algiers		1 18 0
One bar: Anson, August 23rd, 1806		20 0 0

	£. s. d.	£ s. d.
Two bars: Off Rota, April 4th, 1808, Syria, 1st Lieutenant R.N.	. .	15 10 0
One bar: Pompee, June 17th, 1809 (21 issued)	17 0 0 to	26 0 0
One bar: Boat Service, July 29th, 1809 (12 issued)	. .	26 0 0
One bar: Spartan, May 3rd, 1810 (32 issued)	. .	10 0 0
One bar: Banda Neira	. .	17 0 0
One bar: Cherokee, July 10th, 1810 (4 issued)	. .	26 0 0
One bar: "Boadicea," September 12th, 1810 (16 issued)	. .	40 0 0
Two bars: Martinique (523 issued), Guadaloupe (509 issued)	3 0 0 to	12 10 0
One bar: Anhalt, March 27th, 1811	. .	31 0 0
One bar: Java	. .	4 10 0
Two bars: Pelagosa, November 29th, 1811, St. Sebastian	. .	3 10 0
One bar: "Weasel," April 22nd, 1813	. .	36 0 0
Two bars: "Amazon," March 13th, 1806; Boat Service, January 6th, 1813	. .	38 0 0
Two bars: Boat Service, May 4th, 1811; Boat Service, December 14th, 1814, to Lieutenant	. .	12 0 0
One bar: Boat Service, December 14th, 1814	. .	6 10 0
One bar: "Cherub," March 28th, 1814	. .	10 0 0
One bar: "Venerable," January 16th, 1814	16 0 0 to	20 10 0
Three bars: Santa Margarita, June 8th, 1796 (3 issued); "Fisgard," October 20th, 1798 (9 issued); "Eurotas," February 25th, 1814 (29 issued), to Boatswain, a rare combination, very fine	. .	75 0 0
Two bars: San Fiorenzo, February 11th, 1805, March 8th, 1808	. .	40 0 0
Three bars: Basque Roads, Martinique, Guadaloupe	. .	7 10 0
Two bars: Java, St. Sebastian	. .	2 0 0
Three bars: November 4th, 1805; Basque Roads, 1809; Algiers	. .	15 15 0
One bar: Off Tamatave, 1811 (79 issued)	. .	16 10 0
Two bars: Java, Syria	. .	1 10 0
Two bars: "Sapho," Algiers	. .	40 0 0
One bar: St. Domingo	. .	1 10 0
One bar: Algiers (1,320 issued)	1 0 0 to	1 10 0
One bar: Java (715 issued)	1 10 0 to	2 0 0
One bar: St. Sebastian (292 issued)	. .	2 2 0
One bar: Syria, 1840 (7,057 issued)	. .	0 17 6
One bar: Navarino, 1827 (1,123 issued)	. .	2 0 0

Trafalgar Medals

	£. s. d.	£ s. d.
Officer's Gold Medal for Trafalgar (Captain H.M.S. "Prince"), sold in 1903	. .	245 0 0
One bar: Trafalgar, 1805 (1,710 issued)	2 0 0 to	3 10 0
Two bars: Trafalgar, Algiers	. .	2 12 0

	£ s. d.	£ s. d.
Two bars: Egypt, Trafalgar	1 10 0 to	2 10 0
Two bars: "Penelope," March 30th, 1800, Trafalgar, to Midshipman Carter		41 0 0
Two bars: Trafalgar, Boat Service, December 14th, 1814		4 5 0
Three bars: Copenhagen, Trafalgar, Algiers		18 0 0

Davison's Nile and Trafalgar Medals

Battle of Nile, 1798: Davison's Gold Medal, to Officers		100 16 0
Battle of Nile, 1798: Davison's Silver Medal, to Officers		5 10 0
Battle of Nile, 1798: Davison's Bronze Medal Gilt, to Petty Officers	0 12 0 to	1 10 0
Battle of Nile, 1798: Davison's Bronze Medal, to Seamen	0 10 0 to	1 5 0
Battle of Trafalgar, October 21st, 1805: Davison's Pewter	1 1 0 to	1 5 0

Medals for Seringapatam

Gold Medal for Siege of Seringapatam, May 4th, 1799		16 10 0
Officer's Silver Gilt Medal, Seringapatam	3 15 0 to	6 15 0
Officer's Silver Medal, Seringapatam	1 0 0 to	2 2 0
Bronze Medal for Seringapatam, 1799 (awarded to N.C.O.'s	1 1 0 to	1 10 0
Tin Medal, Siege of Seringapatam. Awarded to Private Soldiers	1 0 0 to	1 10 0

Honourable East India Co.'s Medals

H.E.I. Co.'s Small Deccan Medal, 1784		10 0 0
H.E.I. Co.'s Small Deccan Medal, 1784		13 5 0
H.E.I. Co.'s Mysore Medal, 1791-2		10 0 0
H.E.I. Co.'s Large Mysore Medal. Private Sale		20 0 0
H.E.I. Co.'s Java Medal, 1811		11 0 0
H.E.I. Co.'s Medal for Rodriques, Isle of France, 1810		14 10 0
First Burmese War: H.E.I. Co.'s Medal, 1824-6	4 10 0 to	5 5 0

Army of India Medals

Five bars: Allighur, Battle of Delhi, Laswarree, Battle of Deig, Capture of Deig—name impressed (to Officer). Private sale		150 0 0
Four bars: Asseerghur, Argaum, Gawilghur, Ara—name impressed (to Officer). Private sale		100 0 0
One bar: Corygaum		100 0 0

	£	s.	d.		£	s.	d.
Two bars: Laswarree, Battle of Delhi					55	0	0
One bar: Capture of Deig (to Officer of 8th Light Dragoons), only 48 issued to the Regiment. Private sale					50	0	0
Two bars: Laswarree, Capture of Deig. (8th Dragoons)					38	0	0
One bar: Kirkee					20	0	0
Two bars: Assaye and Nagpore					18	0	0
Two bars: Maheidpore and Nagpore					16	0	0
Two bars: Assaye and Asseerghur (Sergeant Irregular Cavalry)					15	10	0
One bar: Laswarree (Sepoy)					15	0	0
22nd Foot. One bar: Capture of Deig					75	0	0
38th Regt. Native Inf. One bar: Seetabuldee and Nagpore					74	0	0
2nd Native Infantry. One bar: Corygaum					70	0	0
94th Foot. Three bars: Asseerghur, Argaum, Gawilghur					65	0	0
Two bars: Kirkee and Poona, Corygaum					51	0	0
1st Foot. One bar: Nagpore					21	0	0
One bar: Poona					17	10	0
Horse Artillery. One bar: Poona					31	0	0
One bar: Nepaul, 1816 (Officers)	6	10	0	to	11	10	0
Native Infantry (Lieutenant). One bar: Nepaul	6	10	0	to	12	0	0
24th Foot. One bar: Nepaul					1	14	0
Sappers and Miners. One bar: Defence of Delhi					1	10	0
87th Foot. One bar: Ava, 1824–6					3	12	6
38th Foot. Two bars: Poona, Ava					23	0	0
89th Foot. One bar: Ava, 1824–6					3	17	6
Armourers' Mate. One bar: Ava (Naval)					4	12	0
Four Bars: Assaye, Argaum, Poona, Bhurtpoor					23	0	0
Three bars: Assaye, Argaum, Bhurtpoor					25	0	0
16th Light Dragoons. One bar: Bhurtpoor					3	12	6
14th Foot. One bar: Bhurtpoor					3	7	6
Native Infantry. One bar: Maheidpoor	12	10	0	to	15	10	0
11th Light Dragoons. One bar: Bhurtpoor					4	2	6
16th Lancers. One bar: Bhurtpoor					3	7	6
16th Lancers. Ghuznee, July 23rd, 1839	1	5	0	to	2	10	0
Ghuznee, Cabul, 1842					2	12	6
4th Com. 2nd Battn. Bengal Foot Artillery. Ghuznee, Cabul, 1842					5	5	0
3rd K.O. Light Dragoons. Cabul, 1842					2	8	0
40th Regt. Cabul, 1842	2	2	0	to	3	10	0
1st Troop Horse Brigade, Bombay Artillery. Ghuznee, Cabul, 1842: two wreaths					4	0	0
H.M. 40th Regt. Candahar, 1842					7	0	0
2nd Battn. Artillery. Candahar, 1842					1	10	0
Kelat-i-Ghilzie, 1842					14	0	0
Jellalabad, April 7th, 1842. First type, Mural crown, XIII. P.A.L.I.	7	15	0	to	13	10	0

	£ s. d.	£ s. d.
Jellalabad, April 7th, 1842. Second type, "Flying Victory," to XIII. P.A.L.I.		15 0 0
40th Regt. Candahar, Ghuznee, Cabul, 1842	2 8 0 to	2 15 0
Meeanee, 1843		7 10 0
Meeanee (Major)		20 0 0
Meeanee (Officer)		23 0 0
22nd Regt. Meeanee. Hyderabad, 1843	3 10 0 to	4 4 0
1st Troop Horse Artillery. Hyderabad, 1843		5 0 0
Sutlej Campaign (Quartermaster-Sergeant). Three bars: Ferozeshuhur, Aliwal, Sobraon, 1845		2 17 6
3rd Light Dragoons. Sutlej Campaign. Two bars: Ferozeshuhur, Sobraon		2 2 0
50th Regt. Sutlej. Three bars: Ferozeshuhur, Sobraon, Aliwal		2 2 0
44th Regt. N.I. (Assistant Surgeon). Ferozeshuhur		2 18 0
3rd Light Dragoons. Moodkee, 1845. Two bars: Ferozeshuhur, Sobraon		3 3 0
31st Regt. Moodkee. Three bars: Ferozeshuhur, Aliwal, Sobraon		3 0 0

Naval Meeanee, 1843

Private Sale		45 0 0

New Zealand Medals

Naval

1845–6	3 10 0 to	4 0 0
1860–1	1 10 0 to	2 10 0
1863–4	1 0 0 to	1 10 0

Military

2nd Coast Brigade R.A. 1860–5	2 10 0 to	3 5 0
12th Regt. 1860–6		8 0 0
18th Royal Irish Rifles. 1863–6	1 0 0 to	1 15 0
2nd 14th Regt. 1861–6	1 0 0 to	1 10 0
40th Regt. 1860–4	3 7 6 to	3 15 0
40th Regt. 1863–4	1 1 0 to	2 10 0
70th Regt. 1863–5		1 5 0
Royal Artillery. 1863–6	1 5 0 to	1 15 0
Military Train. 1864	3 2 6 to	3 15 0
43rd Light Infantry. 1864–6	1 7 0 to	1 10 0
57th Regt. 1865–6	2 0 0 to	2 10 0
65th Regt. 1860–1–5		2 10 0
65th Regt. 1865		6 0 0

China Medals

	£ s. d.	£ s. d.
China, 1842. Lieutenant 26th Regt. Foot	. . .	2 2 0
China, 1842. A.B. Naval	0 10 0 to	0 18 6
China, 1857. Two bars: Unnamed	0 12 6 to	0 15 0
China, 1860. Two bars: Unnamed	0 12 6 to	0 15 0
China, 1860. One bar: Taku Forts, 1860: Acting Master H.M.S. "Bernici," I.N.	. . .	1 10 0
China, 1900. One bar: Defence of Legations. Lance-Corporal R.M.L.I.	. . .	9 10 0
Defence of Legations. Private Sale. (Naval)	. . .	10 5 0
China, 1900. Relief of Pekin. (Naval)	. . .	0 15 0
China, 1900. No bar. (Naval)	. . .	0 7 6
China, 1900. No bar. Nurse	. . .	0 15 0

Crimea Medals

Military

	£ s. d.	£ s. d.
1st Royal Dragoons, Heavy Brigade (Quartermaster). Three bars: Balaklava, Inkermann, Sebastopol, Turkish Crimea medal, and Miniature medals of same	. . .	6 5 0
17th Lancers, Light Brigade. Four bars: Alma, Balaklava, Inkermann, Sebastopol, and Turkish medal	. . .	5 5 0
Grenadier Guards. One bar: Inkermann	. . .	0 8 6
Coldstream Guards (Sergeant). Four bars: Alma, Balaklava, Inkermann, Sebastopol	1 10 0 to	1 15 0
Coldstream Guards (Private). Four bars: Alma, Balaklava, Inkermann, Sebastopol	1 7 0 to	1 12 0
2nd Dragoon Guards. One bar: Balaklava	0 15 0 to	1 5 0
2nd Dragoons, Heavy Brigade. Three bars: Balaklava, Inkermann, Sebastopol; Turkish Crimea and Long Service medal	. . .	2 10 0
8th Hussars. One bar: Inkermann	. . .	0 7 6
18th Regt. (Drum-major). One bar: Sebastopol	. .	0 12 0
Royal Artillery (Gunner). Two bars: Inkermann, Sebastopol	. . .	0 17 0
55th Regt. One bar: Sebastopol	. . .	0 5 0
57th Regt. Two bars: Balaklava, Inkermann	. . .	1 0 0
79th Cameron Highlanders. Two bars: Balaklava, Sebastopol	. . .	1 5 0
Turkish Contingent (English Surgeon). Turkish Crimea	. . .	0 15 0

Naval

	£ s. d.	£ s. d.
One bar: Sebastopol. Schoolmaster	. . .	0 16 0
One bar: Inkermann (A.B.)	. . .	0 12 0
Two bars: Inkermann, Sebastopol (A.B.)	. . .	0 10 0
Two bars: Inkermann, Sebastopol (Boy)	. . .	0 15 0

Indian Mutiny Medals

	£ s. d.	£ s. d.
1st Battn. 5th Fusiliers. Two bars: Defence of Lucknow, 1857	. .	2 4 0
9th Lancers. Three bars: Delhi, Relief of Lucknow, Lucknow	2 2 0 to	2 12 0
32nd Light Infantry. One bar: Defence of Lucknow	. .	1 16 0
78th Highlanders. Defence of Lucknow . .	. .	1 12 0
84th Regt. (Corporal). Defence of Lucknow, Lucknow.	. .	1 15 0
90th Light Infantry. Two bars: Defence of Lucknow	. .	1 15 0
Royal Horse Artillery (Lieutenant). Two bars: Lucknow, Relief of Lucknow . . .	. .	4 12 6
1st Sikhs Infantry (Assistant Surgeon). No bar	. .	2 18 0

Naval Brigade

	£ s. d.	£ s. d.
A. B. "Pearl." No bar	. .	1 16 0
A.B. 1st Naval Brigade H.M.P.V., "Calcutta" .	. .	2 4 0
A. B. "Shannon." No bar.	. .	1 12 0
A.B. "Shannon," 1857. Two bars: Lucknow, Relief of Lucknow	. .	2 14 0
A.B. "Shannon," 1857. One bar: Lucknow, "Shannon"	. .	5 5 0

African Medals

	£ s. d.	£ s. d.
2nd Regt. South Africa, 1853	0 12 6 to	0 15 0
South Africa, 1853. (Naval)	. .	0 15 0
H.M.S. "Bittern." One bar: Commassie, 1873	. .	0 15 0
42nd Black Watch. One bar: Commassie	0 12 6 to	1 0 0
1st Royal Highlanders. Five bars: Egypt, 1882; Tel-el-Kebir; Suakin; 1884; El Teb, The Nile, 1884–5; Kirbekan	. .	3 3 0
Gordon Highlanders. Four bars: Egypt, 1882; Tel-el-Kebir, Suakin, 1884–5; Khedive's Bronze Star. Long Service medal . .	. .	1 8 0
1st Royal Highlanders (Black Watch). Two bars: Tel-el-Kebir, Suakin, 1884	. .	0 10 6
1st Royal Highlanders. Two bars: El Teb, The Nile, 1884–5	. .	0 12 6
1st Royal Highlanders. Two bars: El Teb, Tamaai	. .	0 14 0
1st Royal Highlanders. Five bars: Tel-el-Kebir; Suakin, 1884, El Teb; The Nile, 1884–5; Kirbekan	. .	3 10 0
Royal Artillery (Gunner). Three bars: Suakin, 1884; El Teb; The Nile, 1884–5 . . .	. .	1 0 0
Naval. One bar: Witu, 1890	. .	1 1 0
Victoria Column (Trooper). British South Africa Co.'s medal for Matabeleland, 1893 . .	. .	2 2 0

	£ s. d.	£ s. d.
Raaf's Column (Trooper). Matabeleland, 1893, with bar: Rhodesia, 1896. (In 1901 a medal of this kind realised at auction £10) .	. .	3 5 0
32nd Bengal Infantry (Sepoy). Central Africa, 1893, with swivel and ring suspender . .	. .	3 0 0
Royal Engineers (Sergeant). West Africa. One bar: 1893–4	. .	2 18 0
B.C.A. Rifles. Central Africa. One bar: 1894–8	. .	13 13 0
2nd York and Lancs Regt. British South Africa Co.'s medal for Rhodesia, 1896	. .	1 10 0
Kimberley Regt. (Lieutenant). Cape General Service. One bar: Bechuanaland 1896–7 .	. .	3 0 0
Cape Mounted Rifles. Cape General Service. One bar: Bechuanaland	. .	1 13 0
P.A. Guards. One bar: Basutoland, 1896–7 .	. .	1 16 0
Cape Town Rangers. Cape Colony Service medal. Two bars: Transkei, 1896–7; Basutoland	. .	2 12 0
D.E.O.V.R. Cape General Service. One bar: Basutoland, 1896–7	. .	1 8 0
Raaf's Column. British South Africa Co.'s medal for Matabeleland. One bar: Rhodesia, 1896	. .	4 15 0
Gwelo Volunteers (Trooper). British South Africa's Co.'s medal for Rhodesia, 1896. One bar: Mashonaland, 1897	. .	2 8 0
Transport (Corporal). Two bars: Transkei, Bechuanaland, 1896–7	. .	4 5 0
P.A. Guards. Cape General Service. One bar: Bechuanaland, 1896–7	. .	2 0 0
Niger Coast Protectorate (Lieutenant). West Africa. One bar: Benin, 1897	. .	1 16 0
H.M.S. "St. George" (Gunner R.M.A.). West Africa. One bar: Benin, 1897 . . .	. .	0 15 6
British South African Police. Mashonaland, 1897	. .	2 12 6
27th Bombay Light Infantry. East and Central Africa. One bar: Uganda, 1897–8 . .	. .	1 12 0
Africa Field Force (Doctor). Ashanti, 1900 .	. .	3 17 6
West Africa Field Force. Ashanti, 1900. One bar: Kumassi	2 7 6 to	2 17 6
Nigeria Regt. African General Service medal. One bar: North Nigeria, 1900–1 . .	. .	1 6 0
16th Bombay Light Infantry. One bar: Jubaland, 1900–1	. .	1 6 0
Naval. One bar: Jubaland, 1900–1 . .	0 15 0 to	1 4 0
Lagos Battn. West Africa Field Force. Two bars: Aro, 1901–2; North Nigeria, 1903 .	. .	2 10 0
Royal Warwick Regt. One bar: Somaliland, 1902–4	. .	1 2 0
52nd Sikhs. Africa General Service medal. Two bars: Somaliland, 1902–4; Jidballi . .	0 17 0 to	1 1 0

	£ s. d.	£ s. d.
Nigeria Regt. One bar: North Nigeria, 1902		1 7 0
1/A Northern Nigeria Regt. (Gunner). Two bars: Northern Nigeria, 1903–4; Nigeria, 1904		4 0 0
P.C.E. A.P.F. One bar: East Africa, 1905		3 5 0
3rd King's African Rifles (A.G.S.). One bar: East Africa, 1906		3 0 0
Southern Nigeria Regt. One bar: West Africa, 1906		3 3 0
Northern D. M. Rifles. Natal Rebellion. One bar: 1906		1 0 0
H.M.S. "Hyacinth" (A.B.). One bar: Somaliland, 1908–10		2 4 0
H.M.S. "Fox." One bar: Somaliland, 1908–10		2 0 0
The British Niger Co.'s Bronze medal with bar, Nigeria, 1886–97		2 10 0
H.M.S. "Sparrow" (A.B.). Two bars: Witu, 1891–2; August, 1893		2 10 0
H.M.S. "Swallow" (R.M.). West Africa. One bar: Witu, August 1893		1 16 0
H.M.S. "Swallow." One bar: Witu, August 1893		1 6 0
H.M.S. "Electo." Two bars: Gambia, 1894; Benin, 1897		1 10 0
One bar: Gambia, 1894		0 15 0
H.M.S. "Forte" (A.B.). One bar: Gambia. Only 15 issued to officers, and 22 to men		8 0 0
West India Regt. One bar: Gambia	1 0 0 to	1 5 0
H.M.S. "Barossa." Two bars: Brass River, 1895; Benin, 1897. (The medal with same bars to same ship realised £5 7*s.* 6*d.* in 1901)		1 5 0
H.M.S. "Philomel." Somaliland, 1908–10		3 10 0
3rd King's African Rifles (A.G.S.). Two bars: East Africa, 1906, Somaliland, 1908–10		4 0 0

Canada Medals for 1866–70

25th Regt. One bar: Fenian Raid, 1866.	1 18 0
Light Infantry. One bar: Fenian Raid, 1866	1 15 0
1st Rifle Brigade. Two bars: Fenian Raid, 1866; Fenian Raid, 1870	3 10 0
7th Royal Fusiliers. One bar: Fenian Raid, 1866. Bugler	1 8 0
Montreal G.A. One bar: Fenian Raid, 1870. Gunner	2 0 0

North-West Canada, 1885 Medal

One bar: Saskatchewan (Bugler)	2 0 0
One bar: Saskatchewan (Naval), awarded to men from the steamer "Northcote" who took part in the Battle of Batouche. (Rare)	10 10 0

Boer War Medals

	£ s. d.	£ s. d.
H.M.S. "Powerful" (A.B.). Five bars: Belmont, Modder River, Relief of Kimberley, Paardeberg, Driefontein	. .	2 5 0
H.M.S. "Doris" (A.B.). Five bars: Cape Colony, Paardeberg, Driefontein, Diamond Hill, Belfast	. .	1 18 0
Cape Mounted Rifles. Four bars: Cape Colony, Wepener, Transvaal, Wittebergen . .	. .	1 11 0
Horse Artillery (Gunner). Three bars: Cape Colony, Orange Free State, Transvaal .	. .	1 10 0
Lancs Fusiliers. Seven bars: 1900 . . .	. .	0 17 6
Dublin Fusiliers. Six bars: 1900 . .	0 10 0 to	0 15 0
Rhodesia Regt. (Trooper). Four bars: Rhodesia. Relief of Mafeking, Orange Free State, Transvaal	. .	1 15 0
South African Corps (Captain). Two bars: Relief of Mafeking, Transvaal; and British South Africa Co.'s medal for Rhodesia, 1896	. .	4 5 0
Royal Welsh Fusiliers. Four bars: Relief of Mafeking, Tugela Heights, Relief of Ladysmith, Transvaal	. .	1 16 0
British South African Police (Trooper). One bar: Defence of Mafeking	. .	3 0 0
K.R.R.C. Eight bars: Cape Colony, Tugela Heights, Orange Free State, Relief of Ladysmith, Laing's Nek, Belfast, S. Africa, 1901, S. Africa, 1902	. .	1 6 0
Brabant's Horse. Five bars: Cape Colony, Wepener, Wittebergen, Belfast, South Africa, 1902	. .	1 18 0
Brabant's Horse (Corporal). Four bars: Cape Colony, Wepener, Wittebergen, Belfast .	. .	2 2 0
Brabant's Horse (Lieutenant). Two bars: Cape Colony, Wepener	. .	3 12 0
Two bars: Defence of Ladysmith and Elandslaagte (realised as much as £3 2*s.* 6*d.* in 1905)	1 1 0 to	1 5 0
Leicester Regt. Three bars: Cape Colony, Laing's Nek, Belfast. Egyptian Medal. Two bars: The Nile, 1884–5; Abu Klea and Khedive's Bronze Star	. .	1 6 0

Modern Indian Medals

	£ s. d.	£ s. d.
43rd Regt. Native Infantry. One bar: Naga, 1879–80	. .	1 10 0
44th Regt. Native Infantry	1 14 0 to	2 2 0
One bar: Persia	0 9 6 to	0 10 6
One bar: Jowaki, 1877–8	0 10 0 to	0 12 0
One bar: North-West Frontier	0 10 6 to	1 0 0
One bar: North-West Frontier, 1908 . .	0 9 6 to	0 12 6

	£ s. d.	£ s. d.
4th Battn. K.R.R.C. One bar: Lushai, 1889–92	. .	1 14 0
One bar: Looshai	1 10 0 to	2 2 0
2nd Kashmir Infantry. Hunza, 1891	. .	3 5 0
3rd Kashmir Infantry. Hunza, 1891	. .	3 7 6
Muleteer, Peshawar Transport. Hunza, 1891	. .	3 0 0
Bronze Collar badge for Hunza, Naga Expedition, 1891	. .	1 10 0
Defence of Chitral, 1895. Maharajah of Kashmir's bronze medal for Chitral, 1895	. .	4 0 0
Warwick Regt. One bar: North-West Frontier, 1908	. .	0 14 0
Abor, 1911	3 3 0 to	4 4 0

Victoria Cross

The following are a few prices realised in the sale-room: £43 (Lieutenant); £47 (Gunner), R.A.; £50, Argyll and Sutherlands; £53, Madras Fusiliers; £58, 65th Regt.; £54 (Sergeant-Major), 37th B.N.I.; £61, 2nd Dragoons; £62, with Crimean Medals, Grenadier Guards; £67, and Mutiny Medal, 9th Lancers, £67; £80, 49th Regt. (said to be second issued); £85 with Mutiny Medal and three bars, 9th Lancers; V.C. and South Africa Medal to Rorke's Drift hero, £95, 24th Regt.; 145 guineas for group of seven medals including V.C., 8th Lancers, and a V.C. won before Sebastopol, £175; a Naval V.C. has realised £170 at auction.

Conspicuous Gallantry Medal

Together with Ashantee Medal to Naval Gunner. Private Sale	. .	40 0 0

Distinguished Service Order

2nd W. Riding Regt. (Victorian)	. .	7 10 0

Order of Merit

1st Class (gold, enamelled decoration) for Conspicuous Gallantry in the Field	. .	8 10 0
2nd Class (Gold, Enamelled Centre)	3 10 0 to	4 10 0
3rd Class (entirely silver and centre enamelled)	2 10 0 to	3 10 0

These orders have realised as much as £58, £40, and £35, according to class.

Royal Red Cross

For services in Crimea	. .	10 10 0

The Albert Medal

1st Class, gold, for Gallantry in saving life on land	. .	75 0 0

Groups

	£ s. d.	£ s. d.
Gold Cross for Fuentes d'Onor, Badajoz, Ciudad Rodrigo, Orthes; the Gold Medal with bars for Badajoz and Ciudad Rodrigo, Peninsular Medal with three bars and gold badge of Military Companies of the Order of the Bath, to Lieutenant-Colonel Russell Manners, C.B., 74th Foot	. .	600 0 0
15th Light Dragoons. Waterloo Medal and three-bar Peninsular	20 0 0 to	30 0 0
6th Foot (Lieutenant). Five-bar Peninsular Medal with Gold Order of Knight of Hanover	. .	28 0 0
Two Naval, Mutiny, with two bars: Lucknow, Relief of Lucknow. Baltic, 1854–5. To Midshipman who became M.P.	. .	17 0 0
One bar: Aro, 1901–2: and West Africa medal. One bar: 1897–8. West Africa Field Force	. .	2 0 0
Three to Private, P. A. L. I.: Ghuznee, July 23rd, 1839, Cabul, 1842; Jellalabad, April 7th, 1842, first issue	. .	7 5 0
13th Regt. (Sergeant-Major). Ghuznee, 1839; Cabul, 1842; Jellalabad, 1843; and Meritorious Service Medal	. .	41 0 0
13th Lght Infantry. Jellalabad, 1842, with Mural Crown and Indian Mutiny medal.	. .	5 0 0
Three to Colour-Sergeant, 31st Regt.: Cabul, 1842; Sutlej, three bars: Ferozeshuhur, Aliwal, Sobraon; Meritorious Service	. .	6 10 0
Two to 43rd Regt.: Kelat-i-Ghilzie, 1842; Ghuznee, Cabul, 1842—two wreaths	. .	5 5 0
Two to Gunner, 2nd Battn. Bengal Artillery: Meeanee, Hyderabad, 1843; Indian Mutiny. Three bars: Delhi, Relief of Lucknow, Lucknow	. .	8 5 0
Two to Riding-master, 2nd Brigade Horse Artillery: Punniar Star, December 29th, 1843; Sutlej medal, Ferozeshuhur, 1843: one bar: Sobraon	. .	5 15 0
Two to Lieutenant, 10th Foot: Sobraon, 1846; Punjab, two bars: Goojerat, Mooltan	. .	5 10 0
Three to Sergeant, H.M. 101st Regt.: Indian General Service, one bar: Umbeyla, 1864; Distinguished Conduct in the Field; Meritorious Service	. .	10 10 0
Four to 1st Cameron Highlanders: The Sudan: Two bars: The Atbara, Khartoum; English Sudan; Boer War: Four bars: Cape Colony, Orange Free State, Transvaal; King's medal	. .	1 7 6
Two, 1895, Defence of Chitral: Maharajah of Kashmir's bronze medal, Chitral, 1895	. .	4 0 0

	£ s. d.	£ s. d.
Three to Able Seaman: West African, Gambia, 1894; Benin River, 1894; Brass River, 1895; Boer War; China, 1900	. .	4 0 0
New Zealand Volunteers, Silver medal for 12 years' service to Lieutenant, New Zealand Field Artillery. Boer War medal to same recipient as Private, N.Z. Mounted Rifles .	. .	6 0 0
66th Foot. Afghan, Kandahar, and Distinguished Conduct Medal, July 14th, 1880. For the Battle of Maiwand.	. .	6 15 0
1st Connaught Rangers (Sergeant-Major). Five bars: Cape Colony, Tugela Heights, Orange Free State, Relief of Ladysmith, Transvaal, and Medal for Distinguished Conduct in the Field	. .	6 0 0
Hampshire Regt. (Sergeant). Four bars: Relief of Kimberley, Paardeberg, Johannesburg, Wittebergen, and West Africa Medal, one bar: 1897–8	. .	3 3 0
North-West Frontier, 1908. King George V, Delhi Durbar medal, December 12th, 1911 (both to Gunner)	. .	4 0 0

MISCELLANEOUS MEDALS

Large silver medal. Naval Reward. Struck for distribution amongst those who had commanded the royal forces and fleets which opposed the double invasion of Monmouth and Argyle	. .	13 10 0
Large Silver Medal, Charles II	4 10 0 to	6 6 0
Oval Gold Badge, for taking of Portobello by Admiral Vernon	. .	10 0 0
American Indian Chief's large silver medal, bust of King George III in armour	. .	15 0 0
American Indian Chief's medal. Obverse: Bust of George III, 1794. Reverse: Royal Arms	2 2 0 to	5 0 0
Gold Medal and Chain given by German Emperor to eight officers of the 15th Light Dragoons, for "brilliant and important services" at Villiers-en-Couche (near Cambray), April 24th, 1794	. .	260 0 0
Order of Dooranée. 3rd Class. Private sale .	. .	45 0 0
Highland Society's Silver Medal, for Egypt, 1801	. .	2 7 6
Medal for Coorg, April 1837	. .	26 0 0
Defence of Kars (Turkish)	1 0 0 to	1 5 0
Defence of Silistria (Turkish)	0 17 6 to	1 5 0
George V Special Reserve medal for Long Service and Good Conduct	. .	0 15 0
George V Indian Volunteer Forces Officer's decoration	. .	1 10 0
Naval Best Shot	3 3 0 to	4 10 0

INDEX